Foundations in Signal Processing, Communications and Networking

Series Editors: W. Utschick, H. Boche, R. Mathar

Foundations in Signal Processing, Communications and Networking

Series Editors: W. Utschick, H. Boche, R. Mathar

Vol.1. Dietl, G. K. E.
Linear Estimation and Detection in Krylov Subspaces, 2007
ISBN 978-3-540-68478-7

Guido K. E. Dietl

Linear Estimation and Detection in Krylov Subspaces

With 53 Figures and 11 Tables

 Springer

Series Editors:

Wolfgang Utschick
TU Munich
Associate Institute for Signal
Processing
Arcisstrasse 21
80290 Munich, Germany

Holger Boche
TU Berlin
Dept. of Telecommunication Systems
Heinrich-Hertz-Chair for Mobile
Communications
Einsteinufer 25
10587 Berlin, Germany

Rudolf Mathar
RWTH Aachen University
Institute of Theoretical
Information Technology
52056 Aachen, Germany

Author:

Guido Dietl
Munich, Germany
guido.dietl@mytum.de

ISSN print edition: 1863-8538

ISBN 978-3-540-68478-7 Springer Berlin Heidelberg New York

Library of Congress Control Number: 2007928954

Springer is a part of Springer Science+Business Media

springer.com

© Springer-Verlag Berlin Heidelberg 2007

Typesetting: by the author and Integra using Springer LaTeX package
Cover Design: deblik, Berlin

Printed on acid-free paper   SPIN: 11936770   42/3100/Integra   5 4 3 2 1 0

To Anja

Preface

One major area in the theory of statistical signal processing is reduced-rank estimation where optimal linear estimators are approximated in low-dimensional subspaces, e. g., in order to reduce the noise in overmodeled problems, enhance the performance in case of estimated statistics, and/or save computational complexity in the design of the estimator which requires the solution of linear equation systems. This book provides a comprehensive overview over reduced-rank filters where the main emphasis is put on matrix-valued filters whose design requires the solution of linear systems with multiple right-hand sides. In particular, the multistage matrix Wiener filter, i. e., a reduced-rank Wiener filter based on the multistage decomposition, is derived in its most general form.

In numerical mathematics, iterative block Krylov methods are very popular techniques for solving systems of linear equations with multiple right-hand sides, especially if the systems are large and sparse. Besides presenting a detailed overview of the most important block Krylov methods in Chapter 3, which may also serve as an introduction to the topic, their connection to the multistage matrix Wiener filter is revealed in this book. Especially, the reader will learn the restrictions of the multistage matrix Wiener filter which are necessary in order to end up in a block Krylov method. This relationship is of great theoretical importance because it connects two different fields of mathematics, viz., statistical signal processing and numerical linear algebra.

This book mainly addresses readers who are interested in the theory of reduced-rank signal processing and block Krylov methods. However, it includes also practical issues like efficient algorithms for direct implementation or the exact computational complexity in terms of the required number of floating point operations. If the reader is not interested in these practical aspects, Sections 2.2, 4.3, and 4.4 of this book can be skipped.

Finally, the book covers additionally the application of the proposed linear estimators to a detection problem occurring at the receiver of a digital communication system. An iterative (Turbo) multiuser detector is considered where users are separated via spread spectrum techniques. Besides using

Monte Carlo simulations, the communication system is investigated in terms of the expected iterative estimation error based on extrinsic information transfer charts. It should be mentioned that the extrinsic information transfer characteristics that are shown in these charts, are calculated in a semianalytical way as derived in Section 6.1.2.

This text has been written at the Associate Institute for Signal Processing, Munich University of Technology, Germany, where I was working as a research engineer towards my doctoral degree. I would like to express my deep gratitude to Prof. Wolfgang Utschick who supervised me at the institute. I appreciate his helpful advice and steady support, as well as the numerous discussions which had a great impact on this book. Besides, I thank Prof. Michael L. Honig of the Northwestern University, USA, and Prof. Joachim Hagenauer of the Institute for Communications Engineering, Munich University of Technology, both members of my dissertation committee, for reviewing this manuscript. I also thank Prof. Michael D. Zoltowski of Purdue University, USA, for giving me the opportunity to stay at Purdue University in winter 2000/2001 and summer 2004. In fact, he initiated my research on the multistage Wiener filter and Krylov methods. Further, I would like to thank Prof. Josef A. Nossek of the Institute for Circuit Theory and Signal Processing, Munich University of Technology, for his support.

Finally, many thanks to all the excellent students which I had the chance to supervise at the Munich University of Technology. Their research results deeply influenced this book. Moreover, I thank all my colleagues at the institute for the nice atmosphere and the inspiring discussions.

Munich, March 2007   *Guido Dietl*

Contents

List of Algorithms

List of Figures

List of Tables

1

Introduction

The basic problem in *estimation theory* (e. g., [150, 210]), also known as *parameter estimation*, is to recover an unknown signal from a perturbed observation thereof. In the *Bayesian framework* where the unknown signal is assumed to be random, one of the most popular approach is to design the estimator such that the *Mean Square Error* (MSE) between its output and the unknown signal is minimized, resulting in the *Minimum MSE* (MMSE) *estimator*. Besides non-linear MMSE estimators like, e. g., the *Conditional Mean Estimator* (CME) (cf., e. g., [210, 188]), the *Wiener Filter* (WF) is a linear MMSE estimator, i. e., its output is a linear transformation of the observation. If both the unknown signal and the observation are multivariate non-stationary random sequences, the WF is time-varying and matrix-valued, therefore, in the following denoted as the *Matrix WF* (MWF) (e. g., [210]). The design of this time-variant MWF requires the solution of a system of linear equations with multiple right-hand sides, the so-called *Wiener–Hopf equation*, for each time index which is computationally cumbersome especially if the dimension of the observation is very large. There are two fundamental approaches for reducing this computational burden: the first one is to apply *methods of numerical linear algebra* (e. g., [237]) to solve the Wiener–Hopf equation in a computationally efficient way whereas the second approach exploits the *theory of statistical signal processing* (e. g., [210]) to decrease computational complexity by approximating the estimation problem itself.

First, we briefly discuss the methods of numerical linear algebra. Numerical mathematics distinguish between two different types of algorithms for solving systems of linear equations: *direct* and *iterative methods* (see Table 1.1). Contrary to direct methods which do not provide a solution until all steps of the algorithm are processed, iterative methods yield an approximate solution at each iteration step which improves from step to step. The most famous direct method is *Gaussian elimination*[1] [74] or the *Gauss–Jordan*

---

[1] Note that the principle of Gaussian elimination has been already mentioned in the Chinese book *Jiuzhang Suanshu* (Chinese, The Nine Chapters on the Math-

algorithm[2] [142, 35] both based on the *LU factorization* (e. g., [94]), i. e., the factorization of the system matrix in a lower and an upper triangular matrix. If the system matrix is Hermitian and positive definite as in the case of the Wiener–Hopf equation, the lower and upper triangular matrix are Hermitian versions of each other and the LU factorization can be reformulated as the computationally cheaper *Cholesky factorization* [11, 72, 236, 18]. Two of the first iterative methods are the *Gauss–Seidel* [76, 220] and the *Jacobi* [138, 139] *algorithm* which iteratively update the single entries in the solution assuming that the remaining entries are the values from previous iteration steps. Compared to these methods which need an infinite number of iterations to converge to the optimum solution, iterative *Krylov methods* [204] produce, at least in exact arithmetics, the solution in a finite number of steps. The most general Krylov method is the *Generalized Minimal RESidual* (GMRES) *algorithm* [205] which approximates the exact solution in the Krylov subspace formed by the system matrix and the right-hand side, via the minimization of the residual norm. In each iteration step, the dimension of the Krylov subspace is increased. Whereas the GMRES method can be applied to arbitrarily invertible system matrices, the first Krylov methods, i. e., the *Conjugate Gradient* (CG) [117] and the *Lanczos algorithm* [156, 157] both being mathematically identical, have been derived for Hermitian and positive definite system matrices. Precisely speaking, in [156], the Lanczos algorithm has been also applied to non-Hermitian matrices by using the method of *bi-orthogonalization* but the *Arnoldi algorithm* [4] is the computationally more efficient procedure for arbitrarily invertible system matrices. The extension of Krylov methods to systems of linear equations with multiple right-hand sides result in the *block Krylov methods* like, e. g., the *block Arnoldi algorithm* [208, 207], the *block Lanczos algorithm* [93], and the *Block Conjugate Gradient* (BCG) *algorithm* [175]. For the interested reader, a detailed overview over iterative methods for solving systems of linear equations can be found in [267].

Table 1.1. Different algorithms for solving systems of linear equations with system matrix A

	Direct methods	Iterative Krylov methods
$A \neq A^{\mathrm{H}}$	Gaussian elimination (LU factorization)	Generalized Minimal RESidual (GMRES), Arnoldi, etc.
$A = A^{\mathrm{H}}$	Cholesky factorization (A positive definite)	Conjugate Gradient (CG, A positive definite), Lanczos

---

ematical Art) and in the commentary on this book given by Liu Hui in the year 263 [144].

[2] Gaussian elimination and the Gauss–Jordan algorithm differ only in the economy of data storage [133].

Second, in the theory of statistical signal processing, *reduced-rank estimation*, i. e., the estimation of the unknown signal based on a subspace approximation of the observation, decreases computational complexity by reducing the size and the structure of the Wiener–Hopf equation. In fact, if the computation of the subspace basis and the solution of the reduced-size Wiener–Hopf equation is computationally cheaper than solving the original system of linear equations, the resulting-reduced rank method is a candidate for a low-complexity estimator. Note that besides the complexity issue, reduced-rank methods enhance the robustness against estimation errors of second order statistics due to low sample support. First reduced-rank approaches were based on the approximation of the linear MMSE estimator in eigensubspaces spanned by eigenvectors of the system matrix of the Wiener–Hopf equation. One of them is the *Principal Component* (PC) *method* [132] which chooses the subspace spanned by the eigenvectors corresponding to the largest eigenvalues, i. e., the reduced-dimension eigensubspace with the largest signal energy. Nevertheless, in the case of a mixture of signals, the PC procedure does not distinguish between the signal of interest and the interference signal. Hence, the performance of the algorithm degrades drastically in interference dominated scenarios. This drawback of the PC algorithm lead to the invention of the *Cross-Spectral* (CS) *method* [85] which selects the eigenvectors such that the MSE over all eigensubspace based methods with the same rank is minimal. To do so, it considers additionally the cross-covariance between the observation and the unknown signal, thus, being more robust against strong interference. A third reduced-rank method is based on a *multistage decomposition* of the MMSE estimator whose origin lies in the *Generalized Sidelobe Canceler* (GSC) [3, 98]. The resulting reduced-rank *MultiStage WF* (MSWF) [89], or *MultiStage MWF* [88, 34] if the unknown signal is multivariate, is no longer based on eigensubspaces but on a subspace whose first basis vectors are the columns of the right-hand side matrix of the Wiener–Hopf equation.

Michael L. Honig showed in [130] that a special version of the MSWF is the approximation of the WF in a *Krylov subspace*. This fundamental observation connects the methods of numerical linear algebra with the theory of reduced-rank signal processing. Due to the excellent performance of these special MSWF versions, it follows that Krylov methods which are basically used for solving very large and sparse systems of linear equations [204], are good candidates for solving the Wiener–Hopf equation as well. It remains to discuss the relationship between the general MSWF and Krylov methods as well as the connection between the MSMWF and block Krylov methods which involves a multitude of new problems.

Whereas estimation theory deals with the recovery of a continuous-valued parameter, *detection theory* (e. g., [151, 210]) solves the problem of estimating a signal with a finite set of possible values also known as *hypothesis testing*. One example of hypothesis testing is the receiver of a digital communication system detecting the *binary digits* (bits) of a data source where only the perturbed observation thereof after the transmission over the channel is available.

Note that the mapping of the detector from the observation to the estimated bit which is one of the two possible binary values is clearly *non-linear*. However, a detector can be designed based on a linear estimator or *equalizer* followed by a non-linear operation like, e. g., a *hard decision device* or a *soft-input hard-output decoder*. Recently, *iterative* or *Turbo receivers* [61] have been introduced where the non-linear *soft-input soft-output decoder* which follows after linear estimation is used to compute *a priori information* about the transmitted bits. This *a priori* knowledge can be exploited by the linear estimator to improve the quality of its output. After further decoding, the procedure can be repeated for several iterations until the decoder delivers also the detected data bits. Since the use of the *a priori* information leads to a non-zero mean and non-stationary signal model, the linear estimator has to be designed as a time-varying filter which is not strictly linear but *affine*. If the communication system under consideration has multiple antennas, multiple users which are separated based on long signatures, and channels of high orders, the design of the linear estimator is computationally cumbersome. In such cases, low-complexity solutions based on either numerical linear algebra or statistical signal processing as described above play an important role.

1.1 Overview and Contributions

Chapter 2: Efficient Matrix Wiener Filter Implementations

Section 2.1 briefly reviews the *Matrix Wiener Filter* (MWF) (e. g., [210]) for estimating a non-zero mean and non-stationary signal vector sequence based on an observation vector sequence by minimizing the *Mean Square Error* (MSE) of the Euclidean norm between its output and the unknown signal vector sequence. Before presenting suboptimal reduced-complexity approaches in Section 2.3, Section 2.2 introduces the *Cholesky factorization* [11, 72, 236, 18] as a computationally cheap *direct method* to solve the *Wiener–Hopf equation*. Again, the Cholesky method exploits the fact that the system matrix of the Wiener–Hopf equation, i. e., the auto-covariance matrix of the observation vector, is Hermitian and positive definite.[3] Besides, we derive an optimal reduced-complexity method which can be applied to scenarios where the auto-covariance matrix of the observation vector has a specific time-dependent sub-matrix structure as in the Turbo system of Chapter 5. The resulting algorithm is a generalization of the reduced-complexity method in [240] from a single-user single-antenna to a multiple-user multiple-antenna system. The computational complexities of the Cholesky based and the reduced-complexity MWF are compared based on the exact number of required *FLoating point OPerations* (FLOPs). Finally, in Section 2.3, subspace approximations of the MWF,

---

[3] In this book, we exclude the special case where the system matrix of the Wiener–Hopf equation is rank-deficient and therefore no longer positive definite but semidefinite.

i. e., *reduced-rank MWFs*, are investigated as alternative low-complexity methods. Compared to the Cholesky factorization and its derivatives, reduced-rank MWFs are no longer optimal even if performed in exact arithmetics. Section 2.3 concludes with the brief review of the eigensubspace based reduced-rank MWFs, i. e., the *Principal Component* (PC) [132] and the MSE optimal *Cross-Spectral* (CS) [85] *method.*

Chapter 3: Block Krylov Methods

Iterative *Krylov methods* [204] are used in numerical linear algebra to solve systems of linear equations as well as eigenvalue problems. If the system of linear equations has several right-hand sides, the most important algorithms are the *block Arnoldi* [208, 207], the *block Lanczos* [93], and the *Block Conjugate Gradient* (BCG) [175] *algorithm.* This chapter derives these *block Krylov methods* in Sections 3.2, 3.3, and 3.4, and investigate their properties. The block Krylov methods iteratively improve the approximation of the exact solution from step to step. The *rate of convergence*, i. e., the number of iterations which are necessary to get a sufficient approximation of the exact solution, is derived for the BCG algorithm in Subsection 3.4.2. Compared to the same result as published in [175], the derivation presented in Subsection 3.4.2 is more elegant. Moreover, we analyze in Subsection 3.4.3 the *regularizing characteristic* of the BCG procedure, i. e., the fact that it produces an improved solution if stopped before convergence in cases where the system matrix and/or the right-hand sides are perturbed. For example, this is the case when these algorithms are applied to the Wiener–Hopf equation where second order statistics are not perfectly known due to estimation errors. We investigate the regularizing effect of the BCG algorithm by deriving its *filter factor representation* [110, 43]. So far, *filter factors* have been only presented (e. g., [110]) for the *Conjugate Gradient* (CG) *algorithm* [117], i. e., the special case of the BCG method for solving a system of linear equations with one right-hand side. Note that filter factors are a well-known tool in the theory of *ill-posed* and *ill-conditioned problems* to describe regularization [110]. Eventually, special versions of the block Lanczos as well as the BCG algorithm are derived where the solution matrix is computed columnwise instead of blockwise in order to achieve more flexibility in stopping the iterations. The resulting *block Lanczos–Ruhe algorithm* [200] and the *dimension flexible BCG method* [101] (see Subsections 3.3.2 and 3.4.4) can be used to approximate the exact solution in Krylov subspaces with dimensions which are no longer restricted to integer multiples of the dimension of the unknown signal vector.

Chapter 4: Reduced-Rank Matrix Wiener Filters in Krylov Subspaces

The original *MultiStage MWF* (MSMWF) [88, 34, 45] is a reduced-rank approximation of the MWF based on *orthonormal correlator matrices*, so-called *blocking matrices*, and *quadratic MWFs* of reduced size. In Section 4.1, we derive the MSMWF in its most general form where the orthonormal correlator matrices have been replaced by arbitrary bases of the corresponding correlated subspaces. Here, a *correlated subspace* is the subspace spanned by the columns of the correlator matrix. Besides, Section 4.1 includes a discussion of the MSMWF's fundamental properties. The general derivation of the MSMWF makes it possible to properly investigate its relationship to the Krylov subspace in Section 4.2. In fact, it reveals the restrictions on the MSMWF which are necessary to end up in one of the block Krylov methods (cf. Subsections 4.2.2 and 4.2.3). Whereas the connection between the *MultiStage Wiener Filter* (MSWF) [89] and Krylov subspaces has been already discussed in [128], the relationship between the MSMWF and the Krylov subspace goes back to [45]. After summarizing the resulting Krylov subspace based MSMWF implementations in Section 4.3, Section 4.4 concludes the chapter by revealing their computational efficiency to the reduced-complexity methods of Chapter 2 with respect to the required number of FLOPs.

Chapter 5: System Model for Iterative Multiuser Detection

This chapter defines the system model of a *Multiple-Access* (MA) *communication system* with iterative multiuser detection (cf., e. g., [254, 127]) which is used in Chapter 6 in order to analyze the performance of the algorithms proposed in the previous chapters. In an iterative detection scenario, the transmitter of each user consists of an encoder, interleaver, and mapper (cf. Section 5.1). Here, the MA scheme is realized using *Direct Sequence Code Division Multiple-Access* (DS-CDMA), i. e., the symbol streams of the different users are spread using *Orthogonal Variable Spreading Factor* (OVSF) *codes* as reviewed in Subsection 5.1.4. After the transmission over a frequency-selective channel as described in Section 5.2, the spread signals of the different users are received by multiple antennas. Then, the observation signal is processed by an *iterative* or *Turbo receiver* [61] exchanging soft information between linear estimator and decoder. Section 5.3 explains how the linear estimators of the previous chapters can be used as linear equalizers in the Turbo receiver, i. e., how they exploit the *a priori information* obtained from the *soft-input soft-output decoder* and how they transform the estimate at the output of the equalizer to an *extrinsic information* which is needed by the soft-input soft-output decoder to compute the *a priori* information required by the linear estimator at the next Turbo iteration (cf. [254, 240, 49]). Besides, the decoder provides the detected data bits after the last Turbo step. The *a priori* information is used to adapt the signal model for the linear estimator design. To do so, the transmitted signal must be assumed to be non-zero mean and non-stationary and the resulting estimator is time-varying. Note that the time-varying statistics can be also approximated via a time-average approach leading to further low-complexity solutions if applied to the already mentioned reduced-complexity algorithms except from the one which exploits the time-dependent structure of the auto-covariance matrix. Finally, Subsection 5.3.4 reveals the relationship of the proposed equalizers to *iterative soft interference cancellation* [254, 198] or *parallel decision feedback detection* [62, 127].

Chapter 6: System Performance

The final chapter analyzes the performance of the coded multiuser DS-CDMA system as introduced in Chapter 5. Besides the comparison of the different equalizers in Section 6.3 based on the *Bit Error Rate* (BER) which is obtained via *Monte Carlo simulations*, we investigate their performance in Section 6.2 using *EXtrinsic Information Transfer* (EXIT) *charts* [19] which are briefly reviewed in Section 6.1. Contrary to the simulated calculation of the equalizer's EXIT characteristic, Subsection 6.1.2 proposes a method for their semianalytical computation which is a generalization of the contributions in [195]. Moreover, Section 6.2 answers the question when rank-reduction is preferable to order-reduction of the equalizer filter in order to obtain computationally cheap implementations by introducing *complexity–performance charts*. Eventually, the regularizing property of the MSMWF's BCG implementation as discussed in Subsection 3.4.3 are investigated when the channel is not perfectly known at the receiver but estimated based on the *Least Squares* (LS) method (cf., e. g., [188, 53]).

1.2 Notation

Contrary to mathematical operators which are written based on the standard text font, mathematical symbols are denoted by italic letters where vectors are additionally lower case bold and matrices capital bold. Random variables are written using *sans serif* font and their realizations with the same font but with serifs. Finally, we use calligraphic letters for subspaces and the blackboard notation for sets.

Table 1.2 summarizes frequently used operators and Table 1.3 frequently used symbols. Besides, the matrix

$$S_{(\nu,m,n)} = \begin{bmatrix} \mathbf{0}_{m\times\nu} & I_m & \mathbf{0}_{m\times(n-\nu)} \end{bmatrix} \in \{0,1\}^{m\times(m+n)}, \tag{1.1}$$

is used to select m rows of a matrix beginning from the $(\nu+1)$th row by applying it to the matrix from the left-hand side. Analogous, if its transpose is applied to a matrix from the right hand side, m columns are selected beginning from the $(\nu+1)$th column. Besides, the matrix $S_{(\nu,m,n)}$ can be used to describe convolutional matrices.

An alternative access to submatrices is given by the notation $[A]_{m:n,i:j}$ which denotes the $(n - m + 1) \times (j - i + 1)$ submatrix of A including the elements of A from the $(m+1)$th to the $(n+1)$th row and from the $(i+1)$th to the $(j+1)$th column. Further, if we want access to a single row or column

Table 1.2. Frequently used operators

Operator	Description
$\oplus$, $\diamondsuit$	two general operations defined in a field
$\boxplus$, $\boxdot$	modulo 2 addition and modulo 2 multiplication
$+$, $\cdot$	common addition and multiplication
$\oplus$	direct sum of subspaces
$\oslash$	elementwise division
$\otimes$	Kronecker product
$*$	convolution
$\mathrm{E}\{\cdot\}$	expectation
$(\cdot)^{\circ}$	zero-mean part of a random sequence
$\mathrm{Re}\{\cdot\}$	real part
$\mathrm{Im}\{\cdot\}$	imaginary part
$|\cdot|$	absolute value
$(\cdot)^{*}$	complex conjugate
$(\cdot)^{\mathrm{T}}$	transpose
$(\cdot)^{\mathrm{H}}$	Hermitian, i.e., conjugate transpose
$(\cdot)^{-1}$	inverse
$(\cdot)^{\dagger}$	Moore–Penrose pseudoinverse
$\sqrt{\cdot}$	square root or general square root matrix
$\sqrt[\oplus]{\cdot}$	positive semidefinite square root matrix
$\mathrm{tr}\{\cdot\}$	trace of a matrix
$\mathrm{rank}\{\cdot\}$	rank of a matrix
$\mathrm{range}\{\cdot\}$	range space of a matrix
$\mathrm{null}\{\cdot\}$	null space of a matrix
$\mathrm{vec}\{\cdot\}$	vectorization of a matrix by stacking its columns
$\mathrm{diag}\{\cdot\}$	diagonal matrix with the scalar arguments or the elements of the vector argument on the diagonal
$\mathrm{bdiag}\{\cdot\}$	block diagonal matrix with the matrix arguments on the diagonal
$\lceil\cdot\rceil$	closest integer being larger than or equal to the argument
$\lfloor\cdot\rfloor$	closest integer being smaller than or equal to the argument
$\mathrm{O}(\cdot)$	Landau symbol
$\|\cdot\|$	general norm of a vector or matrix
$\|\cdot\|_{2}$	Euclidean norm of a vector or induced 2-norm of a matrix
$\|\cdot\|_{\mathrm{F}}$	Hilbert–Schmidt or Frobenius norm of a matrix
$\|\cdot\|_{A}$	A-norm of a matrix

of a matrix, respectively, we use the abbreviation 'n' instead of writing '$n:n$'. Note that '$:$' is the abbreviation for '$0:z$' where z is the maximum value of the index under consideration. Eventually, $[a]_{m:n}$ denotes the $(n-m+1)$-dimensional subvector of a including the $(m+1)$th to the $(n+1)$th element.

The *probability* $\mathrm{P}(d=d)$ denotes the likelihood that a realization of the discrete random variable $d \in \mathbb{C}$ is equal to $d \in \mathbb{C}$ and $\mathrm{p}_{\mathsf{x}}(x)$ is the *probability density function* of the continuous random variable $\mathsf{x} \in \mathbb{C}$. The soft

Table 1.3. Frequently used symbols

Symbol	Description
$:=$	definition
$\perp\,\mathcal{S}$	orthogonal subspace complement of $\mathcal{S}$
$\boldsymbol{P}_{\mathcal{S}}$	projector matrix projecting on subspace $\mathcal{S}$
$\hat{(\cdot)}$	estimated value
$\tilde{(\cdot)}$	detected value
$\delta_{i,j}$	Kronecker delta
$\delta[n]$	unit impulse
$\boldsymbol{I}_n$	$n \times n$ identity matrix
$\boldsymbol{e}_i$	ith column of the $n \times n$ identity matrix[4]
$\boldsymbol{1}_n$	n-dimensional all-ones vector
$\boldsymbol{0}_{n \times m}$	$n \times m$ zero matrix
$\boldsymbol{0}_n$	n-dimensional zero vector
$\mathbb{C}$	set of complex numbers
$\mathbb{R}$	set of real numbers
$\mathbb{R}_{0,+}$	set of non-negative real numbers
$\mathbb{R}_+$	set of positive real numbers
$\mathbb{Z}$	set of integers
$\mathbb{N}_0$	set of non-negative integers
$\mathbb{N}$	set of positive integers

information of a binary random variable $b \in \{0,1\}$ is represented by the *Log-Likelihood Ratio* (LLR) [107]

$$l = \ln \frac{\mathrm{P}\,(b = 0)}{\mathrm{P}\,(b = 1)} \in \mathbb{R}. \tag{1.2}$$

Table 1.4 depicts the first and second order statistical moments of the scalar random sequences $x[n] \in \mathbb{C}$ and $y[n] \in \mathbb{C}$, and the vector random sequences $\boldsymbol{u}[n] \in \mathbb{C}^\ell$ and $\boldsymbol{v}[n] \in \mathbb{C}^m$ at time index n. Note that the cross correlations and cross-covariances are functions of the index n and the latency time ν. Since we choose a fixed ν throughout this book, an additional index is omitted in the given notation. Besides, here, the auto-correlations and auto-covariances denote the values of the corresponding auto-correlation and auto-covariance functions evaluated at zero, i.e., for a zero-shift. For example, the auto-correlation $r_x[n]$ of $x[n]$ is equal to the auto-correlation function $r_x[n, \nu] = \mathrm{E}\{x[n]x^*[n - \nu]\}$ at $\nu = 0$, i.e., $r_x[n] = r_x[n, 0]$. The same holds for $c_x[n]$, $\boldsymbol{R}_{\boldsymbol{u}}[n]$, and $\boldsymbol{C}_{\boldsymbol{u}}[n]$.

---

[4] The dimension of the unit vector $\boldsymbol{e}_i$ is defined implicitly via the context.

Table 1.4. Statistical moments at time index n

Moment	Description
$m_x[n] = \mathrm{E}\{x[n]\}$	mean of $x[n]$
$\boldsymbol{m}_u[n] = \mathrm{E}\{\boldsymbol{u}[n]\}$	mean of $\boldsymbol{u}[n]$
$r_x[n] = \mathrm{E}\{|x[n]|^2\}$	auto-correlation of $x[n]$
$c_x[n] = r_x[n] - |m_x[n]|^2$	auto-covariance or variance of $x[n]$
$\sigma_x[n] = \sqrt{c_x[n]}$	standard deviation of $x[n]$
$\boldsymbol{R}_u[n] = \mathrm{E}\{\boldsymbol{u}[n]\boldsymbol{u}^{\mathrm{H}}[n]\}$	auto-correlation matrix of $\boldsymbol{u}[n]$
$\boldsymbol{C}_u[n] = \boldsymbol{R}_u[n] - \boldsymbol{m}_u[n]\boldsymbol{m}_u^{\mathrm{H}}[n]$	auto-covariance matrix of $\boldsymbol{u}[n]$
$r_{x,y}[n] = \mathrm{E}\{x[n]y^*[n-\nu]\}$	cross-correlation between $x[n]$ and $y[n]$
$c_{x,y}[n] = r_{x,y}[n] - m_x[n]m_y^*[n-\nu]$	cross-covariance between $x[n]$ and $y[n]$
$\boldsymbol{r}_{u,x}[n] = \mathrm{E}\{\boldsymbol{u}[n]x^*[n-\nu]\}$	cross-correlation vector between $\boldsymbol{u}[n]$ and $x[n]$
$\boldsymbol{c}_{u,x}[n] = \boldsymbol{r}_{u,x}[n] - \boldsymbol{m}_u[n]m_x^*[n-\nu]$	cross-covariance vector between $\boldsymbol{u}[n]$ and $x[n]$
$\boldsymbol{R}_{u,v}[n] = \mathrm{E}\{\boldsymbol{u}[n]\boldsymbol{v}^{\mathrm{H}}[n-\nu]\}$	cross-correlation matrix between $\boldsymbol{u}[n]$ and $\boldsymbol{v}[n]$
$\boldsymbol{C}_{u,v}[n] = \boldsymbol{R}_{u,v}[n] - \boldsymbol{m}_u[n]\boldsymbol{m}_v^{\mathrm{H}}[n-\nu]$	cross-covariance matrix between $\boldsymbol{u}[n]$ and $\boldsymbol{v}[n]$

Part I

Theory: Linear Estimation in Krylov Subspaces

2

Efficient Matrix Wiener Filter Implementations

Norbert Wiener published in 1949 his book [259] about *linear prediction* and *filtering* of stationary time series, i. e., random sequences and processes. The linear predictor or *extrapolator* estimates the continuation of a time series realization based on the observation thereof up to a specific time instance, by minimizing the *Mean Square Error* (MSE) between the estimate and the exact value. Contrary to that, the linear filtering problem considers the *Minimum MSE* (MMSE) estimation of the time series realization at an arbitrary time index where the observation thereof is additionally randomly perturbed. According to [259], Wiener developed this theory in the years 1940 and 1941. Note that Andrey N. Kolmogorov presented similar results already in 1939 [152] and 1941 [153]. However, his theory has been restricted to *interpolation* and extrapolation of stationary random sequences, i. e., time-discrete time series. The consideration of filtering a time series realization out of a perturbed observation thereof, as well as the time-continuous case is missing in his work. Thus, the optimal linear filter with respect to the MSE criterion is usually denoted as the *Wiener Filter* (WF) (see, e. g., [143] or [210]).

For time-continuous time series, deriving the WF results in an integral equation which has been firstly solved by Norbert Wiener and Eberhard F. F. Hopf in 1931 [260]. In today's literature (see, e. g., [210]), not only this integral equation is denoted as the *Wiener–Hopf equation* but also the system of linear equations which is obtained by deriving the WF for time-discrete time series with a finite number of observation samples which is the focus of this book. Note that from an analytical point of view, the latter problem is trivial compared to the original integral equation solved by Norbert Wiener and Eberhard F. F. Hopf. However, solving the finite time-discrete Wiener Hopf equation can be also very challenging if we are interested in numerically cheap and stable algorithms.

In this chapter, we apply the Wiener theory to non-stationary sequences which are not necessarily zero-mean. Compared to time series where the non-stationarity obeys a deterministic model as investigated in [109, 10], we assume their first and second order statistics to be known at any time instance.

The main focus of this chapter is the derivation of efficient WF algorithms. Therefore, we briefly review an algorithm based on the computationally cheap *Cholesky factorization* [11, 94] which exploits the fact that the system matrix of the Wiener–Hopf equation is Hermitian and positive definite, as well as an algorithm which can be used if the system matrix has a specific time-variant structure. As an alternative reduced-complexity approach, we present *reduced-rank signal processing* by approximating the WF in certain subspaces of the observation space. One of these subspaces has been described by Harold Hotelling who introduced in 1933 the *Principal Component* (PC) *method* [132] which transforms a random vector with correlated elements into a reduced-dimension random vector with uncorrelated entries based on an eigensubspace approximation. Note that although Harold Hotelling introduced the main ideas approximately ten years earlier, this procedure is usually denoted as the *Karhunen–Loève Transform* (KLT) [147, 164] which has been published by Kari Karhunen and Michel Loève in the years 1945 and 1946. Nevertheless, their contribution considers not only vectors of random variables but also random processes.

One of the first applications of the KLT to the filtering problem has been published in the year 1971 [137]. There, the author assumes a circulant system matrix such that the *Fast Fourier Transform* (FFT) [75, 38, 114, 37] can be used for *eigen decomposition* (see, e.g., [131]). However, in [137], no rank reduction is considered, i.e., the complexity reduction is solely based on the application of the computationally efficient FFT. The *MUltiple SIgnal Classification* (MUSIC) *algorithm* [214] introduced by Ralph O. Schmidt in 1979, can be seen as one of the first reduced-rank signal processing technique where the eigen decomposition is used to identify the signal and noise subspace which are needed for parameter estimation. Finally, the origin of reduced-rank filtering as introduced in Section 2.3, can be found in the contributions of Donald W. Tufts et al. [242, 241].

2.1 Matrix Wiener Filter

In this chapter, we discuss the fundamental problem of estimating a signal $x[n] \in \mathbb{C}^M$, $M \in \mathbb{N}$, based on the observation $y[n] \in \mathbb{C}^N$, $N \in \mathbb{N}$, $N \geq M$, which is a perturbed transformation of $x[n]$. One of the main ideas of *Bayes parameter estimation* (e.g., [150, 210]) which lies in our focus, is to model the unknown perturbance as well as the desired signal $x[n]$ as realizations of random sequences with known statistics. As a consequence, the known observation signal $y[n]$ is also a realization of a random sequence. In the following, the realizations $x[n]$ and $y[n]$ correspond to the random sequences $\mathbf{x}[n] \in \mathbb{C}^M$ and $\mathbf{y}[n] \in \mathbb{C}^N$, respectively. If we choose the optimization criterion or *Bayes risk* to be the *Mean Square Error* (MSE) between the estimate and the desired signal, the optimal non-linear filter, i.e., the *Minimum MSE* (MMSE) *estimator* is the well-known *Conditional Mean Estimator* (CME) (cf., e.g., [210, 188])

$$\hat{x}_{\mathrm{CME}}[n] = \mathrm{E}\left\{ x[n] | y[n] = y[n] \right\}. \tag{2.1}$$

Although Bayes parameter estimation is based on non-linear filters in general (cf. the CME as an example), we restrict ourselves to the linear case where the CME reduces to the *linear MMSE estimator* or *Matrix Wiener Filter* (MWF) (see, e.g., [210]). We refer the reader who is interested in non-linear estimation techniques, to one of the books [188, 150, 210, 226]. Note that in the following derivation of the MWF, the random sequences are neither restricted to be stationary nor are they assumed to be zero-mean.

The MWF given by the matrix $W[n] \in \mathbb{C}^{N \times M}$ and the vector $a[n] \in \mathbb{C}^M$ computes the estimate

$$\hat{x}[n] = W^{\mathrm{H}}[n] y[n] + a[n] \in \mathbb{C}^M, \tag{2.2}$$

of the vector random sequence $x[n]$ based on the observed vector random sequence $y[n]$.[1] The MWF is designed to minimize the MSE $\xi_n(W', a') = \mathrm{E}\{\| \varepsilon'[n]\|_2^2\}$ where $\varepsilon'[n] = x[n] - \hat{x}'[n] \in \mathbb{C}^M$ is the error between the estimate $\hat{x}'[n] = W'^{,\mathrm{H}} y[n] + a' \in \mathbb{C}^M$ and the unknown signal $x[n]$,[2] over all linear filters $W' \in \mathbb{C}^{N \times M}$ and $a' \in \mathbb{C}^M$ at each time index n (cf. Figure 2.1).[3] Mathematically speaking,

$$(W[n], a[n]) = \operatorname*{argmin}_{(W',a') \in \mathbb{C}^{N \times M} \times \mathbb{C}^M} \xi_n(W', a'), \tag{2.3}$$

where

$$\begin{aligned}
\xi_n(W', a') &= \mathrm{E}\left\{ \left\| x[n] - W'^{,\mathrm{H}} y[n] - a' \right\|_2^2 \right\} \\
&= \mathrm{tr}\left\{ R_x[n] - R_{y,x}^{\mathrm{H}}[n] W' - m_x[n] a'^{,\mathrm{H}} \right. \\
&\quad - W'^{,\mathrm{H}} R_{y,x}[n] + W'^{,\mathrm{H}} R_y[n] W' + W'^{,\mathrm{H}} m_y[n] a'^{,\mathrm{H}} \\
&\quad \left. - a' m_x^{\mathrm{H}}[n] + a' m_y^{\mathrm{H}}[n] W' + a' a'^{,\mathrm{H}} \right\}.
\end{aligned} \tag{2.4}$$

By partially differentiating $\xi_n(W', a')$ with respect to $W'^{,*}$ and $a'^{,*}$, we get the following systems of linear equations:

---

[1] Note that we apply the MWF $W[n]$ using the Hermitian operator in order to simplify the considerations in the remainder of this book.

[2] The optimal estimate $\hat{x}[n]$ with the error $\varepsilon[n] = x[n] - \hat{x}[n] \in \mathbb{C}^M$ is obtained by applying the optimal filter coefficients $W[n]$ and $a[n]$ whereas $\hat{x}'[n]$ and $\varepsilon'[n]$ denote the estimate and the error, respectively, obtained by applying arbitrary filter coefficients W' and a'.

[3] Note that the considered filters are not *strictly linear* but *affine* due to the shift a'. However, we denote them as *linear* filters because affine filters can always be mapped to linear filters by an adequate coordinate translation without changing the filter coefficients W'.

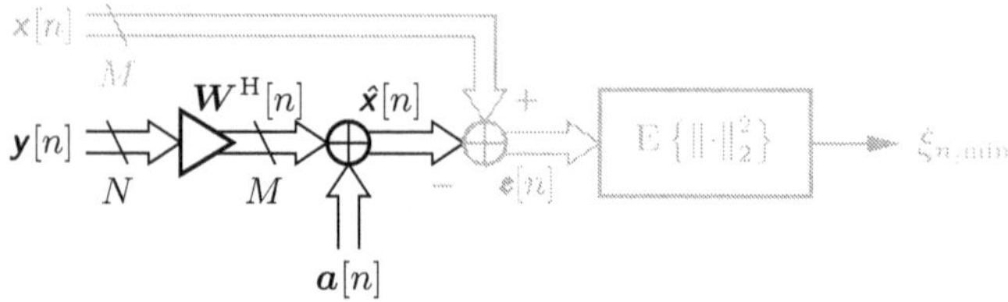

Fig. 2.1. Matrix Wiener Filter (MWF)

$$\left.\frac{\partial \xi_n\left(\boldsymbol{W}', \boldsymbol{a}'\right)}{\partial \boldsymbol{W}'^{,*}}\right|_{\substack{\boldsymbol{W}'=\boldsymbol{W}[n] \\ \boldsymbol{a}'=\boldsymbol{a}[n]}} = \boldsymbol{R}_{\boldsymbol{y}}[n]\boldsymbol{W}[n] + \boldsymbol{m}_{\boldsymbol{y}}[n]\boldsymbol{a}^{\mathrm{H}}[n] - \boldsymbol{R}_{\boldsymbol{y},\boldsymbol{x}}[n]$$

$$\stackrel{!}{=} \boldsymbol{0}_{N \times M}, \tag{2.5}$$

$$\left.\frac{\partial \xi_n\left(\boldsymbol{W}', \boldsymbol{a}'\right)}{\partial \boldsymbol{a}'^{,*}}\right|_{\substack{\boldsymbol{W}'=\boldsymbol{W}[n] \\ \boldsymbol{a}'=\boldsymbol{a}[n]}} = \boldsymbol{a}[n] + \boldsymbol{W}^{\mathrm{H}}[n]\boldsymbol{m}_{\boldsymbol{y}}[n] - \boldsymbol{m}_{\boldsymbol{x}}[n] \stackrel{!}{=} \boldsymbol{0}_M. \tag{2.6}$$

If we solve Equation (2.6) with respect to $\boldsymbol{a}[n]$ and plug the result into Equation (2.5), the solution of the optimization defined in Equation (2.3), computes as

$$\left| \begin{aligned} \boldsymbol{W}[n] &= \boldsymbol{C}_{\boldsymbol{y}}^{-1}[n]\boldsymbol{C}_{\boldsymbol{y},\boldsymbol{x}}[n], \\ \boldsymbol{a}[n] &= \boldsymbol{m}_{\boldsymbol{x}}[n] - \boldsymbol{W}^{\mathrm{H}}[n]\boldsymbol{m}_{\boldsymbol{y}}[n], \end{aligned} \right| \tag{2.7}$$

with the auto-covariance matrix $\boldsymbol{C}_{\boldsymbol{y}}[n] = \boldsymbol{R}_{\boldsymbol{y}}[n] - \boldsymbol{m}_{\boldsymbol{y}}[n]\boldsymbol{m}_{\boldsymbol{y}}^{\mathrm{H}}[n] \in \mathbb{C}^{N \times N}$ and the cross-covariance matrix $\boldsymbol{C}_{\boldsymbol{y},\boldsymbol{x}}[n] = \boldsymbol{R}_{\boldsymbol{y},\boldsymbol{x}}[n] - \boldsymbol{m}_{\boldsymbol{y}}[n]\boldsymbol{m}_{\boldsymbol{x}}^{\mathrm{H}}[n] \in \mathbb{C}^{N \times M}$. Note that the first line of Equation (2.7) is the solution of the *Wiener–Hopf equation* $\boldsymbol{C}_{\boldsymbol{y}}[n]\boldsymbol{W}[n] = \boldsymbol{C}_{\boldsymbol{y},\boldsymbol{x}}[n]$. Finally, with the auto-covariance matrix $\boldsymbol{C}_{\boldsymbol{\varepsilon}}[n] = \boldsymbol{R}_{\boldsymbol{\varepsilon}}[n]$ of the estimation error $\boldsymbol{\varepsilon}[n] = \boldsymbol{x}[n] - \hat{\boldsymbol{x}}[n] \in \mathbb{C}^M$,[4] the MMSE can be written as

$$\begin{aligned} \xi_{n,\min} &= \xi_n\left(\boldsymbol{W}[n], \boldsymbol{a}[n]\right) = \mathrm{tr}\left\{\boldsymbol{C}_{\boldsymbol{\varepsilon}}[n]\right\} \\ &= \mathrm{tr}\left\{\boldsymbol{C}_{\boldsymbol{x}}[n] - \boldsymbol{C}_{\boldsymbol{y},\boldsymbol{x}}^{\mathrm{H}}[n]\boldsymbol{C}_{\boldsymbol{y}}^{-1}[n]\boldsymbol{C}_{\boldsymbol{y},\boldsymbol{x}}[n]\right\}. \end{aligned} \tag{2.8}$$

If we recall Equation (2.2) and replace $\boldsymbol{W}[n]$ and $\boldsymbol{a}[n]$ by the expressions given in Equation (2.7), we get

$$\hat{\boldsymbol{x}}[n] = \boldsymbol{C}_{\boldsymbol{y},\boldsymbol{x}}^{\mathrm{H}}[n]\boldsymbol{C}_{\boldsymbol{y}}^{-1}[n]\left(\boldsymbol{y}[n] - \boldsymbol{m}_{\boldsymbol{y}}[n]\right) + \boldsymbol{m}_{\boldsymbol{x}}[n], \tag{2.9}$$

which gives a good insight in the MWF when applied to non-zero-mean random sequences. First, the sequence $\boldsymbol{y}[n]$ is transformed to the zero-mean sequence $\boldsymbol{y}^\circ[n] = \boldsymbol{y}[n] - \boldsymbol{m}_{\boldsymbol{y}}[n]$ before estimating the zero-mean random sequence $\boldsymbol{x}^\circ[n] = \boldsymbol{x}[n] - \boldsymbol{m}_{\boldsymbol{x}}[n]$ by applying the MWF $\boldsymbol{W}[n]$ with the output

---

[4] Note that the estimation error $\boldsymbol{\varepsilon}[n]$ is zero-mean because $\boldsymbol{m}_{\boldsymbol{\varepsilon}}[n] = \boldsymbol{m}_{\boldsymbol{x}}[n] - \boldsymbol{m}_{\hat{\boldsymbol{x}}}[n] = \boldsymbol{m}_{\boldsymbol{x}}[n] - \boldsymbol{W}^{\mathrm{H}}[n]\boldsymbol{m}_{\boldsymbol{y}}[n] - \boldsymbol{a}[n] = \boldsymbol{0}_M$ (cf. Equation 2.7).

$\hat{x}^{\circ}[n] = W^{\mathrm{H}}[n]y^{\circ}[n]$. To get finally an estimate of the non-zero-mean sequence $x[n]$, the mean $m_x[n]$ has to be added, i.e., $\hat{x}[n] = \hat{x}^{\circ}[n] + m_x[n]$. Again, $\varepsilon^{\circ}[n] = x^{\circ}[n] - \hat{x}^{\circ}[n] = \varepsilon[n]$. Note for later that

$$
\begin{aligned}
C_{y,x}[n] &= R_{y,x}[n] - m_y[n]m_x^{\mathrm{H}}[n] \\
&= C_{y^{\circ},x}[n] = R_{y^{\circ},x}[n] \\
&= C_{y^{\circ},x^{\circ}}[n] = R_{y^{\circ},x^{\circ}}[n], \\
C_y[n] &= C_{y^{\circ}}[n] = R_{y^{\circ}}[n].
\end{aligned}
$$

$$(2.10)$$
$$(2.11)$$

It remains to mention that the time-dependency of the filter coefficients $W[n]$ and $a[n]$ is necessary due to the assumption of non-stationary random sequences. However, at each time index n, the MWF can also be seen as an estimator for a set of random variables. If we consider the special case of stationary random sequences, the filter coefficients are independent of n and can be written as $W[n] = W$ and $a[n] = a$.

Note that we consider a MWF which estimates an M-dimensional signal $x[n]$. If $M = 1$, the cross-covariance matrix $C_{y,x}[n]$ shrinks to a cross-covariance vector and we end up with the *Vector Wiener Filter* (VWF) which produces a scalar output. Considering Equation (2.7), it is obvious that estimating each element of $x[n]$ with M parallel VWFs produces the same estimate $\hat{x}[n]$ as a MWF since the auto-covariance matrix $C_y[n]$ is the same and the cross-covariance vectors of the VWFs are the columns of the cross-covariance matrix $C_{y,x}[n]$.[5]

It remains to point out that the highest computational complexity when computing the MWF, arises from solving the Wiener–Hopf equation $C_y[n]W[n] = C_{y,x}[n]$. Hence, in the sequel of this chapter, we present methods to reduce this computational burden. The computational cost required to calculate $a[n]$ is no longer considered in the following investigations since it is equal for all presented methods.

2.2 Reduced-Complexity Matrix Wiener Filters

In this section, we investigate different methods to reduce the computational complexity of the MWF without loss of optimality. Here, we measure the computational complexity in terms of *FLoating point OPerations* (FLOPs) where an addition, subtraction, multiplication, division, or root extraction is counted as one FLOP, no matter if operating on real or complex values. Note that the number of FLOPs is a theoretical complexity measure which is widely used in numerical mathematics. In practical implementations however, the

---

[5] Throughout this book, the term VWF denotes a MWF with $M = 1$ if it is part of a filter bank estimating a vector signal with $M > 1$. If the VWF is used alone to estimate a single scalar signal, we denote it simply as the *Wiener Filter* (WF).

exact computational burden can differ from the number of FLOPs depending on the given processor architecture. For instance, a root extraction or division can be computationally much more intense than an addition. Anyway, since we want to present general results without any restriction to specific processors, we choose the number of FLOPs as the complexity measure.

The following reduced-complexity methods are all based on exploitations of special structures in the auto-covariance matrix, i. e., the system matrix of the Wiener–Hopf equation. Whereas Subsection 2.2.1 exploits the fact that the auto-covariance matrix is Hermitian and positive definite, Subsection 2.2.2 considers a special time-dependency of its submatrix structure. Besides these two given special structures, there exists auto-covariance matrices with further structural restrictions which can be used to get other reduced-complexity implementations. One example thereof is the *Toeplitz* or *block Toeplitz structure* [94] which can be exploited by using the computationally cheap *Fast Fourier Transform* (FFT) [75, 38, 114, 37] for the solution of the corresponding system of linear equations. The resulting method for Toeplitz matrices has been introduced by William F. Trench in 1964 and is today known as the *Trench algorithm* [238]. Extensions to the block Toeplitz case have been presented in [2, 255]. Since the system matrices in the application of Part II have neither Toeplitz nor block Toeplitz structure, the associated algorithms are no longer discussed here.

2.2.1 Implementation Based on Cholesky Factorization

Since we assume the Hermitian auto-covariance matrix $C_{\mathbf{y}}[n] \in \mathbb{C}^{N \times N}$ to be positive definite,[6] the *Cholesky Factorization* (CF) [11, 18, 94] can be used to solve the Wiener–Hopf equation $C_{\mathbf{y}}[n]W[n] = C_{\mathbf{y},\mathbf{x}}[n] \in \mathbb{C}^{N \times M}$ at each time index n. Due to simplicity, we omit the time index n in the following of this subsection although the matrices under consideration are still time-varying. The procedure to compute the MWF $W \in \mathbb{C}^{N \times M}$ is as follows (cf., e. g., [94]):

1. *Cholesky Factorization* (CF):
   Compute the lower triangular matrix $L \in \mathbb{C}^{N \times N}$ such that $C_{\mathbf{y}} = LL^{\mathrm{H}}$.
2. *Forward Substitution* (FS):
   Solve the system $LV = C_{\mathbf{y},\mathbf{x}} \in \mathbb{C}^{N \times M}$ of linear equations according to $V \in \mathbb{C}^{N \times M}$.
3. *Backward Substitution* (BS):
   Solve the system $L^{\mathrm{H}}W = V$ of linear equations with respect to W.

---

[6] Again, the special case where the auto-covariance matrix is rank-deficient and therefore no longer positive definite but semidefinite, is excluded in the considerations of this book because it has no practical relevance. In the perturbed linear system model of Chapter 5, we will see that this special case corresponds to the marginal case where the perturbance has variance zero.

The CF computes the decomposition $\boldsymbol{C_y} = \boldsymbol{L}\boldsymbol{L}^{\mathrm{H}}$ in an iterative manner. For its derivation, we firstly partition the auto-covariance matrix $\boldsymbol{C_y}$ according to

$$\boldsymbol{C_y} = \begin{bmatrix} d_{0,0} & \boldsymbol{d}_{1,0}^{\mathrm{H}} \\ \boldsymbol{d}_{1,0} & \boldsymbol{D}_{1,1} \end{bmatrix}, \tag{2.12}$$

where $d_{0,0} \in \mathbb{R}_+$, $\boldsymbol{d}_{1,0} \in \mathbb{C}^{N-1}$, and the Hermitian and positive definite matrix $\boldsymbol{D}_{1,1} \in \mathbb{C}^{(N-1)\times(N-1)}$,[7] and the lower triangular matrix $\boldsymbol{L}$ according to

$$\boldsymbol{L} = \begin{bmatrix} \ell_{0,0} & \boldsymbol{0}_{N-1}^{\mathrm{T}} \\ \boldsymbol{l}_{1,0} & \boldsymbol{L}_{1,1} \end{bmatrix}, \tag{2.13}$$

where $\ell_{0,0} \in \mathbb{R}_+$, $\boldsymbol{l}_{1,0} \in \mathbb{C}^{N-1}$, and $\boldsymbol{L}_{1,1} \in \mathbb{C}^{(N-1)\times(N-1)}$ which is a lower triangular matrix of reduced dimension. If we apply the partitioned matrices to $\boldsymbol{C_y} = \boldsymbol{L}\boldsymbol{L}^{\mathrm{H}}$, we get

$$\begin{bmatrix} d_{0,0} & \boldsymbol{d}_{1,0}^{\mathrm{H}} \\ \boldsymbol{d}_{1,0} & \boldsymbol{D}_{1,1} \end{bmatrix} = \begin{bmatrix} \ell_{0,0}^2 & \ell_{0,0}\boldsymbol{l}_{1,0}^{\mathrm{H}} \\ \ell_{0,0}\boldsymbol{l}_{1,0} & \boldsymbol{l}_{1,0}\boldsymbol{l}_{1,0}^{\mathrm{H}} + \boldsymbol{L}_{1,1}\boldsymbol{L}_{1,1}^{\mathrm{H}} \end{bmatrix}, \tag{2.14}$$

which can be rewritten in the following three fundamental equations:

$$\ell_{0,0} = \sqrt{d_{0,0}}, \tag{2.15}$$

$$\boldsymbol{l}_{1,0} = \boldsymbol{d}_{1,0}/\ell_{0,0}, \tag{2.16}$$

$$\boldsymbol{L}_{1,1}\boldsymbol{L}_{1,1}^{\mathrm{H}} = \boldsymbol{D}_{1,1} - \boldsymbol{l}_{1,0}\boldsymbol{l}_{1,0}^{\mathrm{H}}. \tag{2.17}$$

With Equations (2.15) and (2.16), we already computed the first column of the lower triangular matrix $\boldsymbol{L}$. The remaining Equation (2.17) defines the CF of the Hermitian and positive definite $(N-1) \times (N-1)$ matrix $\boldsymbol{D}_{1,1} - \boldsymbol{l}_{1,0}\boldsymbol{l}_{1,0}^{\mathrm{H}} =: \boldsymbol{D}_1$. The principal idea of the Cholesky algorithm is to apply again the procedure explained above in order to compute the CF $\boldsymbol{D}_1 = \boldsymbol{L}_{1,1}\boldsymbol{L}_{1,1}^{\mathrm{H}}$. For $i \in \{1, 2, \ldots, N-2\}$, $N > 2$, the Hermitian and positive definite matrix $\boldsymbol{D}_i \subset \mathbb{C}^{(N-i)\times(N-i)}$ and the lower triangular matrix $\boldsymbol{L}_{i,i} \in \mathbb{C}^{(N-i)\times(N-i)}$ are partitioned according to

$$\boldsymbol{D}_i = \begin{bmatrix} d_{i,i} & \boldsymbol{d}_{i+1,i}^{\mathrm{H}} \\ \boldsymbol{d}_{i+1,i} & \boldsymbol{D}_{i+1,i+1} \end{bmatrix}, \quad \boldsymbol{L}_{i,i} = \begin{bmatrix} \ell_{i,i} & \boldsymbol{0}_{N-i-1}^{\mathrm{T}} \\ \boldsymbol{l}_{i+1,i} & \boldsymbol{L}_{i+1,i+1} \end{bmatrix}, \tag{2.18}$$

where $d_{i,i}, \ell_{i,i} \in \mathbb{R}_+$, $\boldsymbol{d}_{i+1,i}, \boldsymbol{l}_{i+1,i} \in \mathbb{C}^{N-i-1}$, and $\boldsymbol{D}_{i+1,i+1}, \boldsymbol{L}_{i+1,i+1} \in \mathbb{C}^{(N-i-1)\times(N-i-1)}$. The application to the CF $\boldsymbol{D}_i = \boldsymbol{L}_{i,i}\boldsymbol{L}_{i,i}^{\mathrm{H}}$ yields the following three equations:

---

[7] The reason why $d_{0,0}$ is positive real-valued and $\boldsymbol{D}_{1,1}$ is Hermitian and positive definite, is as follows: Any principal submatrix $\boldsymbol{A}' \in \mathbb{C}^{m\times m}$, i.e., a submatrix in the main diagonal, of a Hermitian and positive definite matrix $\boldsymbol{A} \in \mathbb{C}^{n\times n}$, $n \geq m$, is again Hermitian and positive definite. This can be easily verified by replacing $\boldsymbol{x} \in \mathbb{C}^n$ in $\boldsymbol{x}^{\mathrm{H}}\boldsymbol{A}\boldsymbol{x} > 0$ by $[\boldsymbol{0}^{\mathrm{T}}, \boldsymbol{x}'^{,\mathrm{T}}, \boldsymbol{0}^{\mathrm{T}}]^{\mathrm{T}}$ with $\boldsymbol{x}' \in \mathbb{C}^m$, such that $\boldsymbol{x}^{\mathrm{H}}\boldsymbol{A}\boldsymbol{x} = \boldsymbol{x}'^{,\mathrm{H}}\boldsymbol{A}'\boldsymbol{x}' > 0$.

$$\ell_{i,i} = \sqrt{d_{i,i}}, \tag{2.19}$$

$$l_{i+1,i} = d_{i+1,i}/\ell_{i,i}, \tag{2.20}$$

$$L_{i+1,i+1}L_{i+1,i+1}^{\mathrm{H}} = D_{i+1,i+1} - l_{i+1,i}l_{i+1,i}^{\mathrm{H}} =: D_{i+1}. \tag{2.21}$$

Hence, with Equations (2.19) and (2.20), we calculated the $(i+1)$th column of L and Equation (2.21) defines the CF of the Hermitian and positive definite $(N-i-1)\times(N-i-1)$ matrix $D_{i+1} = L_{i+1,i+1}L_{i+1,i+1}^{\mathrm{H}}$. In other words, at each step i, one additional column of the lower triangular matrix L is computed and the CF problem is reduced by one dimension. Finally, at the last step, i. e., for $i = N - 1$, the reduced-dimension CF is $D_{N-1} = L_{N-1,N-1}L_{N-1,N-1}^{\mathrm{H}}$ where both D_{N-1} and $L_{N-1,N-1}$ are real-valued scalars. With $d_{N-1} = D_{N-1} \in \mathbb{R}_+$ and $\ell_{N-1,N-1} = L_{N-1,N-1} \in \mathbb{R}_+$, we get the remaining element in the last column of L, i. e.,

$$\ell_{N-1,N-1} = \sqrt{d_{N-1}}. \tag{2.22}$$

Algorithm 2.1. summarizes Equations (2.15), (2.19), (2.22) in Line 3, Equations (2.16) and (2.20) in Line 5, and Equations (2.17) and (2.21) in the **for**-loop from Line 6 to 8. The computationally efficient **for**-loop implementation of Equations (2.17) and (2.21) exploits the Hermitian property of D_{i+1} by computing only the lower triangular part[8] thereof. Besides, at the end of each step $i \in \{0, 1, \ldots, N-1\}$, the lower triangular part of the matrix D_{i+1} is stored in the matrix $L_{i+1,i+1}$, i. e., the part of L which has not yet been computed. This memory efficient procedure is initiated by storing the lower triangular part of the auto-covariance matrix C_y in L (Line 1 of Algorithm 2.1.). Note that $[A]_{n:m,i:j}$ denotes the $(m-n+1)\times(j-i+1)$ submatrix of A including the elements of A from the $(n+1)$th to the $(m+1)$th row and from the $(i+1)$th to the $(j+1)$th column. Further, if we want access to a single row or column of a matrix, respectively, we use the abbreviation 'n' instead of writing '$n : n$'. Thus, $\ell_{i,i} = [L]_{i,i}$, $l_{i+1,i} = [L]_{i+1:N-1,i}$, and $L_{i+1,i+1} = [L]_{i+1:N-1,i+1:N-1}$.

Algorithm 2.1. Cholesky Factorization (CF)

> Choose L as the lower triangular part of C_y.
> 2: **for** $i = 0, 1, \ldots, N - 1$ **do**
> $\qquad [L]_{i,i} = \sqrt{[L]_{i,i}}$ $\qquad\qquad\langle 1\rangle$
> 4: $\quad$ **if** $i < N - 1$ **then**
> $\qquad\quad [L]_{i+1:N-1,i} \leftarrow [L]_{i+1:N-1,i} / [L]_{i,i}$ $\qquad\langle N - i - 1\rangle$
> 6: $\qquad$ **for** $j = i + 1, i + 2, \ldots, N - 1$ **do**
> $\qquad\qquad [L]_{j:N-1,j} \leftarrow [L]_{j:N-1,j} - [L]_{j,i}^{*}[L]_{j:N-1,i}$ $\qquad\langle 2(N - j)\rangle$
> 8: $\qquad$ **end for**
> $\qquad$ **end if**
> 10: **end for**

---

[8] Here, the lower triangular part of a matrix includes its main diagonal.

By the way, Algorithm 2.1. is the *outer product version* [94] of the CF which is one of its computationally cheapest implementations. In Algorithm 2.1., the number in angle brackets are the FLOPs required to calculate the corresponding line of the algorithm. After summing up, the number of FLOPs required to perform Algorithm 2.1. is

$$\zeta_{\mathrm{CF}}(N) = \sum_{\substack{i=0 \\ N>1}}^{N-2} \left(N - i + \sum_{j=i+1}^{N-1} 2(N-j) \right) + 1 \tag{2.23}$$

$$= \frac{1}{3}N^3 + \frac{1}{2}N^2 + \frac{1}{6}N.$$

The next step is FS, i. e., to compute the solution of the system $LV = C_{y,x}$ of linear equations. For $i \in \{0, 1, \ldots, N-1\}$, the $(i+1)$th row $[L]_{i,:}V = [C_{y,x}]_{i,:}$ of the system can be written as

$$[C_{y,x}]_{i,:} = [L]_{i,0:i}[V]_{0:i,:} \tag{2.24}$$

$$= \begin{cases} [L]_{0,0}[V]_{0,:}, & i = 0, \\ [L]_{i,i}[V]_{i,:} + [L]_{i,0:i-1}[V]_{0:i-1,:}, & i \in \{1, 2, \ldots, N-1\}, \ N > 1, \end{cases}$$

if we exploit the lower triangular structure of L, i. e., the fact that the elements of the $(i+1)$th row with an index higher than i are zero, and where we separated $[L]_{i,i}[V]_{i,:}$ from the given sum. Note that ':' is the abbreviation for '$0:z$' where z is the maximum value of the index under consideration. If we solve Equation (2.24) according to $[V]_{i,:}$, we end up with Algorithm 2.2..

Algorithm 2.2. Forward Substitution (FS)

$\quad [V]_{0,:} \leftarrow [C_{y,x}]_{0,:}/[L]_{0,0}$ $\hfill \langle M \rangle$
2: **for** $i = 1, 2, \ldots, N-1 \land N > 1$ **do**
$\quad [V]_{i,:} \leftarrow \left([C_{y,x}]_{i,:} - [L]_{i,0:i-1}[V]_{0:i-1,:}\right)/[L]_{i,i}$ $\hfill \langle M(2i+1) \rangle$
4: **end for**

For the following complexity investigation, note that the scalar product of two m-dimensional vectors requires m multiplications and $m-1$ additions, thus, $2m-1$ FLOPs. Consequently, the matrix-matrix product of an $n \times m$ matrix by an $m \times p$ matrix has a computational complexity of $np(2m-1)$ FLOPs. Thus, Algorithm 2.2. needs $M(2i+1)$ FLOPs at each step i which results in the total computational complexity of

$$\zeta_{\mathrm{FS}}(N, M) = M + \sum_{\substack{i=1 \\ N>1}}^{N-1} M(2i+1) = MN^2 \quad \text{FLOPs.} \tag{2.25}$$

Finally, the system $L^{\mathrm{H}}W = V$ of linear equations is solved based on BS. The $(i+1)$th column of $L^{\mathrm{H}}W = V$ can be written as

$$[\boldsymbol{V}]_{i,:} = \left[\boldsymbol{L}^{\mathrm{H}}\right]_{i,:}\boldsymbol{W} = \left[\boldsymbol{L}^{\mathrm{H}}\right]_{i,i:N-1}[\boldsymbol{W}]_{i:N-1,:} \tag{2.26}$$

$$= \begin{cases} [\boldsymbol{L}]_{i,i}[\boldsymbol{W}]_{i,:} + [\boldsymbol{L}]_{i+1:N-1,i}^{\mathrm{H}}[\boldsymbol{W}]_{i+1:N-1,:}, & i \in \{0,1,\ldots,N-2\},\ N>1 \\ [\boldsymbol{L}]_{N-1,N-1}[\boldsymbol{W}]_{N-1,:}, & i = N-1, \end{cases}$$

exploiting the upper triangular structure of the matrix $\boldsymbol{L}^{\mathrm{H}}$, i.e., the fact that the non-zero elements in its $(i+1)$th row have an index greater or equal to i. Remember that $[\boldsymbol{L}^{\mathrm{H}}]_{i,i} = [\boldsymbol{L}]_{i,i} \in \mathbb{R}_+$ and $[\boldsymbol{L}^{\mathrm{H}}]_{i,i+1:N-1} = [\boldsymbol{L}]_{i+1:N-1,i}^{\mathrm{H}}$. Solving Equation (2.26) with respect to $[\boldsymbol{W}]_{i,:}$ yields the fundamental line of Algorithm 2.3..

Algorithm 2.3. Backward Substitution (BS)

$$[\boldsymbol{W}]_{N-1,:} \leftarrow [\boldsymbol{V}]_{N-1,:}/[\boldsymbol{L}]_{N-1,N-1} \qquad\qquad \langle M \rangle$$

2: **for** $i = N-2, N-3, \ldots, 0 \wedge N > 1$ **do**

$$[\boldsymbol{W}]_{i,:} \leftarrow \left([\boldsymbol{V}]_{i,:} - [\boldsymbol{L}]_{i+1:N-1,i}^{\mathrm{H}}[\boldsymbol{W}]_{i+1:N-1,:}\right)/[\boldsymbol{L}]_{i,i} \qquad \langle M(2(N-i)-1) \rangle$$

4: **end for**

The resulting number of FLOPs required to perform Algorithm 2.3. is

$$\zeta_{\mathrm{BS}}(N,M) = M + \sum_{\substack{i=N-2 \\ N>1}}^{0} M(2(N-i)-1) = \sum_{\substack{i=1 \\ N>1}}^{N-1} M(2i+1) = MN^2, \tag{2.27}$$

i.e., FS and BS have the same computational complexity.

If we combine Algorithm 2.1., 2.2., and 2.3., we end up with a method for calculating the MWF and the resulting total number of FLOPs $\zeta_{\mathrm{MWF}}(N,M)$ computes as

$$\left| \begin{aligned} \zeta_{\mathrm{MWF}}(N,M) &= \zeta_{\mathrm{CF}}(N) + \zeta_{\mathrm{FS}}(N,M) + \zeta_{\mathrm{BS}}(N,M) \\ &= \frac{1}{3}N^3 + \left(2M + \frac{1}{2}\right)N^2 + \frac{1}{6}N, \end{aligned} \right. \tag{2.28}$$

which is of cubic order if $N \to \infty$. Note that this computational burden occurs at each time index n if the MWF is time-variant. Remember that this is the case in our considerations although we omitted the index n due to simplicity.

2.2.2 Exploitation of Structured Time-Dependency

In this subsection, we consider the case where the time-dependency of the auto-covariance matrix $\boldsymbol{C_y}[n]$ is only restricted to some submatrices thereof. By exploiting the structured time-dependency, we will reduce computational complexity by approximately one order at each time index n.

We assume that the auto-covariance matrices $C_y[i]$ and their inverses $C_y^{-1}[i]$ at time indices $i = n - 1$ and $i = n$ are partitioned according to

$$C_y[i] = \begin{bmatrix} D[i] & E[i] \\ E^{\mathrm{H}}[i] & F[i] \end{bmatrix}, \quad C_y^{-1}[i] = \begin{bmatrix} \bar{D}[i] & \bar{E}[i] \\ \bar{E}^{\mathrm{H}}[i] & \bar{F}[i] \end{bmatrix}, \quad i \in \{n-1, n\}, \quad (2.29)$$

where $D[n]$, $F[n-1]$, $\bar{D}[n]$, and $\bar{F}[n-1]$ are $M \times M$ matrices, $E[n]$, $E^{\mathrm{H}}[n-1]$, $\bar{E}[n]$, and $\bar{E}^{\mathrm{H}}[n-1]$ are $M \times (N-M)$ matrices, and $F[n]$, $D[n-1]$, $\bar{F}[n]$, and $\bar{D}[n-1]$ are $(N-M) \times (N-M)$ matrices. Note that the definitions in Equation (2.29) for $i \in \{n-1, n\}$ is a compact notation for two types of decompositions with matrices of different sizes which is visualized exemplarily for the auto-covariance matrix in Figure 2.2. The inverses of the auto-covariance matrix $C_y[i]$ for $i \in \{n-1, n\}$ have the same structure as given in Figure 2.2 but the submatrices are denoted by an additional bar.

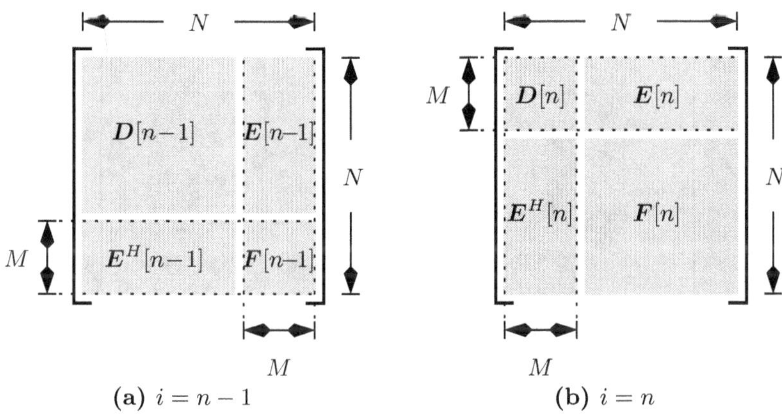

(a) $i = n - 1$ (b) $i = n$

Fig. 2.2. Structure of the auto-covariance matrix $C_y[i]$ for $i \in \{n - 1, n\}$

The following *Reduced Complexity* (RC) implementation of the time variant MWF $W[n]$ which is an extension of the RC method presented in [240], is based on the assumption that the previous auto-covariance matrix $C_y[n-1]$ and the current auto-covariance matrix $C_y[n]$ are related to each other via

$$\boxed{F[n] = D[n - 1].} \quad (2.30)$$

In Chapter 5, we will see that this relation occurs in common system models used for mobile communications. We reduce computational complexity by avoiding the explicit inversion of $C_y[n]$ at each time step. Precisely speaking, we exploit the relation in Equation (2.30) in order compute $C_y^{-1}[n]$, viz., $\bar{D}[n]$, $\bar{E}[n]$, and $\bar{F}[n]$, using the block elements of $C_y[n]$ and the blocks of the already computed inverse $C_y^{-1}[n-1]$ from the previous time index. According to Appendix A.1, we get the following dependencies:[9]

$$\bar{D}[n] = \left(D[n] - E[n]F^{-1}[n]E^{\mathrm{H}}[n]\right)^{-1}, \tag{2.31}$$

$$\bar{E}[n] = -\bar{D}[n]E[n]F^{-1}[n], \tag{2.32}$$

$$\bar{F}[n] = F^{-1}[n] - F^{-1}[n]E^{\mathrm{H}}[n]\bar{E}[n]. \tag{2.33}$$

Furthermore, the inverse of $F[n]$ can be expressed by the *Schur complement* (cf. Appendix A.1) of $\bar{D}[n-1]$ in $C_y^{-1}[n-1]$, i.e.,

$$F^{-1}[n] = D^{-1}[n-1] = \bar{D}[n-1] - \bar{E}[n-1]\bar{F}^{-1}[n-1]\bar{E}^{\mathrm{H}}[n-1]. \tag{2.34}$$

Equations (2.31), (2.32), (2.33), and (2.34) are summarized in Algorithm 2.4. which computes efficiently the inverse of the auto-covariance matrix based on its inverse from the previous time step. Note that we use the following abbreviations:

$$\bar{D}' := \bar{D}[n-1] \in \mathbb{C}^{(N-M)\times(N-M)}, \quad \bar{D} := \bar{D}[n] \in \mathbb{C}^{M\times M}, \tag{2.35}$$

$$\bar{E}' := \bar{E}[n-1] \in \mathbb{C}^{(N-M)\times M}, \quad\quad\;\; \bar{E} := \bar{E}[n] \in \mathbb{C}^{M\times(N-M)}, \tag{2.36}$$

$$\bar{F}' := \bar{F}[n-1] \in \mathbb{C}^{M\times M}, \quad\quad\quad\quad\; \bar{F} := \bar{F}[n] \in \mathbb{C}^{(N-M)\times(N-M)}, \tag{2.37}$$

$$D := D[n] \in \mathbb{C}^{M\times M}, \tag{2.38}$$

$$E := E[n] \in \mathbb{C}^{M\times(N-M)}, \tag{2.39}$$

$$F := F[n] \in \mathbb{C}^{(N-M)\times(N-M)}, \tag{2.40}$$

The matrix $V \in \mathbb{C}^{M\times(N-M)}$ in Line 7 of Algorithm 2.4. has been introduced to further increase computational efficiency.

It remains to investigate the computational complexity of Algorithm 2.4.. Again, the number of FLOPs for each line are given in angle brackets. Apart from the already given complexity of a matrix-matrix product (cf. Subsection 2.2.1), we use the number of FLOPs of the following fundamental operations:

- The Cholesky factorization of a Hermitian and positive definite $n \times n$ matrix requires $\zeta_{\mathrm{CF}}(n) = n^3/3 + n^2/2 + n/6$ FLOPs (cf. Equation 2.23 of Subsection 2.2.1).
- The inversion of a lower triangular $n \times n$ matrix requires $\zeta_{\mathrm{LTI}}(n) = n^3/3 + 2n/3$ FLOPs (cf. Equation B.4 of Appendix B.1).
- The product of a lower triangular $n \times n$ matrix by an arbitrary $n \times m$ matrix requires $m \sum_{i=0}^{n-1}(2(n-i)-1) = mn^2$ FLOPs since the calculation of the $(i+1)$-th row of the result involves the computation of m scalar products between two $(n-i)$-dimensional vectors for all $i \in \{0,1,\ldots,n-1\}$.

---

[9] The solution can be easily derived by rewriting the identities $C_y[i]C_y^{-1}[i] = I_N$, $i \in \{n, n-1\}$, into eight equations using the block structures defined in Equation (2.29).

Algorithm 2.4. Reduced-Complexity (RC) MWF exploiting time-dependency

$$\bar{D}' \leftarrow [C_y^{-1}[n-1]]_{0:N-M-1,0:N-M-1}$$
2: $\bar{E}' \leftarrow [C_y^{-1}[n-1]]_{0:N-M-1,N-M:N-1}$
$$\bar{F}' \leftarrow [C_y^{-1}[n-1]]_{N-M:N-1,N-M:N-1}$$
4: $D \leftarrow [C_y[n]]_{0:M-1,0:M-1}$
$$E \leftarrow [C_y[n]]_{0:M-1,M:N-1}$$
6: $F^{-1} \leftarrow \bar{D}' - \bar{E}'\bar{F}'^{,-1}\bar{E}'^{,\mathrm{H}}$ $\qquad\langle 2M^3/3 + M^2/2 + 5M/6 + M(N-M)(N+1)\rangle$
$\quad V \leftarrow EF^{-1}$ $\qquad\qquad\qquad\qquad\qquad\langle M(N-M)(2(N-M)-1)\rangle$
8: $\bar{D} \leftarrow (D - VE^{\mathrm{H}})^{-1}$ $\qquad\qquad\qquad\langle\zeta_{\mathrm{HI}}(M) + M(M+1)(N-M)\rangle$
$\quad \bar{E} \leftarrow -\bar{D}V$ $\qquad\qquad\qquad\qquad\qquad\langle M(2M-1)(N-M)\rangle$
10: $\bar{F} \leftarrow F^{-1} - V^{\mathrm{H}}\bar{E}$ $\qquad\qquad\qquad\langle M(N-M)(N-M+1)\rangle$
$$C_y^{-1}[n] \leftarrow \begin{bmatrix} \bar{D} & \bar{E} \\ \bar{E}^{\mathrm{H}} & \bar{F} \end{bmatrix}$$
12: $W[n] \leftarrow C_y^{-1}[n]C_{y,x}[n]$ $\qquad\qquad\qquad\qquad\qquad\langle MN(2N-1)\rangle$

- The Hermitian product of an $n \times m$ and $m \times n$ matrix requires $(2m-1)n(n+1)/2$ FLOPs where we exploit the fact that only the lower triangular part of the result has to be computed.[10]
- The subtraction of two Hermitian $n \times n$ matrices requires $n(n+1)/2$ FLOPs for the same reason as above.
- The inversion of a Hermitian and positive definite $n \times n$ matrix requires $\zeta_{\mathrm{HI}}(n) = n^3 + n^2 + n$ FLOPs (cf. Equation B.7 of Appendix B.1).

Note that Line 6 of Algorithm 2.4. is computed in the following computational efficient way: After the Cholesky factorization of $\bar{F}' = LL^{\mathrm{H}} \in \mathbb{C}^{M \times M}$ according to Algorithm 2.1., the resulting lower triangular matrix $L \in \mathbb{C}^{M \times M}$ is inverted using Algorithm B.1., and finally, the Gramian matrix $\bar{E}'\bar{F}'^{,-1}\bar{E}'^{,\mathrm{H}} \in \mathbb{C}^{(N-M) \times (N-M)}$ is generated by multiplying the Hermitian of the product $L^{-1}\bar{E}'^{,\mathrm{H}} \in \mathbb{C}^{M \times (N-M)}$ to itself and exploiting the Hermitian structure of the result. Besides, it should be mentioned that Lines 8 and 9 could also be computed by solving the system $(D - VE^{\mathrm{H}})\bar{E} = -V \in \mathbb{C}^{M \times (N-M)}$ of linear equations according to $\bar{E} \in \mathbb{C}^{M \times (N-M)}$ based on the Cholesky method as described in the previous subsection. However, for $N > 2M^2/3 - M/2 + 5/6$, i. e., for large N, the direct inversion as performed in Line 8 is computationally cheaper (cf. the end of Appendix B.1).

Finally, after summing up, the RC MWF implementation in Algorithm 2.4. has a total computational complexity of

$$\zeta_{\mathrm{RCMWF}}(N, M) = 6MN^2 - 4M^2N + \frac{5}{3}M^3 + \frac{1}{2}M^2 + \frac{11}{6}M$$
$$+ \frac{1}{S}\left(N^3 + N^2 + N\right), \tag{2.41}$$

---

[10] Instead of computing all n^2 elements of the Hermitian $n \times n$ matrix, it suffices to compute the $n(n+1)/2$ elements of its lower triangular part.

if it is applied to estimate a signal of length S. The term with the S-fraction is due to the fact that the inverse of the $N \times N$ auto-covariance matrix $\boldsymbol{C_y}[n-1]$ which requires $\zeta_{\mathrm{HI}}(N) = N^3 + N^2 + N$ FLOPs (cf. Appendix B.1), has to be computed explicitly at the beginning of the signal estimation process since no inverse of the previous auto-covariance matrix is available there. Thus, only for long signals, i.e., large S, the third order term in Equation (2.41) can be neglected and the order of computational complexity for $N \to \infty$ is two, i.e., one smaller than the complexity order of the MWF implementation based on the CF as presented in Subsection 2.2.1.

2.3 Reduced-Rank Matrix Wiener Filters

In the preceding subsection, we discussed RC implementations of the MWF which still perform optimally in the MSE sense. Now, we follow the idea to approximate the MWF with a performance still very close to the optimal linear matrix filter but where the calculation of the filter coefficients is computationally more efficient. In the sequel, we concentrate on MWF approximations based on *reduced-rank signal processing* as introduced in [242, 241, 213, 52]. Besides the computational efficiency of reduced-rank signal processing, it can even increase the performance of the filter if the estimation of the associated statistics has a low sample support, i.e., it has errors since the number of available training samples is too small. This quality of noise reduction is especially apparent if the dimension of the signal subspace is smaller than the dimension of the observation space which is often denoted as an *overmodeled problem* (cf., e.g., [85]). Note that due to the *bias–variance trade-off* [213], the filter performance degrades again if its rank is chosen to be smaller than the optimal rank.

The basic idea of reduced-rank filtering as depicted in Figure 2.3, is to approximate the MWF $\boldsymbol{W}[n]$ and $\boldsymbol{a}[n]$ in a D-dimensional subspace $\mathcal{T}^{(D)}[n] \in G(N, D)$ where $G(N, D)$ denotes the (N, D)-*Grassmann manifold* (cf., e.g., [13]), i.e., the set of all D-dimensional subspaces in $\mathbb{C}^N$. To do so, we firstly prefilter the N-dimensional observed vector random sequence $\boldsymbol{y}[n]$ by the matrix $\boldsymbol{T}^{(D)}[n] \in \mathbb{C}^{N \times D}$, $D \in \{1, 2, \ldots, N\}$, with

$$\mathrm{range}\left\{\boldsymbol{T}^{(D)}[n]\right\} = \mathcal{T}^{(D)}[n] \in G(N, D), \tag{2.42}$$

leading to the reduced-dimension vector random sequence

$$\boldsymbol{y}^{(D)}[n] = \boldsymbol{T}^{(D),\mathrm{H}}[n]\boldsymbol{y}[n] \in \mathbb{C}^D. \tag{2.43}$$

In other words, we apply a coordinate transform where the elements of $\boldsymbol{y}^{(D)}[n]$ are the coordinates of the observed vector random sequence $\boldsymbol{y}[n]$ according to the basis defined by the columns of the prefilter matrix $\boldsymbol{T}^{(D)}[n]$. The second step is to apply the *reduced-dimension MWF* $\boldsymbol{G}^{(D)}[n] \in \mathbb{C}^{D \times M}$ and $\boldsymbol{a}^{(D)}[n] \in \mathbb{C}^M$ to $\boldsymbol{y}^{(D)}[n]$ in order to get the estimate

$$\hat{\boldsymbol{x}}^{(D)}[n] = \boldsymbol{G}^{(D),\mathrm{H}}[n]\boldsymbol{y}^{(D)}[n] + \boldsymbol{a}^{(D)}[n] \in \mathbb{C}^M, \tag{2.44}$$

of the unknown signal $\boldsymbol{x}[n] \in \mathbb{C}^M$, with the error $\boldsymbol{\varepsilon}^{(D)}[n] = \boldsymbol{x}[n] - \hat{\boldsymbol{x}}^{(D)}[n] \in \mathbb{C}^M$. Note that $\boldsymbol{G}^{(D)}[n]$ is a reduced-dimension MWF since its dimension is by a factor of N/D smaller than the dimension of the optimal MWF $\boldsymbol{W}[n] \in \mathbb{C}^{N \times M}$.

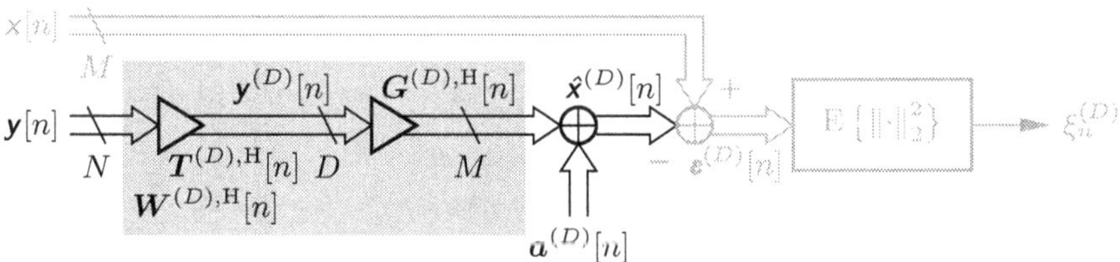

Fig. 2.3. Reduced-rank MWF

Using the MSE criterion as defined in Equation (2.4), the reduced-dimension MWF is the solution of the optimization

$$\left(\boldsymbol{G}^{(D)}[n], \boldsymbol{a}^{(D)}[n]\right) = \underset{(\boldsymbol{G}',\boldsymbol{a}') \in \mathbb{C}^{D \times M} \times \mathbb{C}^M}{\operatorname{argmin}} \xi_n\left(\boldsymbol{T}^{(D)}[n]\boldsymbol{G}', \boldsymbol{a}'\right). \tag{2.45}$$

With a similar derivation as given in Section 2.1, the reduced-dimension MWF computes as

$$\boldsymbol{G}^{(D)}[n] = \boldsymbol{C}_{\boldsymbol{y}^{(D)}}^{-1}[n]\boldsymbol{C}_{\boldsymbol{y}^{(D)},\boldsymbol{x}}[n]$$

$$= \left(\boldsymbol{T}^{(D),\mathrm{H}}[n]\boldsymbol{C}_{\boldsymbol{y}}[n]\boldsymbol{T}^{(D)}[n]\right)^{-1}\boldsymbol{T}^{(D),\mathrm{H}}[n]\boldsymbol{C}_{\boldsymbol{y},\boldsymbol{x}}[n], \tag{2.46}$$

$$\boldsymbol{a}^{(D)}[n] = \boldsymbol{m}_{\boldsymbol{x}}[n] - \boldsymbol{G}^{(D),\mathrm{H}}[n]\boldsymbol{T}^{(D),\mathrm{H}}[n]\boldsymbol{m}_{\boldsymbol{y}}[n]. \tag{2.47}$$

If we combine the prefilter matrix $\boldsymbol{T}^{(D)}[n]$ and the reduced-dimension MWF $\boldsymbol{G}^{(D)}[n]$ (cf. Figure 2.3), we obtain the *reduced-rank MWF* $\boldsymbol{W}^{(D)}[n] = \boldsymbol{T}^{(D)}[n]\boldsymbol{G}^{(D)}[n] \in \mathbb{C}^{N \times M}$ which has the same dimension as the optimal MWF $\boldsymbol{W}[n] \in \mathbb{C}^{N \times M}$. With Equations (2.46) and (2.47), the reduced-rank MWF can finally be written as

$$\left| \begin{aligned} &\boldsymbol{W}^{(D)}[n] - \boldsymbol{T}^{(D)}[n]\left(\boldsymbol{T}^{(D),\mathrm{H}}[n]\boldsymbol{C}_{\boldsymbol{y}}[n]\boldsymbol{T}^{(D)}[n]\right)^{-1}\boldsymbol{T}^{(D),\mathrm{H}}[n]\boldsymbol{C}_{\boldsymbol{y},\boldsymbol{x}}[n], \\ &\boldsymbol{a}^{(D)}[n] = \boldsymbol{m}_{\boldsymbol{x}}[n] - \boldsymbol{W}^{(D),\mathrm{H}}[n]\boldsymbol{m}_{\boldsymbol{y}}[n], \end{aligned} \right. \tag{2.48}$$

and achieves the MMSE

$$\xi_n^{(D)} = \xi_n\left(\boldsymbol{W}^{(D)}[n], \boldsymbol{a}^{(D)}[n]\right) = \operatorname{tr}\left\{\boldsymbol{C}_{\boldsymbol{\varepsilon}^{(D)}}[n]\right\}$$

$$= \operatorname{tr}\left\{\boldsymbol{C}_{\boldsymbol{x}}[n] - \boldsymbol{C}_{\boldsymbol{y},\boldsymbol{x}}^{\mathrm{H}}[n]\boldsymbol{T}^{(D)}[n]\boldsymbol{C}_{\boldsymbol{y}^{(D)}}^{-1}[n]\boldsymbol{T}^{(D),\mathrm{H}}[n]\boldsymbol{C}_{\boldsymbol{y},\boldsymbol{x}}[n]\right\}. \tag{2.49}$$

The solution $\boldsymbol{W}^{(D)}[n]$ and $\boldsymbol{a}^{(D)}[n]$ is called reduced-rank MWF (see, e.g., [85]) since it is based on a reduced-rank signal model [213]. However, some publications [233, 211, 135, 210, 64] refer the reduced-rank MWF to a rank-deficient approximation of the MWF which is included in our definition as the special case where $D < M$.

Note that in general, $\hat{\boldsymbol{x}}^{(D)}[n]$ is unequal to the estimate $\boldsymbol{x}[n]$ if $D < N$. However, in the special case where the prefilter subspace $\mathcal{T}^{(D)}[n]$ is equal to the subspace spanned by the columns of $\boldsymbol{W}[n]$, the reduced-rank MWF is equal to the full-rank or optimal MWF. Since finding such a subspace requires the same computational complexity as calculating the MWF itself, this special case has no practical relevance.

Proposition 2.1. *The reduced-rank MWF $\boldsymbol{W}^{(D)}[n]$ and $\boldsymbol{a}^{(D)}[n]$, thus, also the estimate $\hat{\boldsymbol{x}}^{(D)}[n]$, is invariant with respect to the basis $\boldsymbol{T}^{(D)}[n]$ spanning the subspace $\mathcal{T}^{(D)}[n]$. Hence, the performance of the reduced-rank MWF depends solely on the subspace $\mathcal{T}^{(D)}[n]$ and not on the basis thereof.*

Proof. Proposition 2.1 can be easily verified by replacing $\boldsymbol{T}^{(D)}[n]$ by an alternative basis of $\mathcal{T}^{(D)}[n]$. With $\boldsymbol{T}^{(D),\prime}[n] = \boldsymbol{T}^{(D)}[n]\boldsymbol{\Psi}$ where the matrix $\boldsymbol{\Psi} \in \mathbb{C}^{M \times M}$ is non-singular, i.e., $\mathrm{range}\{\boldsymbol{T}^{(D),\prime}[n]\} = \mathcal{T}^{(D)}[n]$, the resulting reduced-rank MWF $\boldsymbol{W}^{(D),\prime}[n]$ computes as (cf. Equation 2.48)

$$
\begin{aligned}
\boldsymbol{W}^{(D),\prime}[n] &= \boldsymbol{T}^{(D)}[n]\boldsymbol{\Psi} \left(\boldsymbol{\Psi}^{\mathrm{H}}\boldsymbol{T}^{(D),\mathrm{H}}[n]\boldsymbol{C_y}[n]\boldsymbol{T}^{(D)}[n]\boldsymbol{\Psi} \right)^{-1} \boldsymbol{\Psi}^{\mathrm{H}}\boldsymbol{T}^{(D),\mathrm{H}}[n]\boldsymbol{C_{y,x}}[n] \\
&= \boldsymbol{T}^{(D)}[n] \left(\boldsymbol{T}^{(D),\mathrm{H}}[n]\boldsymbol{C_y}[n]\boldsymbol{T}^{(D)}[n] \right)^{-1} \boldsymbol{T}^{(D),\mathrm{H}}[n]\boldsymbol{C_{y,x}}[n] \\
&= \boldsymbol{W}^{(D)}[n].
\end{aligned}
\tag{2.50}
$$

This proves the invariance of the reduced-rank MWF according to the basis of $\mathcal{T}^{(D)}[n]$. Furthermore, it shows that the reduced-rank MWF is the MMSE approximation of the MWF in the D-dimensional subspace $\mathcal{T}^{(D)}[n]$ spanned by the columns of the prefilter matrix $\boldsymbol{T}^{(D)}[n]$. $\qquad\square$

It remains to explain why the reduced-rank MWF is computationally efficient. All reduced-rank MWFs which are considered here, choose the prefilter matrix such that the structure of the auto-covariance matrix of the transformed vector random sequence $\boldsymbol{y}^{(D)}[n]$, i.e., $\boldsymbol{C_{y^{(D)}}}[n] = \boldsymbol{T}^{(D),\mathrm{H}}[n]\boldsymbol{C_y}[n]\boldsymbol{T}^{(D)}[n] \in \mathbb{C}^{D \times D}$, is such that the computational complexity involved in solving the reduced-dimension Wiener–Hopf equation $(\boldsymbol{T}^{(D),\mathrm{H}}[n]\boldsymbol{C_y}[n]\boldsymbol{T}^{(D)}[n])\boldsymbol{G}^{(D)}[n] = \boldsymbol{T}^{(D),\mathrm{H}}[n]\boldsymbol{C_{y,x}}[n]$ (cf. Equation 2.46) can be neglected compared to the computational complexity of the prefilter calculation. If the latter computational complexity is smaller than the one of the optimal MWF implementations as presented in Section 2.2, the reduced-rank MWF can be used as an computationally efficient approximation thereof.

Before we consider the approximation of the MWF in the Krylov subspace composed by the auto-covariance matrix $\boldsymbol{C_y}[n]$ and the cross-covariance matrix $\boldsymbol{C_{y,x}}[n]$ (see Chapter 3) in Chapter 4, we restrict ourselves in the next two subsections on the approximation of the MWF in eigensubspaces, i. e., the columns of the prefilter matrix $\boldsymbol{T}^{(D)}[n]$ are D eigenvectors of the auto-covariance matrix $\boldsymbol{C_y}[n]$.

2.3.1 Principal Component Method

The *Principal Component* (PC) *method* [132, 242, 241] chooses the columns of the prefilter matrix $\boldsymbol{T}^{(D)}[n] \in \mathbb{C}^{N \times D}$ such that the elements of the transformed observation vector random sequence $\boldsymbol{y}^{(D)}[n] \in \mathbb{C}^D$ are uncorrelated, i. e., its auto-covariance matrix $\boldsymbol{C_{y^{(D)}}}[n] = \boldsymbol{T}^{(D),\mathrm{H}}[n]\boldsymbol{C_y}[n]\boldsymbol{T}^{(D)}[n] \in \mathbb{C}^{D \times D}$ is diagonal, and where the limited degrees of freedom are used to preserve as much energy as possible. Due to the diagonal structure of the transformed auto-covariance matrix $\boldsymbol{C_{y^{(D)}}}[n]$, the reduced-dimension MWF decomposes into D scalar WFs whose computational complexity can be neglected compared to the calculation of the prefilter matrix. The PC prefiltering is equal to a reduced-dimension *Karhunen Loève Transform* (KLT) [147, 164] of the observed vector random sequence $\boldsymbol{y}[n] \in \mathbb{C}^N$.

For problems of finite dimension, the KLT can be computed based on the *eigen decomposition* (see, e. g., [131]) of the auto-covariance matrix $\boldsymbol{C_y}[n] \in \mathbb{C}^{N \times N}$. Recalling that $\boldsymbol{C_y}[n]$ is Hermitian and positive definite, we define

$$\boldsymbol{C_y}[n] = \sum_{i=0}^{N-1} \lambda_i[n]\boldsymbol{u}_i[n]\boldsymbol{u}_i^{\mathrm{H}}[n], \qquad (2.51)$$

with the orthonormal *eigenvectors* $\boldsymbol{u}_i[n] \in \mathbb{C}^N$, i. e., $\boldsymbol{u}_i^{\mathrm{H}}[n]\boldsymbol{u}_j[n] = \delta_{i,j}$, $i,j \in \{0, 1, \ldots, N-1\}$, corresponding to the real-valued *eigenvalues* $\lambda_0[n] \geq \lambda_1[n] \geq \ldots \geq \lambda_{N-1}[n] > 0$. Here, $\delta_{i,j}$ denotes the *Kronecker delta* being one for $i = j$ and zero otherwise.

In order to preserve as much signal energy as possible, the PC prefilter matrix $\boldsymbol{T}_{\mathrm{PC}}^{(D)}[n]$ is composed by the D eigenvectors corresponding to the D largest eigenvalues, i. e.,

$$\boldsymbol{T}_{\mathrm{PC}}^{(D)}[n] = [\boldsymbol{u}_0[n]\ \boldsymbol{u}_1[n]\ \cdots\ \boldsymbol{u}_{D-1}[n]] \in \mathbb{C}^{N \times D}. \qquad (2.52)$$

Hence, the transformed auto-covariance matrix computes as

$$\boldsymbol{C_{y_{\mathrm{PC}}^{(D)}}}[n] = \boldsymbol{T}_{\mathrm{PC}}^{(D),\mathrm{H}}[n]\boldsymbol{C_y}[n]\boldsymbol{T}_{\mathrm{PC}}^{(D)}[n] = \boldsymbol{\Lambda}^{(D)}[n] \in \mathbb{R}_{0,+}^{D \times D}, \qquad (2.53)$$

where $\boldsymbol{\Lambda}^{(D)}[n] = \mathrm{diag}\{\lambda_0[n], \lambda_1[n], \ldots, \lambda_{D-1}[n]\}$. If we replace the prefilter matrix in Equation (2.48) by $\boldsymbol{T}_{\mathrm{PC}}^{(D)}[n]$, we get the *PC based rank D MWF* with the filter coefficients $\boldsymbol{W}_{\mathrm{PC}}^{(D)}[n] \in \mathbb{C}^{N \times M}$ and $\boldsymbol{a}_{\mathrm{PC}}^{(D)}[n] \in \mathbb{C}^M$, achieving the MMSE

$$\begin{aligned}
\xi_{n,\mathrm{PC}}^{(D)} &= \xi_n\left(\boldsymbol{W}_{\mathrm{PC}}^{(D)}[n], \boldsymbol{a}_{\mathrm{PC}}^{(D)}[n]\right) \\
&= \mathrm{tr}\left\{\boldsymbol{C}_{\boldsymbol{x}}[n] - \boldsymbol{C}_{\boldsymbol{y},\boldsymbol{x}}^{\mathrm{H}}[n]\boldsymbol{T}_{\mathrm{PC}}^{(D)}[n]\boldsymbol{\Lambda}^{(D),-1}[n]\boldsymbol{T}_{\mathrm{PC}}^{(D),\mathrm{H}}[n]\boldsymbol{C}_{\boldsymbol{y},\boldsymbol{x}}[n]\right\} \\
&= \mathrm{tr}\left\{\boldsymbol{C}_{\boldsymbol{x}}[n]\right\} - \sum_{i=0}^{D-1}\frac{\left\|\boldsymbol{C}_{\boldsymbol{y},\boldsymbol{x}}^{\mathrm{H}}[n]\boldsymbol{u}_i[n]\right\|_2^2}{\lambda_i[n]}.
\end{aligned} \tag{2.54}$$

Since only the N-dimensional eigenvectors corresponding to the D principal eigenvalues must be computed and the inversion of the diagonal matrix $\boldsymbol{\Lambda}^{(D)}[n]$ is computational negligible, the PC based rank D MWF has a computational complexity of $\mathrm{O}(DN^2)$ where '$\mathrm{O}(\cdot)$' denotes the Landau symbol. Note that the D principal eigenvectors of a Hermitian matrix can be computed using the *Lanczos algorithm* (cf. Section 3.3) which is used to firstly tridiagonalize the auto-covariance matrix because the complexity of computing the eigenvectors of a tridiagonal matrix has an negligible order (see, e.g., [237]). The investigation of the exact number of FLOPs required to calculate the PC based rank D MWF is not considered here since we will see in the application of Part II that the eigensubspace based MWFs have no practical relevance because they perform much worse than the Krylov subspace based MWFs introduced in Chapter 4.

Note that the parallel application of reduced-rank VWFs based on the PC method yields again the same estimate as the application of the reduced-rank MWF $\boldsymbol{W}_{\mathrm{PC}}^{(D)}[n]$ and $\boldsymbol{a}_{\mathrm{PC}}^{(D)}[n]$ because the prefilter matrix $\boldsymbol{T}_{\mathrm{PC}}^{(D)}[n]$ does not depend on cross-covariance information and therefore, the term $\boldsymbol{T}_{\mathrm{PC}}^{(D)}[n](\boldsymbol{T}_{\mathrm{PC}}^{(D),\mathrm{H}}[n]\boldsymbol{C}_{\boldsymbol{y}}[n]\boldsymbol{T}_{\mathrm{PC}}^{(D)}[n])^{-1}\boldsymbol{T}_{\mathrm{PC}}^{(D),\mathrm{H}}[n]$ in Equation (2.48) is identical for the equalization of each element of $\boldsymbol{x}[n]$.

2.3.2 Cross-Spectral Method

Whereas the principles of the *Cross-Spectral* (CS) *method* has been introduced by Kent A. Byerly et al. [24] in 1989, the notation is due to J. Scott Goldstein et al. [85, 86, 87]. As the PC method, the CS approach is based on the KLT of the observed vector random sequence $\boldsymbol{y}[n] \in \mathbb{C}^N$. Remember that the PC based reduced-rank MWF is an approximation of the MWF in the D-dimensional eigensubspace with the largest signal energy no matter if the energy belongs to the signal of interest or to an interference signal. Thus, in systems with strong interferers, the performance of the PC based MWF degrades drastically. The CS method solves this main problem of the PC approach by choosing the columns of its prefilter matrix $\boldsymbol{T}_{\mathrm{CS}}^{(D)}[n] \in \mathbb{C}^{N \times D}$ to be the D eigenvectors achieving the smallest MSE of all eigensubspace based reduced-rank MWFs.

To do so, we define the *CS metric* $\vartheta_i[n] \in \mathbb{R}_{0,+}$ corresponding to the eigenvector $\boldsymbol{u}_i[n] \in \mathbb{C}^N$, $i \in \{0, 1, \dots, N-1\}$, as its contribution to minimize the MSE (cf. Equation 2.54), i.e.,

$$\vartheta_i[n] = \frac{\left\| C_{\mathbf{y},\mathbf{x}}^{\mathrm{H}}[n]\mathbf{u}_i[n]\right\|_2^2}{\lambda_i[n]}. \tag{2.55}$$

If we choose the D eigenvectors with the largest CS metric to be the columns of the CS prefilter matrix $\boldsymbol{T}_{\mathrm{CS}}^{(D)}[n]$, i. e.,

$$\left| \boldsymbol{T}_{\mathrm{CS}}^{(D)}[n] = [\boldsymbol{u}_i[n]]_{i\in\mathbb{S}}, \quad \mathbb{S} = \operatorname*{argmax}_{\substack{\mathbb{S}'\subset\{0,1,\ldots,N-1\}\\ |\mathbb{S}'|=D}} \sum_{i\in\mathbb{S}'} \vartheta_i[n], \right| \tag{2.56}$$

the resulting *CS based rank D MWF* with the filter coefficients $\boldsymbol{W}_{\mathrm{CS}}^{(D)}[n] \in \mathbb{C}^{N\times M}$ and $\boldsymbol{a}_{\mathrm{CS}}^{(D)}[n] \in \mathbb{C}^M$ (cf. Equation 2.48) achieves the MMSE (see also Equation 2.54)

$$\xi_{n,\mathrm{CS}}^{(D)} = \xi_n\left(\boldsymbol{W}_{\mathrm{CS}}^{(D)}[n], \boldsymbol{a}_{\mathrm{CS}}^{(D)}[n]\right) = \operatorname{tr}\{\boldsymbol{C}_{\mathbf{x}}[n]\} - \sum_{i\in\mathbb{S}} \frac{\left\| C_{\mathbf{y},\mathbf{x}}^{\mathrm{H}}[n]\mathbf{u}_i[n]\right\|_2^2}{\lambda_i[n]}, \tag{2.57}$$

i. e., the smallest MSE which can be achieved by an approximation of the MWF in a D-dimensional eigensubspace. $|\mathbb{S}|$ denotes the cardinality of the set $\mathbb{S}$ and $[\boldsymbol{a}_i]_{i\in\mathbb{S}}$ is a matrix with the columns $\boldsymbol{a}_i$, $i \in \mathbb{S}$.

Note that we have to compute all eigenvectors in order to determine the corresponding CS metrics which are needed to choose finally the D eigenvectors with the largest sum of metrics. Thus, the complexity order of the CS method is cubic, i. e., there is no reduction in computational complexity compared to the optimal MWF as derived in Subsection 2.2.1. Due to this reason, we present the CS based reduced-rank MWF only as a bound for the best-possible MWF approximation in an eigensubspace.

Although the CS method is not computationally efficient, it remains the superiority of reduced-rank processing compared to full-rank processing in case of estimation errors of statistics due to low sample support (cf., e. g., [87]). In such scenarios, it has been observed that compared to the PC based approach, the rank of the CS based MWF approximation can even be chosen smaller than the dimension of the signal subspace, i. e., the CS based eigensubspace seems to be a compressed version of the signal subspace.

Here, the prefilter matrix $\boldsymbol{T}_{\mathrm{CS}}^{(D)}[n]$ depends on the cross-covariance information and hence, the term $\boldsymbol{T}_{\mathrm{CS}}^{(D)}[n](\boldsymbol{T}_{\mathrm{CS}}^{(D),\mathrm{H}}[n]\boldsymbol{C}_{\mathbf{y}}[n]\boldsymbol{T}_{\mathrm{CS}}^{(D)}[n])^{-1}\boldsymbol{T}_{\mathrm{CS}}^{(D),\mathrm{H}}[n]$ in Equation (2.48) would be different if we estimate each element of $\boldsymbol{x}[n]$ separately. Thus, the CS based reduced-rank MWF is different from the parallel application of M reduced-rank VWFs.

3

Block Krylov Methods

The era of *Krylov subspace methods* [92, 206, 252] started with the invention of the iterative *Lanczos algorithm* [156] in 1950 for solving *eigenvalue problems* (see, e. g., [131]) of symmetric as well as non-symmetric matrices.[1] Cornelius Lanczos proposed a recursive method based on *polynomial expansion* where the combination of previous vectors of the *Krylov subspace* is chosen such that the norm of the resulting vector is minimized. For the application to non-symmetric matrices, Lanczos used *bi-orthogonalization*. One year later, Walter E. Arnoldi simplified this procedure in [4] by finding a way to avoid bi-orthogonalization and obtained the computationally more efficient *Arnoldi algorithm* for non-symmetric matrices.

The first Krylov subspace method for solving systems of linear equations with one right-hand side was the *Conjugate Gradient* (CG) *algorithm* [117] introduced by Magnus R. Hestenes and Eduard Stiefel in 1952.[2] Besides the derivation of the CG method for symmetric and positive definite systems, Hestenes and Stiefel noted that the application to symmetric positive semidefinite matrices can lead to *Least Squares* (LS) *solutions* based on *pseudoinverses* (see also [115]). Moreover, they presented a special type of the CG algorithm for the solution of over-determined systems of linear equations, i. e., the system matrices need to be no longer square. At the same time, Lanczos [157] published independently an algorithm for arbitrary square matrices which is mathematically identical to the CG method if the matrix is restricted to be

---

[1] Unless otherwise stated, the considered matrices are assumed to be square.

[2] According to the article [116] written by Magnus R. Hestenes, he and Eduard Stiefel developed the CG algorithm independently. Stiefel was visiting the *Institute for Numerical Analysis* (INA) of the *University of California, Los Angeles* (UCLA), USA, due to a conference in 1951, where he got a copy of an INA report which Hestenes wrote on the CG method. He recognized that the algorithm which he invented in Zurich, Switzerland, is exactly the same and visited Hestenes in his office to tell him, "This is my talk." Hestenes invited Stiefel to stay for one semester at UCLA and INA where they wrote together their famous paper [117].

symmetric and positive definite. Thus, Hestenes, Stiefel, and Lanczos must be seen as the inventors of the CG algorithm.

Note that the Krylov subspace methods for solving systems of linear equations were firstly seen as *direct methods* like, e.g., *Gaussian elimination* [74] (cf. Chapter 1), since they produce, at least in exact arithmetics, the solution in a finite number of steps.[3] However, already Hestenes and Stiefel mentioned the iterative behavior of the CG procedure since the norm of the residual decreases from step to step, hence, leading to good approximations if stopped before convergence. It took a long time until researchers rediscovered and exploited the potential of the CG algorithm as an iterative method. In 1972, John K. Reid [197] was the first one of these researchers who compared the CG algorithm with iterative methods and showed its superiority.

Although direct methods can be implemented exploiting the sparsity of linear systems in a clever way by avoiding computations with zero-elements as well as zero-element storage, *iterative methods* based on Krylov subspaces have less computational complexity and need a smaller amount of computer storage if the sparse system is very large (see, e. g., [252]). Moreover, compared to most of the classical iterative algorithms like, e.g., the *Gauss–Seidel algorithm* [76, 220] (cf. Chapter 1), Krylov based iterative methods do not suffer from convergence problems and the right selection of iteration parameters.

The second half of the 20th century, mathematicians researched on CG like algorithms for solving symmetric and non-positive definite as well as non-symmetric systems of linear equations. The first attempt in the direction of solving non-symmetric systems was based on bi-orthogonalization, similar to the idea of Lanczos in [156], yielding the *Bi-CG algorithm* [69] published by Roger Fletcher in 1976. There followed many algorithms like, e.g., the *Bi-CG STABilized* (Bi-CGSTAB) *algorithm* [251] developed by Henk A. van der Vorst, which is a combination of the Bi-CG method with residual steps of order one. Probably the most popular method is the *Generalized Minimal RESidual* (GMRES) *algorithm* [205] introduced by Yousef Saad and Martin H. Schultz in 1986. The GMRES method minimizes the residual norm over all vectors of the Krylov subspace, thus, behaving numerically more stable with less computational complexity compared to equivalent methods. Another very important procedure is the *LSQR algorithm* [182] invented by Christopher C. Paige and Michael A. Saunders in 1982 in order to solve numerically efficient LS problems based on normal equations. Note that there exists much more Krylov subspace algorithms for non-symmetric as well as symmetric and non-positive definite systems which are not listed in this book due to its concentration on symmetric (or Hermitian) and positive definite problems. If the reader is interested in solving non-symmetric or symmetric and non-positive definite systems, it is recommended to read the surveys [92, 206, 252] or the books [204, 237].

---

[3] Alston S. Householder, e. g., discusses the CG algorithm in the chapter "Direct Methods of Inversion" of his book [133].

So far, we only discussed methods for solving systems of linear equations with one right-hand side. However, the main focus of this chapter lies in *block Krylov methods* [204, 252] which have been developed in the last decades for solving systems with multiple right-hand sides as well as computing eigenvalues in a more efficient way. Note that we assume that all right-hand sides are *a priori* known. Thus, methods like the *augmented CG algorithm* [66] which computes the columns of the solution matrix consecutively are not discussed here. Before deriving the most important block Krylov methods and discussing their properties, the following section repeats the main principles of block Krylov methods in general and lists the application areas of the most important ones.

3.1 Principles and Application Areas

Similar to Krylov methods, the main application areas of block Krylov algorithms are solving systems of linear equations with multiple right-hand sides and eigenvalue problems. Although these problems can be also solved with Krylov methods, we will see in the next sections that using block versions reduces computational complexity while maintaining the same level of system performance. The most important block Krylov methods are summarized in Table 3.1 where we distinguish between problems with Hermitian matrices, i.e., $A = A^H$, and the non-Hermitian case, i.e., $A \neq A^H$. Throughout this chapter, A is a square matrix.

Table 3.1. Application areas of block Krylov methods

	System of linear equations $AX = B$	Eigenvalue problem $AU = U\Lambda$
$A \neq A^H$	Block Generalized Minimal RESidual (BGMRES) et al.	Block Arnoldi
$A = A^H$	Block Conjugate Gradient (BCG, A positive definite)	Block Lanczos

The *Block Conjugate Gradient* (BCG) *algorithm* [175] for Hermitian and positive definite matrices and the *Block Generalized Minimal RESidual* (BGMRES) *algorithm* [223, 224, 203] for non-Hermitian as well as Hermitian and non-positive definite matrices are the most popular block Krylov procedures for solving systems of linear equations, e.g., $AX = B$ with NM unknowns and M right-hand sides, i.e., $A \in \mathbb{C}^{N \times N}$ and $X, B \in \mathbb{C}^{N \times M}$,

$N, M \in \mathbb{N}$. Note that we assume that N is a multiple integer of M, i.e., $N = LM$, $L \in \mathbb{N}$.[4]

The fundamental idea of block Krylov methods is to approximate each column of the solution $\boldsymbol{X}$ of the linear system $\boldsymbol{AX} = \boldsymbol{B}$ in the D-dimensional Krylov subspace

$$\mathcal{K}^{(D)}(\boldsymbol{A}, \boldsymbol{B}) = \mathrm{range}\left\{ \left[\boldsymbol{B} \; \boldsymbol{AB} \cdots \boldsymbol{A}^{d-1}\boldsymbol{B} \right] \right\} \subseteq \mathbb{C}^N, \tag{3.1}$$

where the dimension D is a multiple integer of M, i.e., $D = dM$, $d \in \{1, 2, \ldots, L\}$. Hence, we approximate $\boldsymbol{X}$ by the matrix polynomial

$$\begin{aligned}\boldsymbol{X} \approx \boldsymbol{X}_d &= \boldsymbol{B}\boldsymbol{\Psi}_0 + \boldsymbol{AB}\boldsymbol{\Psi}_1 + \boldsymbol{A}^2\boldsymbol{B}\boldsymbol{\Psi}_2 + \ldots + \boldsymbol{A}^{d-1}\boldsymbol{B}\boldsymbol{\Psi}_{d-1} \\ &= \sum_{\ell=0}^{d-1} \boldsymbol{A}^\ell \boldsymbol{B}\boldsymbol{\Psi}_\ell \in \mathbb{C}^{N \times M},\end{aligned} \tag{3.2}$$

with the polynomial coefficients $\boldsymbol{B}\boldsymbol{\Psi}_\ell \in \mathbb{C}^{N \times M}$ and the quadratic matrices $\boldsymbol{\Psi}_\ell \in \mathbb{C}^{M \times M}$, $\ell \in \{0, 1, \ldots, d-1\}$.[5] For the latter polynomial approximation, the initial guess of the solution is assumed to be $\boldsymbol{X}_0 = \boldsymbol{0}_{N \times M}$ (cf. Section 3.4). Note that in the special case where $M = 1$, $\boldsymbol{B}$ reduces to the vector $\boldsymbol{b} \in \mathbb{C}^N$, $\boldsymbol{\Psi}_\ell$ to the scalars $\psi_\ell \in \mathbb{C}$, and $D = d$. Thus, the matrix polynomial in Equation (3.2) may be written as $\Psi^{(D-1)}(\boldsymbol{A})\boldsymbol{b} \in \mathbb{C}^N$ where

$$\begin{aligned}\Psi^{(D-1)}(\boldsymbol{A}) &:= \psi_0 \boldsymbol{I}_N + \psi_1 \boldsymbol{A} + \psi_2 \boldsymbol{A}^2 + \ldots + \psi_{D-1}\boldsymbol{A}^{D-1} \\ &= \sum_{\ell=0}^{D-1} \psi_\ell \boldsymbol{A}^\ell \in \mathbb{C}^{N \times N}.\end{aligned} \tag{3.3}$$

In the case where $\boldsymbol{A}$ is Hermitian and positive definite, the scalars ψ_ℓ are real-valued, i.e., $\psi_\ell \in \mathbb{R}$. See Appendix B.2 for a detailed derivation.

If we define $\boldsymbol{B} := [\boldsymbol{b}_0, \boldsymbol{b}_1, \ldots, \boldsymbol{b}_{M-1}]$, it follows for $j \in \{0, 1, \ldots, M-1\}$:

$$\boldsymbol{X}_d \boldsymbol{e}_{j+1} = \sum_{k=0}^{M-1} \Psi_{k,j}^{(d-1)}(\boldsymbol{A})\boldsymbol{b}_k = \sum_{k=0}^{M-1}\sum_{\ell=0}^{d-1} \psi_{k,j,\ell} \boldsymbol{A}^\ell \boldsymbol{b}_k. \tag{3.4}$$

Note that due to the combination of the polynomials $\Psi_{k,j}^{(d-1)}(\boldsymbol{A})$ (cf. Equation 3.4) which depend not only on the right-hand side $\boldsymbol{b}_k$ but also on the current column $j + 1$ of $\boldsymbol{X}_d$ under consideration, the maximal polynomial degree of $d - 1$ is sufficient to obtain an approximation for each column of $\boldsymbol{X}$ in a D-dimensional subspace of $\mathbb{C}^N$ where $D = dM$. Hence, the optimal

---

[4] The matrix $\boldsymbol{B}$ may always be filled with zero columns such that this restriction is true.

[5] Clearly, the quadratic matrices $\boldsymbol{\Psi}_\ell$ depend on d. However, due to simplicity, we introduce only an additional index in the cases where a missing index leads to confusion.

solution is obtained for $d = L = N/M$. Other strategies are to approximate each column $j+1$ of the optimal solution $\boldsymbol{X}$ in a D-dimensional subspace of $\mathbb{C}^N$ considering only the right-hand side $\boldsymbol{b}_j$, i. e.,

$$\boldsymbol{X}\boldsymbol{e}_{j+1} \approx \boldsymbol{X}_D\boldsymbol{e}_{j+1} = \Psi_j^{(D-1)}(\boldsymbol{A})\boldsymbol{b}_j = \sum_{\ell=0}^{D-1} \psi_{j,\ell}\boldsymbol{A}^\ell\boldsymbol{b}_j, \tag{3.5}$$

or by approximating $\boldsymbol{X}$ in a DM-dimensional subspace of $\mathbb{C}^{N\times M}$, i. e.,

$$\boldsymbol{X} \approx \boldsymbol{X}_D = \Psi^{(D-1)}(\boldsymbol{A})\boldsymbol{B} = \sum_{\ell=0}^{D-1} \psi_\ell\boldsymbol{A}^\ell\boldsymbol{B}. \tag{3.6}$$

In both cases, we need a maximal polynomial degree of $D-1 = N-1$ in order to achieve the optimal solution $\boldsymbol{X}$. In other words, the latter two strategies do not exploit the subspace in an efficient way, thus, leading to methods which achieve the same performance as the approximation in $\mathcal{K}^{(D)}(\boldsymbol{A},\boldsymbol{B})$ only with an increase in computational complexity. Since we are interested in computationally cheap approaches, we concentrate in this book on the approximation method presented in Equation (3.4).

The *block Lanczos algorithm* [93, 175] as well as its derivatives [200, 201], and the *block Arnoldi algorithm* [208, 207, 219], respectively, are methods for the computation of eigenvalues of Hermitian and non-Hermitian matrices, respectively. Compared to the Lanczos or Arnoldi procedure, the block versions compute several eigenvalues at once which results in improved rates of convergence. All (block) Krylov procedures are used as a preprocessing step for the *eigen decomposition* (see, e. g., [131]) $\boldsymbol{A} = \boldsymbol{U}\boldsymbol{\Lambda}\boldsymbol{U}^{-1}$ with the diagonal matrix $\boldsymbol{\Lambda} \in \mathbb{C}^{N\times N}$ of *eigenvalues*. Whereas the Lanczos or Arnoldi algorithm transform the matrix $\boldsymbol{A} \in \mathbb{C}^{N\times N}$ to a matrix with *tridiagonal*[6] or *Hessenberg structure*[7], the block variants produce a $(2M+1)$-*diagonal* or *band Hessenberg structure*[8], respectively. Note that good eigenvalue approximations of matrices with such structures can be found in less steps than by directly computing the eigenvalues of the original matrix $\boldsymbol{A}$ (cf., e. g., [204]).

In the sequel of this chapter, we review the mentioned block Krylov methods in more detail. Note that the BGMRES algorithm is no longer considered since we are mainly interested in systems with Hermitian and positive definite matrices. See [223, 204] for detailed derivations and investigations of the BGMRES procedure.

---

[6] An $(2n+1)$-diagonal matrix has $2n$ non-zero subdiagonals besides the main diagonal.

[7] A Hessenberg matrix is an upper triangular matrix with additional non-zero entries in the first subdiagonal.

[8] A band Hessenberg matrix is an upper triangular matrix with additional non-zero entries in the first M subdiagonals.

3.2 Block Arnoldi Algorithm

In this section, we consider the *block Arnoldi procedure* [208, 207, 204] as an algorithm which computes iteratively an orthonormal basis

$$Q = \begin{bmatrix} Q_0 \ Q_1 \ \cdots \ Q_{d-1} \end{bmatrix} \in \mathbb{C}^{N \times D} \quad \text{with } Q^{\mathrm{H}} Q = I_D, \tag{3.7}$$

of the D-dimensional Krylov subspace

$$\mathcal{K}^{(D)}(A, Q_0) = \mathrm{range}\left\{ \begin{bmatrix} Q_0 \ AQ_0 \ \cdots \ A^{d-1}Q_0 \end{bmatrix} \right\} \subseteq \mathbb{C}^N, \tag{3.8}$$

where $A \in \mathbb{C}^{N \times N}$, $Q_0 \in \mathbb{C}^{N \times M}$ has orthonormal columns, and D is a multiple integer of M, i.e., $D = dM$, $d \in \{1, 2, \ldots L\}$. Remember that $N = LM$.

First, we derive the fundamental iteration formula of the block Arnoldi algorithm. Assume that we have already calculated the $N \times M$ matrices Q_0, Q_1, ..., and Q_ℓ, $\ell \in \{0, 1, \ldots, d - 2\}$, such that they are mutually orthonormal,[9] i.e.,

$$Q_i^{\mathrm{H}} Q_j = \begin{cases} I_M, & i = j, \\ 0_{M \times M}, & i \neq j, \end{cases} \quad i, j \in \{0, 1, \ldots, \ell\}, \tag{3.9}$$

and form a basis of the $(\ell + 1)M$-dimensional Krylov subspace $\mathcal{K}^{((\ell+1)M)}(A, Q_0)$, i.e., the matrix Q_ℓ can be expressed by the matrix polynomial $Q_\ell = \sum_{i=0}^{\ell} A^i Q_0 \Psi_{i,\ell} \in \mathbb{C}^{N \times M}$ (cf. Equation 3.2). Then, we compute $Q_{\ell+1} \in \mathbb{C}^{N \times M}$ by orthonormalizing a matrix whose columns are elements of the Krylov subspace $\mathcal{K}^{((\ell+2)M)}(A, Q_0)$, with respect to the matrices Q_0, Q_1, ..., and Q_ℓ. Although the obvious choice of this matrix is $A^{\ell+1}Q_0$ (cf. Equation 3.8), we choose AQ_ℓ due to numerical issues. Note that the columns of AQ_ℓ are elements of $\mathcal{K}^{((\ell+2)M)}(A, Q_0)$. Orthonormalization of the matrix AQ_ℓ with respect to the matrices Q_i, $i \in \{0, 1, \ldots, \ell\}$, is done blockwise based on the *classical Gram–Schmidt procedure* [94, 237, 204], i.e.,

$$Q_{\ell+1} H_{\ell+1,\ell} = A Q_\ell - \sum_{i=0}^{\ell} Q_i H_{i,\ell}. \tag{3.10}$$

The left-hand side of Equation (3.10) is obtained by QR factorization of its right-hand side (see Appendix B.3). Thus, the columns of $Q_{\ell+1}$ are mutually orthonormal, i.e., $Q_{\ell+1}^{\mathrm{H}} Q_{\ell+1} = I_M$, and $H_{\ell+1,\ell}$ is an upper triangular matrix. Further, the coefficients $H_{i,\ell} \in \mathbb{C}^{M \times M}$ are chosen such that $Q_{\ell+1}$ is orthogonal, i.e.,

$$Q_i^{\mathrm{H}} Q_{\ell+1} H_{\ell+1,\ell} = Q_i^{\mathrm{H}} A Q_\ell - \sum_{j=0}^{\ell} Q_i^{\mathrm{H}} Q_j H_{j,\ell} \overset{!}{=} 0_{M \times M}. \tag{3.11}$$

---

[9] Two matrices are denoted to be orthogonal/orthonormal if each column of the one matrix is orthogonal/orthonormal to all the columns of the second matrix.

If we recall that the matrices Q_0, Q_1, ..., and Q_ℓ are already orthonormal to each other (cf. Equation 3.9), the coefficients $H_{i,\ell}$ to be used by the classical Gram–Schmidt procedure in Equation (3.10), compute as

$$H_{i,\ell} = Q_i^{\mathrm{H}} A Q_\ell, \quad i \in \{0, 1, \ldots, \ell\}. \tag{3.12}$$

Now, we derive an alternative iteration formula to the one in Equation (3.10) which is more stable in the presence of rounding errors. Replacing $H_{i,\ell}$ in Equation (3.10) using Equation (3.12), yields

$$Q_{\ell+1} H_{\ell+1,\ell} = \left(I_N - \sum_{i=0}^{\ell} Q_i Q_i^{\mathrm{H}} \right) A Q_\ell. \tag{3.13}$$

With Equation (3.9) and the definition of the matrices $P_{\perp Q_i} := I_N - Q_i Q_i^{\mathrm{H}}$ projecting onto the space orthogonal to the one spanned by the columns of Q_i, we get

$$I_N - \sum_{i=0}^{\ell} Q_i Q_i^{\mathrm{H}} = \prod_{i=0}^{\ell} \left(I_N - Q_i Q_i^{\mathrm{H}} \right) = \prod_{i=0}^{\ell} P_{\perp Q_i}, \tag{3.14}$$

and Equation (3.13) can be rewritten as

$$\left| Q_{\ell+1} H_{\ell+1,\ell} = \left(\prod_{i=0}^{\ell} P_{\perp Q_i} \right) A Q_\ell, \right| \tag{3.15}$$

which is the fundamental iteration formula of the so-called *modified Gram–Schmidt version* [94, 237, 204] of the block Arnoldi algorithm. Again, $Q_{\ell+1}$ and $H_{\ell+1,\ell}$ are obtained by the QR factorization of the right-hand side of Equation (3.15).

In exact arithmetics, the two versions of the Gram–Schmidt procedure are identical. Nevertheless, we prefer the Arnoldi algorithm based on the modified Gram–Schmidt procedure because it is much more reliable in the presence of rounding errors [237]. In order to further improve the reliability, one can introduce an additional reorthogonalization step or use *Householder reflection matrices* [253, 237] to produce a Hessenberg matrix with a slightly increase in computational complexity. Although this high reliability is reasonable for solving eigenvalue problems, the modified Gram–Schmidt with optional reorthogonalization is adequate in most cases if we are interested in solving systems of linear equations.

To end up in a computationally efficient algorithm, we also compute the right-hand side of Equation (3.15) in an iterative manner. Starting with $V_0 = A Q_\ell \in \mathbb{C}^{N \times M}$, we calculate iteratively

$$V_{i+1} = P_{\perp Q_i} V_i = V_i - Q_i Q_i^{\mathrm{H}} V_i \in \mathbb{C}^{N \times M}, \quad i \in \{0, 1, \ldots, \ell\}, \tag{3.16}$$

ending up with $\boldsymbol{V}_{\ell+1} = (\prod_{i=0}^{\ell} \boldsymbol{P}_{\perp \boldsymbol{Q}_i}) \boldsymbol{A} \boldsymbol{Q}_\ell$. Using Equations (3.9) and (3.12), we further get

$$\boldsymbol{Q}_i^{\mathrm{H}} \boldsymbol{V}_i = \boldsymbol{Q}_i^{\mathrm{H}} \left(\prod_{j=0}^{i-1} \boldsymbol{P}_{\perp \boldsymbol{Q}_j} \right) \boldsymbol{A} \boldsymbol{Q}_\ell = \boldsymbol{Q}_i^{\mathrm{H}} \boldsymbol{A} \boldsymbol{Q}_\ell = \boldsymbol{H}_{i,\ell}, \tag{3.17}$$

since $\boldsymbol{Q}_i^{\mathrm{H}} \boldsymbol{P}_{\perp \boldsymbol{Q}_j} = \boldsymbol{Q}_i^{\mathrm{H}}$ for all $j \in \{0, 1, \ldots, i-1\}$ because $\boldsymbol{Q}_i^{\mathrm{H}} \boldsymbol{Q}_j = \boldsymbol{0}_{M \times M}$ for $i \neq j$ (cf. Equation 3.9). Finally, we compute the QR factorization of $\boldsymbol{V}_{\ell+1}$ in order to obtain $\boldsymbol{Q}_{\ell+1}$ and $\boldsymbol{H}_{\ell+1,\ell}$. This implementation of Equation (3.15) is summarized in Algorithm 3.1. where Equation (3.17) can be found in Line 5, the iteration formula of Equation (3.16) in Line 6 with the initial value computed in Line 3, and the QR decomposition in Line 8 (cf. Appendix B.3). Note that we omitted the index at $\boldsymbol{V}$ since it can be overwritten to reduce computer storage.

Algorithm 3.1. Block Arnoldi

    Choose a matrix $\boldsymbol{Q}_0 \in \mathbb{C}^{N \times M}$ with orthonormal columns.
2: **for** $\ell = 0, 1, \ldots, d-2 \wedge d > 1$ **do**
        $\boldsymbol{V} \leftarrow \boldsymbol{A} \boldsymbol{Q}_\ell$
4:     **for** $i = 0, 1, \ldots, \ell$ **do**
            $\boldsymbol{H}_{i,\ell} \leftarrow \boldsymbol{Q}_i^{\mathrm{H}} \boldsymbol{V}$
6:          $\boldsymbol{V} \leftarrow \boldsymbol{V} - \boldsymbol{Q}_i \boldsymbol{H}_{i,\ell}$
    end for
8:     $(\boldsymbol{Q}_{\ell+1}, \boldsymbol{H}_{\ell+1,\ell}) \leftarrow \mathrm{QR}(\boldsymbol{V})$
    end for
10: $\boldsymbol{Q} \leftarrow [\boldsymbol{Q}_0, \boldsymbol{Q}_1, \ldots, \boldsymbol{Q}_{d-1}]$

We conclude this section with the following property of the orthonormal basis $\boldsymbol{Q}$ of the Krylov subspace $\mathcal{K}^{(D)}(\boldsymbol{A}, \boldsymbol{Q}_0)$.

Proposition 3.1. *The transformed matrix* $\boldsymbol{H} := \boldsymbol{Q}^{\mathrm{H}} \boldsymbol{A} \boldsymbol{Q} \in \mathbb{C}^{D \times D}$ *has band Hessenberg structure.*

Proof. The structure of a band Hessenberg matrix is shown in Figure 3.1. If we define $\boldsymbol{H}_{m,\ell}$, $m, \ell \in \{0, 1, \ldots, d-1\}$, to be the $M \times M$ matrix in the $(m+1)$th block row and the $(\ell+1)$th block column of the $D \times D$ matrix $\boldsymbol{H}$, i.e., $\boldsymbol{H}_{m,\ell} = \boldsymbol{Q}_m^{\mathrm{H}} \boldsymbol{A} \boldsymbol{Q}_\ell \in \mathbb{C}^{M \times M}$ if we recall that $\boldsymbol{Q} = [\boldsymbol{Q}_0, \boldsymbol{Q}_1, \ldots, \boldsymbol{Q}_{d-1}]$, Proposition 3.1 is proven if the last two cases of the following equation are true:

$$\boldsymbol{Q}_m^{\mathrm{H}} \boldsymbol{A} \boldsymbol{Q}_\ell = \begin{cases} \boldsymbol{H}_{m,\ell}, & m < \ell + 1, \\ \boldsymbol{H}_{\ell+1,\ell} \text{ (upper triangular matrix)}, & m = \ell + 1, \\ \boldsymbol{0}_{M \times M}, & m > \ell + 1. \end{cases} \tag{3.18}$$

Multiply Equation (3.15) on the left-hand side by $\boldsymbol{Q}_m^{\mathrm{H}}$. If $m > \ell$, we get

$$Q_m^H Q_{\ell+1} H_{\ell+1,\ell} = Q_m^H A Q_\ell, \tag{3.19}$$

since again $Q_m^H P_{\perp Q_i} = Q_m^H$ for all $i \in \{0, 1, \ldots, \ell\}$. Thus, for $m = \ell + 1$, $Q_m^H A Q_\ell = H_{\ell+1,\ell}$ which is an upper triangular matrix due to the QR factorization, and for $m > \ell + 1$, $Q_m^H A Q_\ell = 0_{M \times M}$, due to Equation (3.9). $\qquad \square$

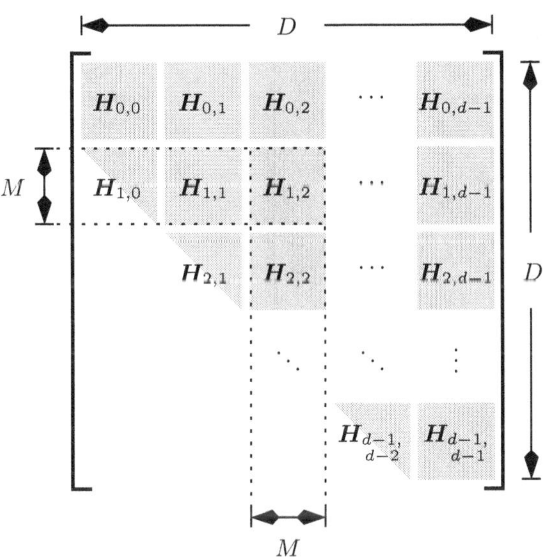

Fig. 3.1. Structure of the $D \times D$ band Hessenberg matrix H with M subdiagonals

Note that the main purpose of Algorithm 3.1. is to compute Q and not the band Hessenberg matrix H. However, with a slightly increase in computational complexity, Algorithm 3.1. can be enhanced to compute also the last block column of H which is missing so far.

3.3 Block Lanczos Algorithm

3.3.1 Original Version

The *block Lanczos algorithm* [93, 175, 94, 204] is a simplified block Arnoldi procedure for the particular case where the matrix A is Hermitian.

Proposition 3.2. *If the block Arnoldi algorithm is applied to a Hermitian matrix A, the generated block matrices $H_{m,\ell}$ are such that $H_{m,\ell} = 0_{M \times M}$ for $m < \ell - 1$ and $H_{\ell,\ell+1} = H_{\ell+1,\ell}^H$.*

Proof. If $\boldsymbol{A}$ is Hermitian, then $\boldsymbol{H} = \boldsymbol{Q}^{\mathrm{H}}\boldsymbol{A}\boldsymbol{Q}$ is Hermitian, i.e., $\boldsymbol{H}_{m,\ell} = \boldsymbol{H}_{\ell,m}^{\mathrm{H}}$ and especially, $\boldsymbol{H}_{\ell,\ell+1} = \boldsymbol{H}_{\ell+1,\ell}^{\mathrm{H}}$. Since $\boldsymbol{H}_{m,\ell} = \boldsymbol{0}_{M\times M}$ for $m > \ell + 1$ due to Proposition 3.1, $\boldsymbol{H}_{m,\ell} = \boldsymbol{0}_{M\times M}$ is also true for $m < \ell - 1$. $\qquad\square$

Thus, $\boldsymbol{H}$ is a Hermitian band Hessenberg matrix with bandwidth $2M + 1$ as shown in Figure 3.2 if we define $\boldsymbol{\Xi}_\ell := \boldsymbol{H}_{\ell,\ell} \in \mathbb{C}^{M\times M}$ and $\boldsymbol{\Upsilon}_{\ell+1} := \boldsymbol{H}_{\ell,\ell+1} \in \mathbb{C}^{M\times M}$. Note that if $M = 1$, the Hermitian band Hessenberg matrix is equal to a Hermitian *tridiagonal matrix*.

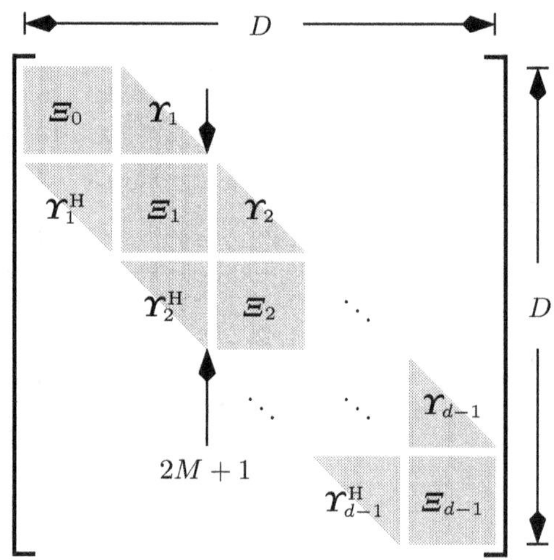

Fig. 3.2. Structure of the Hermitian $D \times D$ band Hessenberg matrix $\boldsymbol{H}$ with bandwidth $2M + 1$

As already mentioned, Lanczos [156] suggested an algorithm to transform a matrix $\boldsymbol{A}$ to a tridiagonal matrix. Besides the derivation of an algorithm for symmetric matrices, he presented already an extended procedure for the non-symmetric case based on bi-orthogonalization which was working on two Krylov subspaces, viz., $\mathcal{K}^{(D)}(\boldsymbol{A}, \boldsymbol{q}_0)$ and $\mathcal{K}^{(D)}(\boldsymbol{A}^{\mathrm{H}}, \boldsymbol{q}_0)$. Note that the matrix $\boldsymbol{Q}_0 \in \mathbb{C}^{N\times M}$ reduces to the vector $\boldsymbol{q}_0 \in \mathbb{C}^N$ if $M = 1$. Arnoldi simplified this method for non-symmetric matrices by using only $\mathcal{K}^{(D)}(\boldsymbol{A}, \boldsymbol{q}_0)$ to generate a Hessenberg matrix instead of a tridiagonal one.

Since $\boldsymbol{H}_{i,\ell} = \boldsymbol{0}_{M\times M}$ for $i < \ell - 1$ (cf. Proposition 3.2), the fundamental iteration formula in Equation (3.10) simplifies to

$$\begin{aligned}
\boldsymbol{Q}_{\ell+1}\boldsymbol{H}_{\ell+1,\ell} &= \boldsymbol{A}\boldsymbol{Q}_\ell - \boldsymbol{Q}_\ell\boldsymbol{H}_{\ell,\ell} - \boldsymbol{Q}_{\ell-1}\boldsymbol{H}_{\ell-1,\ell} \\
&= \left(\boldsymbol{I}_N - \boldsymbol{Q}_\ell\boldsymbol{Q}_\ell^{\mathrm{H}} - \boldsymbol{Q}_{\ell-1}\boldsymbol{Q}_{\ell-1}^{\mathrm{H}}\right)\boldsymbol{A}\boldsymbol{Q}_\ell,
\end{aligned} \tag{3.20}$$

or to the modified Gram–Schmidt version, i.e.,

$$\left| Q_{\ell+1} H_{\ell+1,\ell} = P_{\perp Q_{\ell-1}} P_{\perp Q_\ell} A Q_\ell, \right| \tag{3.21}$$

which are both the fundamental three-term iteration formulas of the block Lanczos procedure. Again, the left-hand sides are obtained by QR factorization of the right-hand sides.

Finally, the block Lanczos method [93] given in Algorithm 3.2. is obtained by modifying the block Arnoldi procedure given in Algorithm 3.1. in an adequate way. For the same reason as above, the inner **for**-loop of Algorithm 3.1. has to be executed only for $i \in \{\ell - 1, \ell\}$ and Lines 3 to 7 of Algorithm 3.1. may be replaced by Lines 5 to 7 of Algorithm 3.2.. Therefore, the amount of computer storage is reduced tremendously compared to the block Arnoldi algorithm. Again, Algorithm 3.2. is designed to compute the basis Q. If one is interested in the Hessenberg matrix H, the given implementation of the block Lanczos procedure has to be extended by the computation of Ξ_{d-1}.

Algorithm 3.2. Block Lanczos

$\quad\ \Upsilon_0 \leftarrow 0_{M \times M}$
2: $\ Q_{-1} \leftarrow 0_{N \times M}$
$\quad\ $ Choose a matrix $Q_0 \in \mathbb{C}^{N \times M}$ with orthonormal columns.
4: **for** $\ell = 0, 1, \ldots, d-2 \wedge d > 1$ **do**
$\quad\quad V \leftarrow A Q_\ell - Q_{\ell-1} \Upsilon_\ell$
6: $\quad\ \Xi_\ell \leftarrow Q_\ell^{\mathrm{H}} V$
$\quad\quad V \leftarrow V - Q_\ell \Xi_\ell$
8: $\quad\ (Q_{\ell+1}, \Upsilon_{\ell+1}^{\mathrm{H}}) \leftarrow \mathrm{QR}(V)$
$\quad\ $ **end for**
10: $Q \leftarrow [Q_0, Q_1, \ldots, Q_{d-1}]$

It is really astonishing that the simple procedure in Algorithm 3.2. guarantees, at least in exact arithmetic, that the matrix Q has orthonormal columns. Nevertheless, in the presence of rounding errors, the global orthonormality may be lost after a few iteration steps. Hence, additional reorthogonalization steps are necessary [94].

Note that in the special case $M = 1$, the Lanczos algorithm is nothing but the *Stieltjes algorithm* [77] for computing a sequence of orthogonal polynomials [204]. Moreover, the characteristic polynomial of the tridiagonal matrix produced by the Lanczos algorithm minimizes the norm $\|P(A)q_0\|_2$ over the *monic polynomials* $P(A)$ (cf., e. g., [202, 237]).[10]

3.3.2 Dimension-Flexible Ruhe Version

Ruhe derived in [200] a version of the block Lanczos procedure where the matrix $Q \in \mathbb{C}^{N \times D}$ is computed columnwise instead of blockwise. First, we

---

[10] Monic polynomials are the polynomials where the coefficient of the term with maximal degree is one.

derive this version for the case where D is still an integer multiple of M, i. e., $D = dM$, $d \in \{1, 2, \ldots, L\}$. Then, we observe that the resulting algorithm can also be used for an arbitrary $D \in \{M, M+1, \ldots, N\}$.[11]

Let $D = dM$. Recall the fundamental iteration formula of the Lanczos algorithm in Equation (3.20). Since $\boldsymbol{Q} = [\boldsymbol{Q}_0, \boldsymbol{Q}_1, \ldots, \boldsymbol{Q}_{d-1}] = [\boldsymbol{q}_0, \boldsymbol{q}_1, \ldots, \boldsymbol{q}_{dM-1}]$, we get for $\ell \in \{0, 1, \ldots, d-2\}$,

$$
\begin{aligned}
\big[\boldsymbol{q}_{(\ell+1)M} \ \boldsymbol{q}_{(\ell+1)M+1} \ \cdots \ \boldsymbol{q}_{(\ell+2)M-1}\big] \boldsymbol{H}_{\ell+1,\ell} \\
= \boldsymbol{A} \big[\boldsymbol{q}_{\ell M} \ \boldsymbol{q}_{\ell M+1} \ \cdots \ \boldsymbol{q}_{(\ell+1)M-1}\big] \\
- \big[\boldsymbol{q}_{\ell M} \ \boldsymbol{q}_{\ell M+1} \ \cdots \ \boldsymbol{q}_{(\ell+1)M-1}\big] \boldsymbol{H}_{\ell,\ell} \\
- \big[\boldsymbol{q}_{(\ell-1)M} \ \boldsymbol{q}_{(\ell-1)M+1} \ \cdots \ \boldsymbol{q}_{\ell M-1}\big] \boldsymbol{H}_{\ell-1,\ell}.
\end{aligned}
\tag{3.22}
$$

Now consider the $(k+1)$th column of Equation (3.22), $k \in \{0, 1, \ldots, M-1\}$, and remember that $\boldsymbol{H}_{\ell+1,\ell}$ is an upper triangular matrix whereas $\boldsymbol{H}_{\ell-1,\ell} = \boldsymbol{H}_{\ell,\ell-1}^{\mathrm{H}}$ is lower triangular. Besides, we define $h_{mM+j,\ell M+k}$ to be the entry in the $(j+1)$th row and $(k+1)$th column, $j, k \in \{0, 1, \ldots, M-1\}$, of the matrix $\boldsymbol{H}_{m,\ell}$, $m, \ell \in \{0, 1, \ldots, d-1\}$, and the $(mM+j+1)$th row and the $(\ell M + k + 1)$th column of $\boldsymbol{H}$, respectively. It follows

$$
\sum_{\substack{i=(\ell+1)M}}^{(\ell+1)M+k} \boldsymbol{q}_i h_{i,\ell M+k} = \boldsymbol{A}\boldsymbol{q}_{\ell M+k} - \sum_{\substack{i=(\ell-1)M+k \\ i \in \mathbb{N}_0}}^{(\ell+1)M-1} \boldsymbol{q}_i h_{i,\ell M+k},
\tag{3.23}
$$

or

$$
\boldsymbol{q}_{(\ell+1)M+k} h_{(\ell+1)M+k,\ell M+k} = \boldsymbol{A}\boldsymbol{q}_{\ell M+k} - \sum_{\substack{i=(\ell-1)M+k \\ i \in \mathbb{N}_0}}^{(\ell+1)M+k-1} \boldsymbol{q}_i h_{i,\ell M+k},
\tag{3.24}
$$

by bringing $\boldsymbol{q}_i h_{i,\ell M+k}$ for $i \in \{(\ell+1)M, (\ell+1)M+1, \ldots, (\ell+1)M+k-1\}$ on the right-hand side. If we define $j+1 := (\ell+1)M + k$ and $m := \ell M + k = j - M + 1$, i. e., $j \in \{M-1, M, \ldots, D-2\}$ and $m \in \{0, 1, \ldots, D-M-1\}$, Equation (3.24) may be rewritten as

$$
\boldsymbol{q}_{j+1} h_{j+1,m} = \boldsymbol{A}\boldsymbol{q}_m - \sum_{\substack{i=m-M \\ i \in \mathbb{N}_0}}^{m+M-1} \boldsymbol{q}_i h_{i,m} \in \mathbb{C}^N,
\tag{3.25}
$$

where $h_{i,m} = \boldsymbol{q}_i^{\mathrm{H}} \boldsymbol{A}\boldsymbol{q}_m$ and $h_{j+1,m} = \boldsymbol{q}_{j+1}^{\mathrm{H}} \boldsymbol{A}\boldsymbol{q}_m$ is chosen to ensure that $\|\boldsymbol{q}_{j+1}\|_2 = 1$. Note that Equation (3.25) is the fundamental iteration formula of the *block Lanczos–Ruhe algorithm* [200, 204].

Starting with the orthonormal vectors $\boldsymbol{q}_0$, $\boldsymbol{q}_1$, ..., and $\boldsymbol{q}_{M-1}$, all elements of $\mathbb{C}^N$, the block Lanczos–Ruhe algorithm multiplies in a first step $\boldsymbol{q}_0$ by $\boldsymbol{A}$

---

[11] Note that N can be also chosen as an arbitrary integer as long as $N \geq M$.

on the left-hand side and orthonormalizes the resulting vector with respect to the vectors q_0, q_1, ..., and q_{M-1} based on a Gram–Schmidt procedure. The resulting vector is defined to be q_M. In the second step, the vector q_{M+1} is obtained by orthonormalizing Aq_1 with respect to all available vectors, viz., q_0, q_1, ..., q_{M-1}, and q_M. At the $(m+1)$th step, the vector q_{m+M} is the part of Aq_m orthonormal to the last at most $2M$ vectors q_i, $i \in \{m - M, m - M + 1, \ldots, m + M - 1\} \cap \mathbb{N}_0$. This procedure is repeated until the vector q_{D-1} is obtained. The resulting block Lanczos–Ruhe method is depicted in Algorithm 3.3.. Note that we exploit the fact that A is Hermitian in Lines 6 to 10 in order to reduce computational complexity. There, we compute $h_{i,m}$ only for $i \geq m$ and use the already computed $h_{m,i}$ to get $h_{i,m} = h_{m,i}^*$ in the cases where $i < m$.

Algorithm 3.3. Block Lanczos–Ruhe

    Choose a matrix $[q_0, q_1, \ldots, q_{M-1}] \in \mathbb{C}^{N \times M}$ with orthonormal columns.
2: **for** $j = M - 1, M, \ldots, D - 2 \wedge D > M$ **do**
        $m \leftarrow j - M + 1$
4:     $v \leftarrow Aq_m$
        for $i = j - 2M + 1, j - 2M + 2, \ldots, j \wedge i \in \mathbb{N}_0$ **do**
6:         **if** $i < m$ **then**
            $h_{i,m} \leftarrow h_{m,i}^*$
8:         **else**
            $h_{i,m} \leftarrow q_i^{\mathrm{H}} v$
10:       **end if**
        $v \leftarrow v - q_i h_{i,m}$
12:     **end for**
        $h_{j+1,m} \leftarrow \|v\|_2$
14:     $q_{j+1} \leftarrow v / h_{j+1,m}$
    end for
16: $Q \leftarrow [q_0, q_1, \ldots, q_{D-1}]$

Note that compared to the block Arnoldi or Lanczos procedure (cf. Algorithm 3.1. and 3.2.), the dimension D of the Krylov subspace is no longer restricted to be an integer multiple of M (cf. Proposition 3.3), resulting in a more flexible iterative solver for systems of linear equations. Nevertheless, it gives up the potential of a parallel implementation since the vectors q_{j+1}, $j \in \{M - 1, M, \ldots, D - 2\}$ are computed sequentially.

Proposition 3.3. *For an arbitrary $D \in \{M, M + 1, \ldots, N\}$, the columns of the matrix $Q = [q_0, q_1, \ldots, q_{D-1}]$ of the block Lanczos–Ruhe algorithm form an orthonormal basis of the sum of Krylov subspaces*

$$\mathcal{K}_{\mathrm{Ruhe}}^{(D)}\left(A, q_0, q_1, \ldots, q_{M-1}\right) := \sum_{k=0}^{\mu-1} \mathcal{K}^{(d)}\left(A, q_k\right) + \sum_{\substack{k=\mu \\ \mu < M}}^{M-1} \mathcal{K}^{(d-1)}\left(A, q_k\right), \quad (3.26)$$

where $d = \lceil D/M \rceil \in \{1, 2, \ldots, \lceil N/M \rceil\}$ and $\mu = D - (d-1)M \in \{1, 2, \ldots, M\}$.

Proof. See Appendix B.4. $\qquad\qquad\square$

It remains to mention that all reviewed algorithms, viz., the block Arnoldi, Lanczos, and Lanczos–Ruhe algorithm, can be extended to solve the system of linear equations $\boldsymbol{AX} = \boldsymbol{B}$. In this case, the initial matrix $\boldsymbol{Q}_0 \in \mathbb{C}^{N \times M}$ has to be chosen to be an orthonormal basis of the subspace $\text{range}\{\boldsymbol{B}\}$,[12] e.g., by QR factorization $(\boldsymbol{Q}_0, \boldsymbol{H}_{0,-1}) = \text{QR}(\boldsymbol{B})$ (cf. Appendix B.3) or by $\boldsymbol{Q}_0 = \boldsymbol{B}(\boldsymbol{B}^{\mathrm{H}}\boldsymbol{B})^{-1/2}$. Hence, $\mathcal{K}^{(D)}(\boldsymbol{A}, \boldsymbol{B}) = \mathcal{K}^{(D)}(\boldsymbol{A}, \boldsymbol{Q}_0)$ since the polynomial coefficients $\boldsymbol{B\Psi}_\ell$, $\ell \in \{0, 1, \ldots, d-1\}$, of the matrix polynomial representing an element of the Krylov subspace (cf. Equation 3.2) can be replaced by $\boldsymbol{BH}_{0,-1}^{-1}\boldsymbol{\Psi}'_\ell = \boldsymbol{Q}_0\boldsymbol{\Psi}'_\ell$ where $\boldsymbol{\Psi}_\ell = \boldsymbol{H}_{0,-1}^{-1}\boldsymbol{\Psi}'_\ell$, or by $\boldsymbol{B}(\boldsymbol{B}^{\mathrm{H}}\boldsymbol{B})^{-1/2}\boldsymbol{\Psi}'_\ell = \boldsymbol{Q}_0\boldsymbol{\Psi}'_\ell$ where $\boldsymbol{\Psi}_\ell = (\boldsymbol{B}^{\mathrm{H}}\boldsymbol{B})^{-1/2}\boldsymbol{\Psi}'_\ell$, respectively. Nevertheless, the block Krylov method usually used for solving systems of linear equations with multiple right-hand sides where $\boldsymbol{A}$ is Hermitian, is the *Block Conjugate Gradient* (BCG) *algorithm* [175, 204] which is discussed in the next section.

3.4 Block Conjugate Gradient Algorithm

3.4.1 Original Version

The *Block Conjugate Gradient* (BCG) *algorithm* [175, 204] is an iterative method to solve the system $\boldsymbol{AX} = \boldsymbol{B}$ of linear equations which we already discussed in the introduction of this chapter. Again, $\boldsymbol{A} \in \mathbb{C}^{N \times N}$, $\boldsymbol{X}, \boldsymbol{B} \in \mathbb{C}^{N \times M}$, $N = LM$, and $L \in \mathbb{N}$.

The first step in the derivation of the BCG algorithm is to interpret the solution of the system of linear equations as a solution of the optimization problem

$$\boldsymbol{X} = \underset{\boldsymbol{X}' \in \mathbb{C}^{N \times M}}{\text{argmin}}\ \varepsilon_{\mathrm{F}/\boldsymbol{A}}\left(\boldsymbol{X}'\right). \tag{3.27}$$

Here, the cost function can be the squared Frobenius norm (cf. Appendix A.2) of the *residuum matrix* $\boldsymbol{R}' = \boldsymbol{B} - \boldsymbol{AX}' \in \mathbb{C}^{N \times M}$, given by

$$\begin{aligned}\varepsilon_{\mathrm{F}}\left(\boldsymbol{X}'\right) &= \|\boldsymbol{R}'\|_{\mathrm{F}}^2 = \|\boldsymbol{B} - \boldsymbol{AX}'\|_{\mathrm{F}}^2 \\ &= \text{tr}\left\{\boldsymbol{X}'^{,\mathrm{H}}\boldsymbol{A}^{\mathrm{H}}\boldsymbol{AX}' - \boldsymbol{X}'^{,\mathrm{H}}\boldsymbol{A}^{\mathrm{H}}\boldsymbol{B} - \boldsymbol{B}^{\mathrm{H}}\boldsymbol{AX}' + \boldsymbol{B}^{\mathrm{H}}\boldsymbol{B}\right\},\end{aligned} \tag{3.28}$$

also used for the derivation of the *Block Generalized Minimal RESidual* (BGMRES) *algorithm* [223, 224, 203]. However, if $\boldsymbol{A}$ is Hermitian and positive definite, the $\boldsymbol{A}$-norm exists (cf. Appendix A.2) and minimizing the squared Frobenius norm over all $\boldsymbol{X}'$ is equal to minimizing the squared $\boldsymbol{A}$-norm of the error matrix $\boldsymbol{E}' = \boldsymbol{X} - \boldsymbol{X}' \in \mathbb{C}^{N \times M}$ written as

---

[12] Again, we assume the initial value of the approximation of $\boldsymbol{X}$ to be $\boldsymbol{X}_0 = \boldsymbol{0}_{N \times M}$.

$$\begin{aligned}
\varepsilon_A\left(X'\right) &= \left\|E'\right\|_A^2 = \left\|X - X'\right\|_A^2 \\
&= \operatorname{tr}\left\{X'^{,\mathrm{H}}AX' - X'^{,\mathrm{H}}B - B^{\mathrm{H}}X' + X^{\mathrm{H}}B\right\},
\end{aligned} \tag{3.29}$$

since both minima compute as $X = A^\dagger B = (A^{\mathrm{H}}A)^{-1}A^{\mathrm{H}}B = A^{-1}B$. The matrix $A^\dagger$ is the left-hand side *Moore–Penrose pseudoinverse* [167, 168, 184, 97] of A. In the sequel of this section, we restrict ourselves to Hermitian and positive definite matrices A and use the A-norm of the error matrix E' as the cost function of the optimization.

In a further step, we solve the optimization in an iterative manner. Starting from an initial approximation $X_0 \in \mathbb{C}^{N\times M}$, the new approximate solution $X_{\ell+1} \in \mathbb{C}^{N\times M}$, $\ell \in \{0, 1, \ldots, L-1\}$, at iteration step $\ell+1$ is obtained by improving the previous approximation $X_\ell \in \mathbb{C}^{N\times M}$ based on an M-dimensional search in the subspace spanned by the columns of the *direction matrix* $D_\ell \in \mathbb{C}^{N\times M}$, i.e.,

$$\boxed{X_{\ell+1} = X_\ell + D_\ell \Phi_\ell.} \tag{3.30}$$

The optimal *step size matrix* $\Phi_\ell \in \mathbb{C}^{M\times M}$ is obtained by minimizing the chosen cost function $\varepsilon_A(X')$ where $X' = X_\ell + D_\ell\Phi$, i.e.,

$$\Phi_\ell = \underset{\Phi \in \mathbb{C}^{M\times M}}{\operatorname{argmin}}\ \varepsilon_A\left(X_\ell + D_\ell\Phi\right). \tag{3.31}$$

Partially differentiating $\varepsilon_A(X_\ell + D_\ell\Phi)$ according to Φ^*, and setting each element of the result equal to zero, yields the optimal step size matrix

$$\left|\ \Phi_\ell = \left(D_\ell^{\mathrm{H}}AD_\ell\right)^{-1}D_\ell^{\mathrm{H}}\left(B - AX_\ell\right) = \left(D_\ell^{\mathrm{H}}AD_\ell\right)^{-1}D_\ell^{\mathrm{H}}R_\ell,\ \right| \tag{3.32}$$

with the residuum matrix $R_\ell = B - AX_\ell \in \mathbb{C}^{N\times M}$. Note that the repeated application of the update formula in Equation (3.30) to X_i for $i \in \{0, 1, \ldots, \ell\}$, yields

$$X_{\ell+1} = X_0 + \sum_{i=0}^{\ell} D_i\Phi_i. \tag{3.33}$$

As we will see, the BCG algorithm is a special case of a *Block Conjugate Direction* (BCD) *method*,[13] i.e., an algorithm where each column of one direction matrix is A-conjugate[14] to all columns of all the remaining direction matrices. It holds for $i, j \subset \{0, 1, \ldots, \ell\}$ that

$$D_i^{\mathrm{H}}AD_j = 0_{M\times M} \quad \text{for all } i \neq j. \tag{3.34}$$

In the following, we denote search direction matrices with this property as A-conjugate or A-orthogonal. This special choice of search direction matrices

---

[13] The first algorithm based on *conjugate directions* has been presented in [72] which denoted the authors as the *method of orthogonal vectors*.

[14] Two vectors x and y are A-*conjugate* or A-*orthogonal* if $x^{\mathrm{H}}Ay = 0$.

makes BCD methods, and therefore, the BCG algorithm, so powerful because it splits the $(\ell+1)M$-dimensional optimization problem at iteration step $\ell+1$, into $\ell+1$ times M-dimensional optimization problems without losing optimality. This fundamental characteristic of BCD algorithms can be formulated more detailed as the following proposition.

Proposition 3.4. *The fact that the search direction matrices are mutually A-conjugate ensures that the choice of $\boldsymbol{\Phi}_\ell$ given by Equation (3.32) minimizes the cost function $\varepsilon_A(\boldsymbol{X}')$ not only in the subspace spanned by the columns of the direction matrix $\boldsymbol{D}_\ell$ (cf. Equation 3.30), but also in the whole subspace spanned by the columns of the direction matrices $\boldsymbol{D}_0$, $\boldsymbol{D}_1$, ..., and $\boldsymbol{D}_\ell$. In other words, with $\boldsymbol{X}_\ell$ according to Equation (3.33), it holds*

$$\varepsilon_A\left(\boldsymbol{X}_{\ell+1}\right) = \min_{\boldsymbol{\Phi}\in\mathbb{C}^{M\times M}} \varepsilon_A\left(\boldsymbol{X}_\ell + \boldsymbol{D}_\ell\boldsymbol{\Phi}\right) = \min_{\substack{\left(\boldsymbol{F}_0',\boldsymbol{F}_1',\dots,\boldsymbol{F}_\ell'\right)\\ \boldsymbol{F}_i'\in\mathbb{C}^{M\times M}}} \varepsilon_A\left(\boldsymbol{X}_0 + \sum_{i=0}^{\ell} \boldsymbol{D}_i\boldsymbol{F}_i'\right). \quad (3.35)$$

Proof. We prove Proposition 3.4 by computing the optimal weights

$$\left(\boldsymbol{F}_0, \boldsymbol{F}_1, \dots, \boldsymbol{F}_\ell\right) = \operatorname*{argmin}_{\substack{\left(\boldsymbol{F}_0',\boldsymbol{F}_1',\dots,\boldsymbol{F}_\ell'\right)\\ \boldsymbol{F}_i'\in\mathbb{C}^{M\times M}}} \varepsilon_A\left(\boldsymbol{X}_0 + \sum_{i=0}^{\ell} \boldsymbol{D}_i\boldsymbol{F}_i'\right), \quad (3.36)$$

of the right-hand side of Equation (3.35), and showing that they are identical to $\boldsymbol{\Phi}_i$, $i \in \{0,1,\dots,\ell\}$, given by Equation (3.32).

With the definition of the cost function in Equation (3.29), we get

$$\varepsilon_A\left(\boldsymbol{X}_0 + \sum_{i=0}^{\ell} \boldsymbol{D}_i\boldsymbol{F}_i'\right) = \operatorname{tr}\left\{\boldsymbol{X}_0^{\mathrm{H}}\boldsymbol{A}\boldsymbol{X}_0 + \boldsymbol{X}_0^{\mathrm{H}}\boldsymbol{A}\sum_{i=0}^{\ell}\boldsymbol{D}_i\boldsymbol{F}_i' + \sum_{i=0}^{\ell}\boldsymbol{F}_i'^{,\mathrm{H}}\boldsymbol{D}_i^{\mathrm{H}}\boldsymbol{A}\boldsymbol{X}_0\right.$$
$$+ \left(\sum_{i=0}^{\ell}\boldsymbol{F}_i'^{,\mathrm{H}}\boldsymbol{D}_i^{\mathrm{H}}\right)\boldsymbol{A}\left(\sum_{i=0}^{\ell}\boldsymbol{D}_i\boldsymbol{F}_i'\right) - \boldsymbol{X}_0^{\mathrm{H}}\boldsymbol{B} - \boldsymbol{B}^{\mathrm{H}}\boldsymbol{X}_0$$
$$\left. - \sum_{i=0}^{\ell}\boldsymbol{F}_i'^{,\mathrm{H}}\boldsymbol{D}_i^{\mathrm{H}}\boldsymbol{B} - \boldsymbol{B}^{\mathrm{H}}\sum_{i=0}^{\ell}\boldsymbol{D}_i\boldsymbol{F}_i' + \boldsymbol{X}^{\mathrm{H}}\boldsymbol{B}\right\}. \quad (3.37)$$

Due to the $\boldsymbol{A}$-conjugacy of the search direction matrices (cf. Equation 3.34), all the summands with the term $\boldsymbol{D}_i^{\mathrm{H}}\boldsymbol{A}\boldsymbol{D}_j$, $i,j \in \{0,1,\dots,\ell\}$, $i \neq j$, vanish and Equation (3.37) simplifies to

$$\varepsilon_A\left(\boldsymbol{X}_0 + \sum_{i=0}^{\ell} \boldsymbol{D}_i\boldsymbol{F}_i'\right) = \operatorname{tr}\left\{\boldsymbol{X}_0^{\mathrm{H}}\boldsymbol{A}\boldsymbol{X}_0 - \boldsymbol{X}_0^{\mathrm{H}}\boldsymbol{B} - \boldsymbol{B}^{\mathrm{H}}\boldsymbol{X}_0 + \boldsymbol{X}^{\mathrm{H}}\boldsymbol{B}\right.$$
$$+ \sum_{i=1}^{\ell}\left(\boldsymbol{F}_i'^{,\mathrm{H}}\boldsymbol{D}_i^{\mathrm{H}}\boldsymbol{A}\boldsymbol{D}_i\boldsymbol{F}_i' - \boldsymbol{F}_i'^{,\mathrm{H}}\boldsymbol{D}_i^{\mathrm{H}}\boldsymbol{B} - \boldsymbol{B}^{\mathrm{H}}\boldsymbol{D}_i\boldsymbol{F}_i'\right.$$
$$\left.\left. + \boldsymbol{X}_0^{\mathrm{H}}\boldsymbol{A}\boldsymbol{D}_i\boldsymbol{F}_i' + \boldsymbol{F}_i'^{,\mathrm{H}}\boldsymbol{D}_i^{\mathrm{H}}\boldsymbol{A}\boldsymbol{X}_0\right)\right\}. \quad (3.38)$$

By partially differentiating with respect to $\boldsymbol{F}_i'^{,*}$, $i \in \{0, 1, \ldots, \ell\}$, and setting the result equal to $\boldsymbol{0}_{M \times M}$ for $\boldsymbol{F}_i' = \boldsymbol{F}_i$, we get the following system of linear equations:

$$\left. \frac{\partial \varepsilon_A \left(\boldsymbol{X}_0 + \sum_{i=0}^{\ell} \boldsymbol{D}_i \boldsymbol{F}_i' \right)}{\boldsymbol{F}_i'^{,*}} \right|_{\boldsymbol{F}_i' = \boldsymbol{F}_i} = \boldsymbol{D}_i^{\mathrm{H}} \boldsymbol{A} \boldsymbol{D}_i \boldsymbol{F}_i - \boldsymbol{D}_i^{\mathrm{H}} \left(\boldsymbol{B} - \boldsymbol{A} \boldsymbol{X}_0 \right)$$

$$\overset{!}{=} \boldsymbol{0}_{M \times M}, \tag{3.39}$$

with the solutions

$$\boldsymbol{F}_i = \left(\boldsymbol{D}_i^{\mathrm{H}} \boldsymbol{A} \boldsymbol{D}_i \right)^{-1} \boldsymbol{D}_i^{\mathrm{H}} \boldsymbol{R}_0, \quad i \in \{0, 1, \ldots, \ell\}. \tag{3.40}$$

Consequently, $\boldsymbol{F}_i = \boldsymbol{\Phi}_i$ for all $i \in \{0, 1, \ldots, \ell\}$ (cf. Equation 3.32) if we can show that $\boldsymbol{D}_i^{\mathrm{H}} \boldsymbol{R}_0 = \boldsymbol{D}_i^{\mathrm{H}} \boldsymbol{R}_i$. Use the definition of the residuum matrix and Equation (3.33) to get

$$\boldsymbol{D}_i^{\mathrm{H}} \boldsymbol{R}_i = \boldsymbol{D}_i^{\mathrm{H}} \boldsymbol{B} - \boldsymbol{D}_i^{\mathrm{H}} \boldsymbol{A} \left(\boldsymbol{X}_0 + \sum_{j=0}^{i-1} \boldsymbol{D}_j \boldsymbol{\Phi}_j \right) = \boldsymbol{D}_i^{\mathrm{H}} \left(\boldsymbol{B} - \boldsymbol{A} \boldsymbol{X}_0 \right)$$

$$= \boldsymbol{D}_i^{\mathrm{H}} \boldsymbol{R}_0, \tag{3.41}$$

where we exploited again the fact that the search direction matrices are mutually $\boldsymbol{A}$-conjugate. $\qquad\square$

Based on Proposition 3.4, we can also conclude that iterative BCD methods compute the exact solution $\boldsymbol{X}$ in less than or equal to $L = N/M$ steps because at the final step, the solution obtained from the iterative M-dimensional optimizations is identical to the solution of the optimization problem in the whole $\mathbb{C}^N$ which is clearly $\boldsymbol{X}_L = \boldsymbol{X} = \boldsymbol{A}^{-1} \boldsymbol{B}$.[15]

Now, since we have seen that the $\boldsymbol{A}$-conjugate search direction matrices simplify tremendously the iterative solving of a system of linear equations, the following question remains: How can we generate the $\boldsymbol{A}$-conjugate search direction matrices in an efficient way?

For this aim, we choose arbitrary $N \times M$ matrices $\boldsymbol{J}_0$, $\boldsymbol{J}_1$, ..., and $\boldsymbol{J}_{L-1}$, $L = N/M$, whose columns are linearly independent, i. e., the matrix $\boldsymbol{J} = [\boldsymbol{J}_0, \boldsymbol{J}_1, \ldots, \boldsymbol{J}_{L-1}] \in \mathbb{C}^{N \times N}$ has full rank. Besides, we assume that we have already computed the mutually $\boldsymbol{A}$-conjugate search direction matrices $\boldsymbol{D}_0$, $\boldsymbol{D}_1$, ..., and $\boldsymbol{D}_\ell$ based on the matrices $\boldsymbol{J}_0$, $\boldsymbol{J}_1$, ..., and $\boldsymbol{J}_\ell$ starting with $\boldsymbol{D}_0 = \boldsymbol{J}_0$. For the next iteration step, we are interested in the search direction matrix $\boldsymbol{D}_{\ell+1}$ which has to be $\boldsymbol{A}$-conjugate to all previous search direction matrices. The *conjugate Gram–Schmidt procedure* similar to the one used in

---

[15] Note that if $\boldsymbol{A} = \bar{\boldsymbol{A}} + \alpha \boldsymbol{I}_N$ is Hermitian and positive definite and $\bar{\boldsymbol{A}}$ has rank r, then the optimum is even achieved after at most $\lceil r/M \rceil + 1$ steps (cf. [94], Theorem 10.2.5).

Section 3.2,[16] can be used to $\boldsymbol{A}$-orthogonalize $\boldsymbol{J}_{\ell+1} \in \mathbb{C}^{N \times M}$ with respect to the previous search direction matrices. We obtain

$$\boxed{\boldsymbol{D}_{\ell+1} = \boldsymbol{J}_{\ell+1} + \sum_{i=0}^{\ell} \boldsymbol{D}_i \boldsymbol{\Theta}_{i,\ell},} \tag{3.42}$$

where $\boldsymbol{\Theta}_{i,\ell} \in \mathbb{C}^{M \times M}$ is chosen such that $\boldsymbol{D}_{\ell+1}$ is $\boldsymbol{A}$-conjugate to all previous search direction matrices, i. e., it holds for $i \in \{0, 1, \ldots, \ell\}$:

$$\boldsymbol{D}_i^{\mathrm{H}} \boldsymbol{A} \boldsymbol{D}_{\ell+1} = \boldsymbol{D}_i^{\mathrm{H}} \boldsymbol{A} \boldsymbol{J}_{\ell+1} + \sum_{j=0}^{\ell} \boldsymbol{D}_i^{\mathrm{H}} \boldsymbol{A} \boldsymbol{D}_j \boldsymbol{\Theta}_{j,\ell} \overset{!}{=} \boldsymbol{0}_{M \times M}. \tag{3.43}$$

Remembering the fact that the previous search direction matrices $\boldsymbol{D}_j$, $j \in \{0, 1, \ldots, \ell\}$, are already $\boldsymbol{A}$-orthogonal, all the summands over j except for $j = i$ vanish, and the matrix $\boldsymbol{\Theta}_{i,\ell}$ computes finally as

$$\left| \boldsymbol{\Theta}_{i,\ell} = - \left(\boldsymbol{D}_i^{\mathrm{H}} \boldsymbol{A} \boldsymbol{D}_i \right)^{-1} \boldsymbol{D}_i^{\mathrm{H}} \boldsymbol{A} \boldsymbol{J}_{\ell+1}. \right| \tag{3.44}$$

Before simplifying Equation (3.42) by choosing the matrices $\boldsymbol{J}_i$, $i \in \{0, 1, \ldots, \ell + 1\}$, in a special way, we list the additional properties of the BCD method defined by Equations (3.30), (3.32), (3.42), and (3.44), besides the main characteristic of $\boldsymbol{A}$-conjugate search direction matrices ensured by the recursion formula in Equation (3.42) with the special choice of $\boldsymbol{\Theta}_{i,\ell}$ in Equation (3.44).

Proposition 3.5. *The $\boldsymbol{A}$-conjugate search direction matrices $\boldsymbol{D}_0$, $\boldsymbol{D}_1$, $\ldots$, and $\boldsymbol{D}_{\ell+1}$, generated using Equations (3.42) and (3.44), the matrices $\boldsymbol{J}_0$, $\boldsymbol{J}_1$, $\ldots$, and $\boldsymbol{J}_\ell$, and the residuum matrix $\boldsymbol{R}_{\ell+1}$ satisfy the following relations:*

$$\boldsymbol{D}_j^{\mathrm{H}} \boldsymbol{R}_{\ell+1} = \boldsymbol{0}_{M \times M}, \qquad\qquad j < \ell + 1, \tag{3.45}$$

$$\boldsymbol{J}_j^{\mathrm{H}} \boldsymbol{R}_{\ell+1} = \boldsymbol{0}_{M \times M}, \qquad\qquad j < \ell + 1, \tag{3.46}$$

$$\boldsymbol{J}_{\ell+1}^{\mathrm{H}} \boldsymbol{R}_{\ell+1} = \boldsymbol{D}_{\ell+1}^{\mathrm{H}} \boldsymbol{R}_{\ell+1}. \tag{3.47}$$

Proof. For the proof of Equation (3.45), we define the error matrix $\boldsymbol{E}_{\ell+1}$ at the $(\ell + 1)$th iteration step to be

$$\boldsymbol{E}_{\ell+1} = \boldsymbol{X} - \boldsymbol{X}_{\ell+1} = \boldsymbol{X}_0 + \sum_{i=0}^{L-1} \boldsymbol{D}_i \boldsymbol{\Phi}_i - \boldsymbol{X}_0 - \sum_{i=0}^{\ell} \boldsymbol{D}_i \boldsymbol{\Phi}_i = \sum_{i=\ell+1}^{L-1} \boldsymbol{D}_i \boldsymbol{\Phi}_i, \tag{3.48}$$

---

[16] Compared to the Gram–Schmidt procedure used in Section 3.2 for orthogonalization, the conjugate Gram–Schmidt method is used for $\boldsymbol{A}$-orthogonalization. Note that the sign before the sum in Equation (3.42) can be changed without loss of generality if the matrix coefficients $\boldsymbol{\Theta}_{i,\ell}$ are adjusted accordingly.

where we used Equation (3.33) to replace $\boldsymbol{X}_{\ell+1}$ as well as $\boldsymbol{X} = \boldsymbol{X}_L$. Thus, the residuum matrix $\boldsymbol{R}_{\ell+1}$ can be written as

$$\boldsymbol{R}_{\ell+1} = \boldsymbol{B} - \boldsymbol{A}\boldsymbol{X}_{\ell+1} = \boldsymbol{A}\left(\boldsymbol{X} - \boldsymbol{X}_{\ell+1}\right) = \boldsymbol{A}\boldsymbol{E}_{\ell+1}, \tag{3.49}$$

and we get for $j \in \{0, 1, \ldots, \ell\}$:

$$\boldsymbol{D}_j^{\mathrm{H}} \boldsymbol{R}_{\ell+1} = \boldsymbol{D}_j^{\mathrm{H}} \boldsymbol{A}\boldsymbol{E}_{\ell+1} = \sum_{i=\ell+1}^{L-1} \boldsymbol{D}_j^{\mathrm{H}} \boldsymbol{A}\boldsymbol{D}_i \boldsymbol{\Phi}_i = \boldsymbol{0}_{M \times M}, \tag{3.50}$$

due to the mutual $\boldsymbol{A}$-conjugacy of the search direction matrices. Note that $i > j$ for all summands. Thus, Equation (3.45) of Proposition 3.5 is verified.

In order to verify the last two Equations (3.46) and (3.47), we take Equation (3.42) Hermitian, replace $\ell+1$ by j, and multiply $\boldsymbol{R}_{\ell+1}$ on the right-hand side. We get

$$\boldsymbol{D}_j^{\mathrm{H}} \boldsymbol{R}_{\ell+1} = \boldsymbol{J}_j^{\mathrm{H}} \boldsymbol{R}_{\ell+1} + \sum_{i=0}^{j-1} \boldsymbol{\Theta}_{i,j-1}^{\mathrm{H}} \boldsymbol{D}_i^{\mathrm{H}} \boldsymbol{R}_{\ell+1}. \tag{3.51}$$

It holds finally due to Equation (3.45) that

$$\boldsymbol{J}_j^{\mathrm{H}} \boldsymbol{R}_{\ell+1} = \begin{cases} \boldsymbol{0}_{M \times M}, & j < \ell+1, \\ \boldsymbol{D}_j^{\mathrm{H}} \boldsymbol{R}_{\ell+1}, & j = \ell+1, \end{cases} \tag{3.52}$$

which proves the remaining equations of Proposition 3.5. $\qquad\square$

In the following, we present the BCG algorithm as a computationally more efficient BCD method. The crucial idea is to choose the basis $\boldsymbol{J}$ such that most of the summands in Equation (3.42) vanish. At iteration step $\ell + 1$, the BCG algorithm sets $\boldsymbol{J}_{\ell+1} = \boldsymbol{R}_{\ell+1}$ beginning with $\boldsymbol{J}_0 = \boldsymbol{R}_0 = \boldsymbol{B} - \boldsymbol{A}\boldsymbol{X}_0$. First, this choice makes the residuum matrices mutually orthogonal due to Equation (3.45), i.e.,

$$\boldsymbol{R}_j^{\mathrm{H}} \boldsymbol{R}_{\ell+1} = \boldsymbol{0}_{M \times M} \quad \text{for all } j < \ell+1. \tag{3.53}$$

Thus, as far as the residuum matrix is unequal to $\boldsymbol{0}_{M \times M}$ which is only the case when the exact solution is reached, the columns of the residuum matrices are linearly independent. Remember that this was a necessary requirement of the basis $\boldsymbol{J}$.

In order to show an even more important consequence of using the residuum matrix $\boldsymbol{R}_{\ell+1}$ as $\boldsymbol{J}_{\ell+1}$ for the conjugate Gram–Schmidt procedure in Equation (3.42), we firstly reformulate the residuum matrix $\boldsymbol{R}_{\ell+1}$ as follows:

$$\left| \boldsymbol{R}_{\ell+1} = \boldsymbol{B} - \boldsymbol{A}\boldsymbol{X}_{\ell+1} = \boldsymbol{B} - \boldsymbol{A}\left(\boldsymbol{X}_\ell + \boldsymbol{D}_\ell \boldsymbol{\Phi}_\ell\right) = \boldsymbol{R}_\ell - \boldsymbol{A}\boldsymbol{D}_\ell \boldsymbol{\Phi}_\ell, \right| \tag{3.54}$$

by using the update formula in Equation (3.30). Then, if we replace ℓ by i and multiply its Hermitian on the left-hand side on $\boldsymbol{R}_{\ell+1}$, we get

$$R_{i+1}^{\mathrm{H}} R_{\ell+1} = R_i^{\mathrm{H}} R_{\ell+1} - \boldsymbol{\Phi}_i^{\mathrm{H}} \boldsymbol{D}_i^{\mathrm{H}} \boldsymbol{A} R_{\ell+1}. \tag{3.55}$$

With Equation (3.53) and the fact that $\boldsymbol{J}_{\ell+1} = \boldsymbol{R}_{\ell+1}$, it holds for $i < \ell+1$ that

$$\boldsymbol{D}_i^{\mathrm{H}} \boldsymbol{A} \boldsymbol{J}_{\ell+1} = \begin{cases} -\boldsymbol{\Phi}_\ell^{\mathrm{H},-1} \boldsymbol{R}_{\ell+1}^{\mathrm{H}} \boldsymbol{R}_{\ell+1}, & i = \ell, \\ \boldsymbol{0}_{M \times M}, & i < \ell, \end{cases} \tag{3.56}$$

and the matrices $\boldsymbol{\Theta}_{i,\ell}$ in Equation (3.44) compute as

$$\boldsymbol{\Theta}_{i,\ell} = \begin{cases} \left(\boldsymbol{\Phi}_\ell^{\mathrm{H}} \boldsymbol{D}_\ell^{\mathrm{H}} \boldsymbol{A} \boldsymbol{D}_\ell\right)^{-1} \boldsymbol{R}_{\ell+1}^{\mathrm{H}} \boldsymbol{R}_{\ell+1}, & i = \ell, \\ \boldsymbol{0}_{M \times M}, & i < \ell, \end{cases} \tag{3.57}$$

i.e., they vanish for $i < \ell$. Hence, we can rewrite Equations (3.42) and (3.44) as

$$\boxed{\boldsymbol{D}_{\ell+1} = \boldsymbol{R}_{\ell+1} + \boldsymbol{D}_\ell \boldsymbol{\Theta}_\ell,} \tag{3.58}$$

and

$$\boldsymbol{\Theta}_\ell := \boldsymbol{\Theta}_{\ell,\ell} = \left(\boldsymbol{\Phi}_\ell^{\mathrm{H}} \boldsymbol{D}_\ell^{\mathrm{H}} \boldsymbol{A} \boldsymbol{D}_\ell\right)^{-1} \boldsymbol{R}_{\ell+1}^{\mathrm{H}} \boldsymbol{R}_{\ell+1}, \tag{3.59}$$

respectively, which is computationally more efficient since only the $\boldsymbol{A}$-conjugacy to $\boldsymbol{D}_\ell$ has to be constructed. Note that this simplification is not possible for non-Hermitian matrices $\boldsymbol{A}$. Thus, there exists no BCG-like algorithm for non-Hermitian matrices [249, 68].

Proposition 3.6. *Each column of the approximation $\boldsymbol{X}_{\ell+1} - \boldsymbol{X}_0$ of $\boldsymbol{X} - \boldsymbol{X}_0$ at iteration step $\ell+1$ lies in the subspace spanned by the columns of the search direction matrices $\boldsymbol{D}_0$, $\boldsymbol{D}_1$, ..., and $\boldsymbol{D}_\ell$ which is identical to the subspace spanned by the columns of the residuum matrices $\boldsymbol{R}_0$, $\boldsymbol{R}_1$, ..., and $\boldsymbol{R}_\ell$ as well as the Krylov subspace $\mathcal{K}^{((\ell+1)M)}(\boldsymbol{A}, \boldsymbol{D}_0) = \mathcal{K}^{((\ell+1)M)}(\boldsymbol{A}, \boldsymbol{R}_0)$.*

Proof. The fact that each column of $\boldsymbol{X}_{\ell+1} - \boldsymbol{X}_0$ lies in the subspace spanned by the columns of the search direction matrices, i.e.,

$$\boldsymbol{X}_{\ell+1} - \boldsymbol{X}_0 = \mathrm{range}\left\{\begin{bmatrix} \boldsymbol{D}_0 & \boldsymbol{D}_1 & \cdots & \boldsymbol{D}_\ell \end{bmatrix}\right\}, \tag{3.60}$$

is already proven since $\boldsymbol{X}_{\ell+1}$ can be expressed as a linear combination of the columns of the search direction matrices $\boldsymbol{D}_0$, $\boldsymbol{D}_1$, ..., and $\boldsymbol{D}_\ell$ (cf. Equation 3.33).

It remains to show that the subspace spanned by the columns of the search direction matrices is identical to the subspace spanned by the columns of the residuum matrices which has to be the Krylov subspace, i.e.,

$$\mathrm{range}\left\{\begin{bmatrix} \boldsymbol{D}_0 & \boldsymbol{D}_1 & \cdots & \boldsymbol{D}_\ell \end{bmatrix}\right\} = \mathrm{range}\left\{\begin{bmatrix} \boldsymbol{R}_0 & \boldsymbol{R}_1 & \cdots & \boldsymbol{R}_\ell \end{bmatrix}\right\} \tag{3.61}$$

$$= \mathcal{K}^{((\ell+1)M)}(\boldsymbol{A}, \boldsymbol{D}_0) = \mathcal{K}^{((\ell+1)M)}(\boldsymbol{A}, \boldsymbol{R}_0).$$

First of all, $\mathcal{K}^{((\ell+1)M)}(\boldsymbol{A}, \boldsymbol{D}_0) = \mathcal{K}^{((\ell+1)M)}(\boldsymbol{A}, \boldsymbol{R}_0)$ since $\boldsymbol{D}_0 = \boldsymbol{J}_0 = \boldsymbol{R}_0 = \boldsymbol{B} - \boldsymbol{A} \boldsymbol{X}_0$. Further on, we prove by induction on ℓ. Consider the case $\ell = 0$.

Clearly, range$\{\boldsymbol{D}_0\}$ = range$\{\boldsymbol{R}_0\}$ = $\mathcal{K}^{(M)}(\boldsymbol{A}, \boldsymbol{R}_0)$ by using again $\boldsymbol{D}_0 = \boldsymbol{R}_0$. Now, assume that Equation (3.61) holds for $\ell = i$, i.e., with the polynomial representation defined in Equation (3.2),

$$\boldsymbol{D}_i = \sum_{\ell=0}^{i} \boldsymbol{A}^\ell \boldsymbol{R}_0 \boldsymbol{\Psi}_{\ell,i} \quad \text{and} \quad \boldsymbol{R}_i = \sum_{\ell=0}^{i} \boldsymbol{A}^\ell \boldsymbol{R}_0 \boldsymbol{\Psi}'_{\ell,i}. \tag{3.62}$$

Then, use Equations (3.54) and (3.58) to get

$$\boldsymbol{R}_{i+1} = \sum_{\ell=0}^{i} \boldsymbol{A}^\ell \boldsymbol{R}_0 \boldsymbol{\Psi}'_{\ell,i} - \boldsymbol{A} \sum_{\ell=0}^{i} \boldsymbol{A}^\ell \boldsymbol{R}_0 \boldsymbol{\Psi}_{\ell,i} \boldsymbol{\Phi}_i =: \sum_{\ell=0}^{i+1} \boldsymbol{A}^\ell \boldsymbol{R}_0 \boldsymbol{\Psi}'_{\ell,i+1}, \tag{3.63}$$

$$\boldsymbol{D}_{i+1} = \sum_{\ell=0}^{i+1} \boldsymbol{A}^\ell \boldsymbol{R}_0 \boldsymbol{\Psi}'_{\ell,i+1} + \sum_{\ell=0}^{i} \boldsymbol{A}^\ell \boldsymbol{R}_0 \boldsymbol{\Psi}_{\ell,i} \boldsymbol{\Phi}_i =: \sum_{\ell=0}^{i+1} \boldsymbol{A}^\ell \boldsymbol{R}_0 \boldsymbol{\Psi}_{\ell,i+1}, \tag{3.64}$$

which shows that Equation (3.61) holds also for $\ell = i + 1$ and the proof of Proposition 3.6 is completed. $\qquad\square$

The fact that the columns of the search direction matrices as well as the columns of the residuum matrices lie in a Krylov subspace is actually the reason why the new residuum matrix $\boldsymbol{R}_{\ell+1}$ has to be only $\boldsymbol{A}$-orthogonalized with respect to the search direction matrix $\boldsymbol{D}_\ell$. Krylov subspaces have the property that $\boldsymbol{A}\mathcal{K}^{(\ell M)}(\boldsymbol{A}, \boldsymbol{R}_0)$ is included in $\mathcal{K}^{((\ell+1)M)}(\boldsymbol{A}, \boldsymbol{R}_0)$.[17] Therefore, since $\boldsymbol{R}_{\ell+1}$ is orthogonal to $\mathcal{K}^{((\ell+1)M)}(\boldsymbol{A}, \boldsymbol{R}_0)$ due to Equation (3.53), it is already $\boldsymbol{A}$-orthogonal to a matrix build up by linearly independent elements of $\mathcal{K}^{(\ell M)}(\boldsymbol{A}, \boldsymbol{R}_0)$. That is the reason why the matrices $\boldsymbol{\Theta}_{i,\ell}$ in Equation (3.42) vanish for $i < \ell$ which results in the computationally efficient Equation (3.58). Note that only the BCG algorithm with $\boldsymbol{J}_{\ell+1} = \boldsymbol{R}_{\ell+1}$ approximates each column of $\boldsymbol{X} - \boldsymbol{X}_0$ in a Krylov subspace. This is not true for BCD methods in general.

Before presenting the BCG algorithm, we further simplify some of its equations. With Equation (3.47) and the fact that $\boldsymbol{J}_{\ell+1} = \boldsymbol{R}_{\ell+1}$, Equation (3.32) may be rewritten as

$$\boxed{\boldsymbol{\Phi}_\ell = \left(\boldsymbol{D}_\ell^{\mathrm{H}} \boldsymbol{A} \boldsymbol{D}_\ell\right)^{-1} \boldsymbol{R}_\ell^{\mathrm{H}} \boldsymbol{R}_\ell.} \tag{3.65}$$

Plugging Equation (3.65) in Equation (3.59), yields

$$\boxed{\boldsymbol{\Theta}_\ell = \left(\boldsymbol{R}_\ell^{\mathrm{H}} \boldsymbol{R}_\ell\right)^{-1} \boldsymbol{R}_{\ell+1}^{\mathrm{H}} \boldsymbol{R}_{\ell+1}.} \tag{3.66}$$

Finally, Equations (3.30), (3.54), (3.58), (3.65), and (3.66) are summarized in Algorithm 3.4. where we adjusted the formulas such that matrix-matrix products which occur several times, are computed only once. Starting from $\boldsymbol{X}_0$,

---

[17] The term $\boldsymbol{A}\mathcal{S}$ denotes a vector subspace where the elements are obtained by the multiplication of the matrix $\boldsymbol{A}$ with each element of the vector subspace $\mathcal{S}$.

Algorithm 3.4. computes the approximation X_d, $d \in \{1, 2, \ldots, L\}$, of the exact solution X of the system $AX = B$ of linear equations such that each column of $X_d - X_0$ lies in the D-dimensional Krylov subspace $\mathcal{K}^{(D)}(A, B - AX_0)$ (cf. Proposition 3.6). Remember that the dimension D is restricted to integer multiples of the number M of right-hand sides, i.e., $D = dM$. At the last iteration step, i.e., for $\ell = d - 1$, Algorithm 3.4. is stopped after executing Line 8 since the desired approximation X_d has been already computed and the matrices R_d, G_d, Θ_{d-1}, and D_d are no longer needed.

Algorithm 3.4. Block Conjugate Gradient (BCG)

$\quad$ Choose an arbitrary matrix $X_0 \in \mathbb{C}^{N \times M}$ (e.g., $X_0 \leftarrow 0_{N \times M}$).
2: $\ R_0 \leftarrow B - AX_0$
$\quad\ \ G_0 \leftarrow R_0^{\mathrm{H}} R_0$
4: $\ D_0 \leftarrow R_0$
$\quad\ \ $**for** $\ell = 0, 1, \ldots, d - 1$ **do**
6: $\quad V \leftarrow AD_\ell$
$\quad\quad\ \ \Phi_\ell \leftarrow (D_\ell^{\mathrm{H}} V)^{-1} G_\ell$
8: $\quad X_{\ell+1} \leftarrow X_\ell + D_\ell \Phi_\ell$
$\quad\quad\ \ R_{\ell+1} \leftarrow R_\ell - V \Phi_\ell$
10: $\quad G_{\ell+1} \leftarrow R_{\ell+1}^{\mathrm{H}} R_{\ell+1}$
$\quad\quad\ \ \Theta_\ell \leftarrow G_\ell^{-1} G_{\ell+1}$
12: $\quad D_{\ell+1} \leftarrow R_{\ell+1} + D_\ell \Theta_\ell$
$\quad\ \ $**end for**

3.4.2 Rate of Convergence

In this subsection, we analyze the convergence of the BCG algorithm by investigating the behavior of the squared A-norm of the error matrix $E_{\ell+1} = X - X_{\ell+1} \in \mathbb{C}^{N \times M}$ (cf. Equation 3.48) between the exact solution $X \in \mathbb{C}^{N \times M}$ and the approximate BCG solution $X_{\ell+1} \in \mathbb{C}^{N \times M}$, in terms of the number $\ell + 1$ of performed iterations. Note that this error norm, i.e., $\|E_{\ell+1}\|_A^2$, is equal to the cost function $\varepsilon_A(X_{\ell+1})$ which is minimized by the BCG method at the $(\ell + 1)$th iteration step (see Equation 3.29 and Proposition 3.4). If $\varepsilon_{\ell+1,j} = E_{\ell+1} e_{j+1} \in \mathbb{C}^N$, $j \in \{0, 1, \ldots, M - 1\}$, denotes the $(j + 1)$th column of the error matrix $E_{\ell+1}$, its squared A-norm reads as (cf. Appendix A.2)

$$\|E_{\ell+1}\|_A^2 = \mathrm{tr}\left\{E_{\ell+1}^{\mathrm{H}} A E_{\ell+1}\right\} = \sum_{j=0}^{M-1} \varepsilon_{\ell+1,j}^{\mathrm{H}} A \varepsilon_{\ell+1,j}$$

$$= \sum_{j=0}^{M-1} \|\varepsilon_{\ell+1,j}\|_A^2 . \tag{3.67}$$

In the following, we investigate the squared $\boldsymbol{A}$-norm of the $(j+1)$th column of the error matrix $\boldsymbol{E}_{\ell+1}$, i.e., $\|\boldsymbol{\varepsilon}_{\ell+1,j}\|_{\boldsymbol{A}}^2$, and use the result to get an expression for $\|\boldsymbol{E}_{\ell+1}\|_{\boldsymbol{A}}^2$ based on Equation (3.67).

From Proposition 3.6, it follows that the approximate solution $\boldsymbol{X}_{\ell+1}$ at the $(\ell+1)$th iteration step of the BCG algorithm, can be expressed by a polynomial in $\boldsymbol{A} \in \mathbb{C}^{N \times N}$ of degree ℓ, i.e.,

$$\boldsymbol{X}_{\ell+1} = \sum_{i=0}^{\ell} \boldsymbol{A}^i \boldsymbol{R}_0 \boldsymbol{\Psi}_{i,\ell} + \boldsymbol{X}_0. \tag{3.68}$$

If we choose the origin as the initial BCG approximation of the solution, i.e., $\boldsymbol{X}_0 = \boldsymbol{0}_{N \times M}$, which we assume in the sequel of this subsection, it holds $\boldsymbol{R}_0 = \boldsymbol{B} - \boldsymbol{A}\boldsymbol{X}_0 = \boldsymbol{B} \in \mathbb{C}^{N \times M}$, and the $(j+1)$th column of $\boldsymbol{X}_{\ell+1}$, i.e., $\boldsymbol{x}_{\ell+1,j} = \boldsymbol{X}_{\ell+1}\boldsymbol{e}_{j+1} \in \mathbb{C}^N$, can be written as (cf. also Equation 3.4)

$$\boldsymbol{x}_{\ell+1,j} = \sum_{k=0}^{M-1} \boldsymbol{\Psi}_{k,j}^{(\ell)}(\boldsymbol{A})\boldsymbol{b}_k = \sum_{k=0}^{M-1} \sum_{i=0}^{\ell} \psi_{k,j,i,\ell} \boldsymbol{A}^i \boldsymbol{b}_k, \tag{3.69}$$

where $\boldsymbol{b}_k = \boldsymbol{B}\boldsymbol{e}_{k+1} \in \mathbb{C}^N$, $k \in \{0, 1, \dots, M-1\}$, is the $(k+1)$th column of the right-hand side $\boldsymbol{B}$. With $\boldsymbol{\varepsilon}_{0,j} = (\boldsymbol{X} - \boldsymbol{X}_0)\boldsymbol{e}_{j+1} = \boldsymbol{X}\boldsymbol{e}_{j+1} =: \boldsymbol{x}_j \in \mathbb{C}^N$, the $(\ell+1)$th column of the error matrix $\boldsymbol{E}_{\ell+1}$ computes as

$$\boldsymbol{\varepsilon}_{\ell+1,j} = \boldsymbol{x}_j - \boldsymbol{x}_{\ell+1,j} = \boldsymbol{\varepsilon}_{0,j} - \sum_{k=0}^{M-1} \boldsymbol{\Psi}_{k,j}^{(\ell)}(\boldsymbol{A})\boldsymbol{b}_k, \tag{3.70}$$

i.e., all right-hand sides are taken into account in order to achieve the minimum error with respect to the $\boldsymbol{A}$-norm. As discussed in Section 3.1 (cf. Equation 3.5), another strategy is to approximate the $(j+1)$th column of $\boldsymbol{X}$, i.e., $\boldsymbol{x}_j$, considering only the corresponding column $\boldsymbol{b}_j$ of the right-hand side. This is done if we apply the BCG algorithm with $M = 1$, which is equal to the *Conjugate Gradient* (CG) *algorithm* [117], separately to each system $\boldsymbol{A}\boldsymbol{x}_j = \boldsymbol{b}_j$, $j \in \{0, 1, \dots, M-1\}$, of linear equations with one right-hand side. The resulting error vector of the $(j+1)$th system can be written as (cf. Equation 3.5)

$$\boldsymbol{\varepsilon}_{\ell+1,j}^{\mathrm{CG}} = \boldsymbol{\varepsilon}_{0,j} - \boldsymbol{\Psi}_j^{(\ell)}(\boldsymbol{A})\boldsymbol{b}_j \in \mathbb{C}^N, \tag{3.71}$$

where $\boldsymbol{\varepsilon}_{0,j} = \boldsymbol{\varepsilon}_{0,j}^{\mathrm{CG}} = \boldsymbol{x}_j$. Clearly, Equation (3.71) is a special case of Equation (3.70) where $\boldsymbol{\Psi}_{k,j}^{(\ell)}(\boldsymbol{A}) = \boldsymbol{0}_{N \times N}$ for $k \neq j$ and $\boldsymbol{\Psi}_{j,j}^{(\ell)}(\boldsymbol{A}) = \boldsymbol{\Psi}_j^{(\ell)}(\boldsymbol{A})$. Thus, the $\boldsymbol{A}$-norm of the error vector $\boldsymbol{\varepsilon}_{\ell+1,j}$ of the BCG algorithm is smaller than or at least equal to the $\boldsymbol{A}$-norm of the error vector $\boldsymbol{\varepsilon}_{\ell+1,j}^{\mathrm{CG}}$ of the CG algorithm for each $j \in \{0, 1, \dots, M-1\}$. Hence, the squared $\boldsymbol{A}$-norm of the BCG error at iteration step $\ell+1$ can be upper bounded by

$$\|\boldsymbol{E}_{\ell+1}\|_{\boldsymbol{A}}^2 = \sum_{j=0}^{M-1} \|\boldsymbol{\varepsilon}_{\ell+1,j}\|_{\boldsymbol{A}}^2 \leq \sum_{j=0}^{M-1} \|\boldsymbol{\varepsilon}_{\ell+1,j}^{\mathrm{CG}}\|_{\boldsymbol{A}}^2 = \|\boldsymbol{E}_{\ell+1}^{\mathrm{CG}}\|_{\boldsymbol{A}}^2, \tag{3.72}$$

where we used Equation (3.67) and the definition of the error matrix $\boldsymbol{E}_{\ell+1}^{\mathrm{CG}} = [\boldsymbol{\varepsilon}_{\ell+1,0}^{\mathrm{CG}}, \boldsymbol{\varepsilon}_{\ell+1,1}^{\mathrm{CG}}, \dots, \boldsymbol{\varepsilon}_{\ell+1,M-1}^{\mathrm{CG}}] \in \mathbb{C}^{N \times M}$.

Further on, we investigate $\|\boldsymbol{\varepsilon}_{\ell+1,j}^{\mathrm{CG}}\|_{\boldsymbol{A}}^2$ in more detail. With $\boldsymbol{b}_j = \boldsymbol{A}\boldsymbol{x}_j = \boldsymbol{A}\boldsymbol{\varepsilon}_{0,j}$, we get

$$\boldsymbol{\varepsilon}_{\ell+1,j}^{\mathrm{CG}} = \left(\boldsymbol{I}_N - \boldsymbol{\Psi}_j^{(\ell)}(\boldsymbol{A})\,\boldsymbol{A} \right) \boldsymbol{\varepsilon}_{0,j} =: P_j^{(\ell+1)}(\boldsymbol{A})\,\boldsymbol{\varepsilon}_{0,j}, \tag{3.73}$$

where $P_j^{(\ell+1)}(\boldsymbol{A}) \in \mathbb{C}^{N \times N}$ is a polynomial in $\boldsymbol{A}$ of degree $\ell+1$ with $P_j^{(\ell+1)}(0) = 1$.[18] Let the *eigen decomposition* (see, e.g., [131]) of the system matrix $\boldsymbol{A}$ be given by

$$\boldsymbol{A} = \sum_{n=0}^{N-1} \lambda_i \boldsymbol{u}_i \boldsymbol{u}_i^{\mathrm{H}} = \boldsymbol{U}\boldsymbol{\Lambda}\boldsymbol{U}^{\mathrm{H}}, \tag{3.74}$$

with the unitary matrix

$$\boldsymbol{U} = \begin{bmatrix} \boldsymbol{u}_0 & \boldsymbol{u}_1 & \cdots & \boldsymbol{u}_{N-1} \end{bmatrix} \in \mathbb{C}^{N \times N}, \tag{3.75}$$

of *eigenvectors* $\boldsymbol{u}_n \in \mathbb{C}^N$, $n \in \{0, 1, \dots, N-1\}$, and the diagonal matrix

$$\boldsymbol{\Lambda} = \mathrm{diag}\,\{\lambda_0, \lambda_1, \dots, \lambda_{N-1}\} \in \mathbb{R}_{0,+}^{N \times N}, \tag{3.76}$$

comprising the N *eigenvalues* $\lambda_0 \geq \lambda_1 \geq \cdots \geq \lambda_{N-1} > 0$. Recall that we restricted the system matrix $\boldsymbol{A}$ to be Hermitian and positive definite (cf. Subsection 3.4.1). Then, with the transformation $\boldsymbol{\varepsilon}_{0,j} = \boldsymbol{U}\boldsymbol{\xi}_j$ and the fact that $P_j^{(\ell+1)}(\boldsymbol{A}) = \boldsymbol{U}P_j^{(\ell+1)}(\boldsymbol{\Lambda})\boldsymbol{U}^{\mathrm{H}}$, we obtain from Equation (3.73)

$$\boldsymbol{\varepsilon}_{\ell+1,j}^{\mathrm{CG}} = \boldsymbol{U}P_j^{(\ell+1)}(\boldsymbol{\Lambda})\,\boldsymbol{\xi}_j = \sum_{n=0}^{N-1} \xi_{j,n} P_j^{(\ell+1)}(\lambda_n)\,\boldsymbol{u}_n, \tag{3.77}$$

with $\xi_{j,n} \in \mathbb{C}$ being the $(n+1)$th element of $\boldsymbol{\xi}_j \in \mathbb{C}^N$. Finally, the squared error norm $\|\boldsymbol{\varepsilon}_{\ell+1,j}^{\mathrm{CG}}\|_{\boldsymbol{A}}^2$ can be written as

$$\|\boldsymbol{\varepsilon}_{\ell+1,j}^{\mathrm{CG}}\|_{\boldsymbol{A}}^2 = \boldsymbol{\xi}_j^{\mathrm{H}} \left(P_j^{(\ell+1)}(\boldsymbol{\Lambda}) \right)^2 \boldsymbol{\Lambda}\boldsymbol{\xi}_j = \sum_{n=0}^{N-1} |\xi_{j,n}|^2 \lambda_n \left(P_j^{(\ell+1)}(\lambda_n) \right)^2, \tag{3.78}$$

if we remember that diagonal matrices can be permuted arbitrarily. As already mentioned, the CG algorithm chooses the polynomial function $P_j^{(\ell+1)}(\lambda)$ such that it minimizes the cost function $\|\boldsymbol{\varepsilon}_{\ell+1,j}^{\mathrm{CG}}\|_{\boldsymbol{A}}^2$ over all polynomial functions $P^{(\ell+1)} : \mathbb{R}_{0,+} \to \mathbb{R}, \lambda \mapsto P^{(\ell+1)}(\lambda)$, with the same degree $\ell+1$ and $P^{(\ell+1)}(0) = 1$, i.e.,

---

[18] Note that the argument of the polynomial $P_j^{(\ell+1)}$ can be either a matrix or a scalar.

$$\left\|\varepsilon_{\ell+1,j}^{\mathrm{CG}}\right\|_{\boldsymbol{A}}^{2} = \min_{P^{(\ell+1)}} \sum_{n=0}^{N-1} |\xi_{j,n}|^{2} \lambda_{n} \left(P^{(\ell+1)}\left(\lambda_{n}\right)\right)^{2} \quad \text{s.\,t.} \quad P^{(\ell+1)}(0) = 1. \quad (3.79)$$

If we replace $|P^{(\ell+1)}\left(\lambda_{n}\right)|$ by its maximum in $[\lambda_{N-1}, \lambda_{0}]$, we obtain finally the upper bound

$$\left\|\varepsilon_{\ell+1,j}^{\mathrm{CG}}\right\|_{\boldsymbol{A}}^{2} \leq \min_{P^{(\ell+1)}} \max_{\lambda \in [\lambda_{N-1},\lambda_{0}]} \left(P^{(\ell+1)}\left(\lambda\right)\right)^{2} \sum_{n=0}^{N-1} |\xi_{j,n}|^{2} \lambda_{n} \quad (3.80)$$

$$= \min_{P^{(\ell+1)}} \max_{\lambda \in [\lambda_{N-1},\lambda_{0}]} \left(P^{(\ell+1)}\left(\lambda\right)\right)^{2} \left\|\varepsilon_{0,j}\right\|_{\boldsymbol{A}}^{2} \quad \text{s.\,t.} \quad P^{(\ell+1)}(0) = 1,$$

where we used $\left\|\varepsilon_{0,j}\right\|_{\boldsymbol{A}}^{2} = \varepsilon_{0,j}^{\mathrm{H}} \boldsymbol{A} \varepsilon_{0,j} = \boldsymbol{\xi}_{j}^{\mathrm{H}} \boldsymbol{\Lambda} \boldsymbol{\xi}_{j} = \sum_{n=0}^{N-1} |\xi_{j,n}|^{2} \lambda_{n}$. Note that in general, the polynomial which fulfills the optimization in Equation (3.80) is not equal to the CG polynomial $P_{j}^{(\ell+1)}(\lambda)$ obtained from the optimization in Equation (3.79).

Proposition 3.7. *Of all polynomials $P^{(\ell+1)}(\lambda)$ of degree $\ell + 1$ and $P^{(\ell+1)}(0) = 1$, the polynomial (cf. Figure 3.3)*

$$\bar{P}^{(\ell+1)}\left(\lambda\right) = \frac{T^{(\ell+1)}\left(\dfrac{\lambda_{0} + \lambda_{N-1} - 2\lambda}{\lambda_{0} - \lambda_{N-1}}\right)}{T^{(\ell+1)}\left(\dfrac{\lambda_{0} + \lambda_{N-1}}{\lambda_{0} - \lambda_{N-1}}\right)}, \quad (3.81)$$

is the best approximation of the polynomial $f(\lambda) = 0$ in the region $[\lambda_{N-1}, \lambda_{0}]$ with respect to the supremum norm[19], i. e.,

$$\bar{P}^{(\ell+1)} = \min_{P^{(\ell+1)}} \max_{\lambda \in [\lambda_{N-1},\lambda_{0}]} \left|P^{(\ell+1)}\left(\lambda\right)\right| \quad \text{s.\,t.} \quad P^{(\ell+1)}(0) = 1. \quad (3.82)$$

Here, $T^{(\ell+1)}(x)$ denotes the Chebyshev polynomial of the first kind and degree $\ell + 1$ as defined in Appendix A.3.

Proof. We prove by contradiction similar to the optimality proof of the Chebyshev polynomials given in [9]. Assume that there exists a polynomial $Q^{(\ell+1)}(\lambda)$ of degree $\ell + 1$ and $Q^{(\ell+1)}(0) = 1$, which approximates $f(\lambda) = 0$ for $\lambda \in [\lambda_{N-1}, \lambda_{0}]$ better than $\bar{P}^{(\ell+1)}(\lambda)$, i. e.,

$$\max_{\lambda \in [\lambda_{N-1},\lambda_{0}]} \left|Q^{(\ell+1)}(\lambda)\right| < \max_{\lambda \in [\lambda_{N-1},\lambda_{0}]} \left|\bar{P}^{(\ell+1)}(\lambda)\right|. \quad (3.83)$$

Recall that the $\ell + 2$ extrema of the Chebyshev polynomial $T^{(\ell+1)}(x)$ for $x \in [-1,1]$ lie at $x_{i} = \cos(i\pi/(\ell + 1))$, $i \in \{0, 1, \ldots, \ell + 1\}$, and alternate

---

[19] The *supremum norm* of the function $f : [a,b] \rightarrow \mathbb{C}, x \mapsto f(x)$, is defined as $\|f\|_{\infty} = \sup_{x \in [a,b]} |f(x)|$ (cf., e.\,g., [268]).

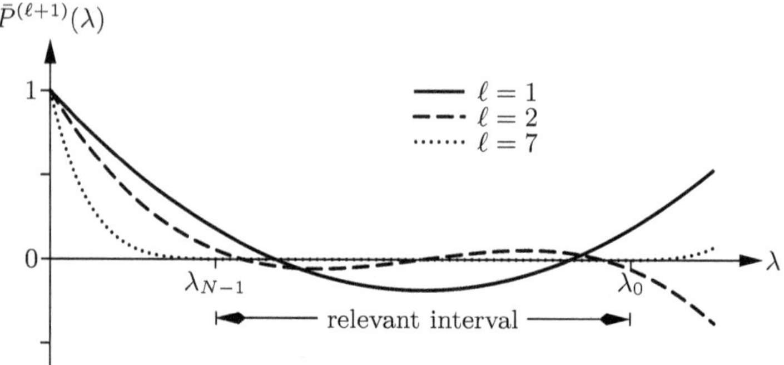

Fig. 3.3. Polynomial approximation $\bar{P}^{(\ell+1)}(\lambda)$ of degree $\ell \in \{1, 2, 7\}$

between ± 1 for increasing x_i, i.e., decreasing i (cf. Equation A.20).[20] Thus, the polynomial $\bar{P}^{(\ell+1)}(\lambda)$ of Equation (3.81) has at

$$\lambda_i' = \frac{1}{2}\left(\lambda_0 + \lambda_{N-1} - (\lambda_0 - \lambda_{N-1})\cos\left(\frac{i\pi}{\ell+1}\right)\right), \qquad (3.84)$$

$i \in \{0, 1, \ldots, \ell+1\}$, the $\ell+2$ extrema

$$\bar{P}^{(\ell+1)}(\lambda_i') = (-1)^i \left(T^{(\ell+1)}\left(\frac{\lambda_0 + \lambda_{N-1}}{\lambda_0 - \lambda_{N-1}}\right)\right)^{-1}, \qquad (3.85)$$

in the region $[\lambda_{N-1}, \lambda_0]$. Hence, for increasing $\lambda = \lambda_i'$, i.e., increasing i, the polynomial $\bar{P}^{(\ell+1)}(\lambda)$ alternates between the positive and negative extrema $\ell+1$ times, and we get with Equation (3.83) and the fact that $T^{(\ell+1)}((\lambda_0 + \lambda_{N-1})/(\lambda_0 - \lambda_{N-1})) \in \mathbb{R}_+$ (cf. Equation 3.88), that

$$Q^{(\ell+1)}(\lambda) > \bar{P}^{(\ell+1)}(\lambda), \quad \text{for } i \text{ odd}, \qquad (3.86)$$

$$Q^{(\ell+1)}(\lambda) < \bar{P}^{(\ell+1)}(\lambda), \quad \text{for } i \text{ even}. \qquad (3.87)$$

Thus, the polynomial $Q^{(\ell+1)}(\lambda) - \bar{P}^{(\ell+1)}(\lambda)$ which is also of degree $\ell+1$, has $\ell+1$ zeros in the region $[\lambda_{N-1}, \lambda_0]$. Besides, since $Q^{(\ell+1)}(0) = \bar{P}^{(\ell+1)}(0) = 1$, the polynomial $Q^{(\ell+1)}(\lambda) - \bar{P}^{(\ell+1)}(\lambda)$ has an additional zero at $\lambda = 0$ which disagrees with its degree. Remember that a polynomial of degree $\ell+1$ can have at most $\ell+1$ zeros. Thus, the polynomial $Q^{(\ell+1)}(\lambda)$ assumed above, does not exist and Proposition 3.7 is proven. $\qquad \square$

With the extrema of $\bar{P}^{(\ell+1)}(\lambda)$ given in Equation (3.85), Equation (3.80) reads as (see also [145, 39, 36, 165])

---

[20] Note that $\cos(x)$ is monotonic decreasing in $x \in [0, \pi]$.

$$\left\| \varepsilon_{\ell+1,j}^{\mathrm{CG}} \right\|_A^2 \le \left(T^{(\ell+1)} \left(\frac{\lambda_0 + \lambda_{N-1}}{\lambda_0 - \lambda_{N-1}} \right) \right)^{-1} \left\| \varepsilon_{0,j} \right\|_A^2$$

$$= 2 \left(\left(\frac{\sqrt{\kappa}+1}{\sqrt{\kappa}-1} \right)^{\ell+1} + \left(\frac{\sqrt{\kappa}-1}{\sqrt{\kappa}+1} \right)^{\ell+1} \right)^{-1} \left\| \varepsilon_{0,j} \right\|_A^2 , \tag{3.88}$$

where we applied additionally Equation (A.14). Here, $\kappa = \lambda_0/\lambda_{N-1} \in \mathbb{R}_+$ denotes the condition number of the system matrix A. If we neglect the second addend in Equation (3.88) because it converges to zero for increasing $\ell + 1$, and recall Equation (3.72), we get finally the upper bound

$$\left\| E_{\ell+1} \right\|_A^2 \le 2 \left(\frac{\sqrt{\kappa}-1}{\sqrt{\kappa}+1} \right)^{\ell+1} \left\| E_0 \right\|_A^2 , \tag{3.89}$$

for the squared A norm of the error matrix $E_{\ell+1}$ of the BCG algorithm. Thus, we have shown that the A-norm of the BCG error matrix converges at least *linearly*[21] to zero with an increasing number of iterations. Note that Diane P. O'Leary derived in [175] the upper bound of $\left\| E_{\ell+1} \right\|_A^2$ directly based on the BCG method without using the CG procedure. However, the resulting upper bound is the same as the one in Equation (3.89). Therefore, with the given convergence analysis, it cannot be shown that the BCG algorithm converges faster than the CG method, which has been observed in many applications (see, e. g., [45]).

The upper bound given by Equation (3.89) matches very well the convergence behavior of the BCG or CG algorithm, respectively, if the eigenvalues of the system matrix A are distributed homogeneously (cf., e. g., Example 3.8). If their distribution is uneven, a closer bound can be found based on a product of suitably chosen Chebyshev polynomials [6]. However, all these upper bounds do not reveal the *superlinear convergence* (cf., e. g., [165]) which means that the rate of convergence increases during the iteration process, of the iterative (B)CG procedure as often observed in numerical applications. Already in the early 1950s, proofs for the superlinear convergence of the CG method have been published for certain linear operators on a *Hilbert space* (cf., e. g., [149]). Nevertheless, these proofs do not give a useful estimate of the rate of convergence for a small number of iterations, which is of great interest because the superlinear convergence behavior of the CG algorithm takes place in early phases of the iteration process. Paul Concus et al. explained in [36] that the maximal eigenvalues of the system matrix A are already approximated very well for a small number of iterations. Roughly speaking, the eigenmodes corresponding to the maximal eigenvalues are switched off one by one during

---

[21] A sequence $\{X_k\}_{k=0}^\infty$ converges *linearly* to X if it converges at least as fast as the geometric sequence $\{ct^k\}_{k=0}^\infty$ for some constant $c \ge 0$, i. e., $\| X_k - X \| \le ct^k$, where $\| \cdot \|$ can be an arbitrary matrix norm (cf. Appendix A.2) and $t \in [0,1)$ denotes the *rate of convergence* (see, e. g., [165]). Note that linear convergence is sometimes also denoted as *geometric convergence*.

the iteration process leading to a decreasing condition number and therefore, to an improving rate of convergence. This observation has been quantitatively proven in [225]. Using finite precision arithmetic, the convergence of the Krylov subspace algorithms is delayed but they still produce acceptable approximations of the solution [227, 96].

Note that the convergence speed of the BCG procedure can be improved by *preconditioning* (cf., e. g., [175]), i. e., a method where the system of linear equations is transformed to a system with a smaller condition number. Although this approach increases computational complexity in general, preconditioning can be implemented very efficiently (see, e. g., [228, 26]) for system matrices with *Toeplitz structure* [94] if we exploit the computationally cheap *Fast Fourier Transform* (FFT) [75, 38, 114, 37]. Since the system matrices in the application part of this book (cf. Part II) have no Toeplitz structure, and due to the fact that the improvement of the approximate solution given by the preconditioned BCG procedure, is very small if applied to communications scenarios [59, 100], preconditioning is no longer considered in this book.

3.4.3 Regularizing Effect

As we will see in the simulation results of Chapter 6, stopping the BCG iterations before the algorithm converges, i. e., choosing $d < L$, reduces not only computational complexity but also the performance loss in case of errors in the system matrix $\boldsymbol{A} \in \mathbb{C}^{N \times N}$ or/and the right-hand side $\boldsymbol{B} \in \mathbb{C}^{N \times M}$, respectively. These errors can be due to the limited accuracy of floating or fixed point arithmetic or due to an estimation process involved in determining $\boldsymbol{A}$ and $\boldsymbol{B}$. Decreasing the effect of these errors on the solution is called *regularization* (see Appendix A.4). Note that the *regularizing effect* of the CG algorithm has been observed very early. Already Cornelius Lanczos mentioned in his paper [156] published in 1950, that his procedure which is an alternative implementation of the CG method, suppresses eigenmodes corresponding to small eigenvalues if stopped before convergence. Although there has been much progress (e. g., [108]), the reason why the CG method regularizes the solution is still an open problem in the theory of *ill-posed* and *ill-conditioned* problems [110]. However, the *filter factors* which have only been derived for the CG method so far, are a well-known tool to analyze and characterize the quality of the regularized solution (see Appendix A.4). In this subsection, we investigate the regularizing effect of the BCG algorithm by deriving the corresponding *filter factor matrix* [43].

Let the eigen decomposition of the Hermitian and positive definite system matrix $\boldsymbol{A} = \boldsymbol{U}\boldsymbol{\Lambda}\boldsymbol{U}^{\mathrm{H}}$ be given as defined in Subsection 3.4.2 (cf. Equations 3.74 to 3.76). Then, the exact solution of the system $\boldsymbol{A}\boldsymbol{X} = \boldsymbol{B}$ of linear equations can be written as

$$\boldsymbol{X} = \boldsymbol{A}^{-1}\boldsymbol{B} = \boldsymbol{U}\boldsymbol{\Lambda}^{-1}\boldsymbol{U}^{\mathrm{H}}\boldsymbol{B}. \tag{3.90}$$

From Appendix A.4, we know that the exact solution $\boldsymbol{X} \in \mathbb{C}^{N \times M}$ can be regularized by introducing the diagonal filter factor matrix $\boldsymbol{F} \in \mathbb{R}_{0,+}^{N \times N}$ such that

$$X_{\mathrm{reg}} = U F \Lambda^{-1} U^{\mathrm{H}} B \in \mathbb{C}^{N \times M}. \tag{3.91}$$

Now, if we find such a filter factor representation of the approximate solution $X_d \in \mathbb{C}^{N \times M}$ obtained from the BCG method summarized in Algorithm 3.4., we can discuss its intrinsic regularizing effect.

Later on, we will see that the *filter factors*, i.e., the diagonal elements of the filter factor matrix, of the BCG algorithm are not only complex-valued but also different for each column of the right-hand side matrix B, i.e.,

$$x_{d,j} = U F_{d,j} \Lambda^{-1} U^{\mathrm{H}} b_j, \quad j \in \{0, 1, \ldots, M-1\}, \tag{3.92}$$

where $F_{d,j} \in \mathbb{C}^{N \times N}$ is diagonal and $F_{d,j} \neq F_{d,i}$ for $j \neq i \in \{0, 1, \ldots, M-1\}$. The vector $x_{d,j} = X_d e_{j+1}$ and $b_j = B e_{j+1}$ denotes the $(j+1)$th column of X_d and B, respectively. From Equation (3.92) follows that the inverse of the system matrix A mapping b_j to $x_{d,j}$, is approximated differently for each column index j. Note that although the matrix A is assumed to be Hermitian and positive definite, the approximation of its inverse depending on the column index j, is generally non-Hermitian because of the complex-valued filter factors in $F_{d,j}$. However, the approximation of the inverse is still a *normal matrix*.[22] The traditional representation of a regularized solution as given in Equation (3.91) where the filter factor matrix F is positive real-valued and independent of the column index j, cannot be derived for the approximate solution obtained from the BCG procedure. However, if $M = 1$, i.e., the system of linear equations under consideration is $A x = b$, the BCG algorithm reduces to the CG method and the traditional filter factor representation of the approximate solution $x_D \approx x$ ($D = dM = d$ if $M = 1$) according to Equation (3.91) exists (cf., e.g., [110]). We define the diagonal filter factor matrix $F_D \in \mathbb{R}_{0,+}^{N \times N}$ of the CG algorithm such that

$$x_D = U F_D \Lambda^{-1} U^{\mathrm{H}} b \in \mathbb{C}^N. \tag{3.93}$$

In the following, we generalize the filter factor representation of the CG method to the one of the BCG procedure corresponding to Equation (3.92) which holds for $M \geq 1$.

With the definition of the block diagonal matrix

$$\check{F}_d = \mathrm{bdiag}\left\{ F_{d,0}, F_{d,1}, \ldots, F_{d,M-1} \right\} \in \mathbb{C}^{NM \times NM}, \tag{3.94}$$

and the vectors $\check{x}_d = \mathrm{vec}\{X_d\} \in \mathbb{C}^{NM}$ and $\check{b} = \mathrm{vec}\{B\} \in \mathbb{C}^{NM}$ (cf. Appendix A.5), Equation (3.92) can be rewritten as

$$\left| \check{x}_d = (I_M \otimes U) \, \check{F}_d \left(I_M \otimes \Lambda^{-1} U^{\mathrm{H}} \right) \check{b}. \right| \tag{3.95}$$

Note that the $N \times N$ blocks $F_{d,j}$ in the diagonal of $\check{F}_d$ are all diagonal, i.e., $\check{F}_d$ is a diagonal matrix. In the remainder of this subsection, we derive the diagonal entries of $\check{F}_d$.

---

[22] The matrix A is normal if $A^{\mathrm{H}} A = A A^{\mathrm{H}}$.

From Proposition 3.6, it follows that the approximate solution $\boldsymbol{X}_d$ of the BCG algorithm can be expressed by a polynomial in $\boldsymbol{A}$ of degree $d-1$ (cf. also Equation 3.68), i.e.,

$$\boldsymbol{X}_d = \sum_{\ell=0}^{d-1} \boldsymbol{A}^\ell \boldsymbol{R}_0 \boldsymbol{\Psi}_\ell + \boldsymbol{X}_0 \in \mathbb{C}^{N \times M}. \tag{3.96}$$

Again, if we choose $\boldsymbol{X}_0 = \boldsymbol{0}_{N \times M}$, it holds $\boldsymbol{R}_0 = \boldsymbol{B} - \boldsymbol{A}\boldsymbol{X}_0 = \boldsymbol{B}$ and Equation (3.96) reads as (cf. also Equation 3.2)

$$\boldsymbol{X}_d = \sum_{\ell=0}^{d-1} \boldsymbol{A}^\ell \boldsymbol{B} \boldsymbol{\Psi}_\ell. \tag{3.97}$$

Applying the vec-operation as defined in Appendix A.5, on Equation (3.97), and exploiting Equations (A.33) and (A.32), yields

$$\begin{aligned}
\check{\boldsymbol{x}}_d &= \sum_{\ell=0}^{d-1} \left(\boldsymbol{\Psi}_\ell^{\mathrm{T}} \otimes \boldsymbol{A}^\ell \right) \check{\boldsymbol{b}} = \sum_{\ell=0}^{d-1} \left(\boldsymbol{\Psi}_\ell^{\mathrm{T}} \otimes \boldsymbol{I}_N \right) \left(\boldsymbol{I}_M \otimes \boldsymbol{A} \right)^\ell \check{\boldsymbol{b}} \\
&= \sum_{\ell=0}^{d-1} \check{\boldsymbol{\Psi}}_\ell \check{\boldsymbol{A}}^\ell \check{\boldsymbol{b}} =: \check{\boldsymbol{\Psi}}^{(d-1)} \left(\check{\boldsymbol{A}} \right) \check{\boldsymbol{b}},
\end{aligned} \tag{3.98}$$

i.e., the vector $\check{\boldsymbol{x}}_d$ can be expressed by a polynomial $\check{\boldsymbol{\Psi}}^{(d-1)}(\check{\boldsymbol{A}}) \in \mathbb{C}^{NM \times NM}$ in $\check{\boldsymbol{A}} = \boldsymbol{I}_M \otimes \boldsymbol{A} \in \mathbb{C}^{NM \times NM}$ of degree $d-1$ with the polynomial coefficients $\check{\boldsymbol{\Psi}}_\ell = \boldsymbol{\Psi}_\ell^{\mathrm{T}} \otimes \boldsymbol{I}_N \in \mathbb{C}^{NM \times NM}$.[23] With the eigen decomposition defined in Equation (3.74) and Equation (A.32), we get

$$\begin{aligned}
\check{\boldsymbol{A}}^\ell &= (\boldsymbol{I}_M \otimes \boldsymbol{A})^\ell = \boldsymbol{I}_M \otimes \boldsymbol{A}^\ell = \boldsymbol{I}_M \otimes \boldsymbol{U}\boldsymbol{\Lambda}^\ell \boldsymbol{U}^{\mathrm{H}} \\
&= (\boldsymbol{I}_M \otimes \boldsymbol{U}) \left(\boldsymbol{I}_M \otimes \boldsymbol{\Lambda}^\ell \right) \left(\boldsymbol{I}_M \otimes \boldsymbol{U}^{\mathrm{H}} \right) \\
&= (\boldsymbol{I}_M \otimes \boldsymbol{U}) (\boldsymbol{I}_M \otimes \boldsymbol{\Lambda})^\ell \left(\boldsymbol{I}_M \otimes \boldsymbol{U}^{\mathrm{H}} \right),
\end{aligned} \tag{3.99}$$

and if we exploit additionally that $(\boldsymbol{\Psi}_\ell^{\mathrm{T}} \otimes \boldsymbol{I}_N)(\boldsymbol{I}_M \otimes \boldsymbol{U}) = \boldsymbol{\Psi}_\ell^{\mathrm{T}} \otimes \boldsymbol{U} = (\boldsymbol{I}_M \otimes \boldsymbol{U})(\boldsymbol{\Psi}_\ell^{\mathrm{T}} \otimes \boldsymbol{I}_N)$ due to Equation (A.32), the polynomial $\check{\boldsymbol{\Psi}}^{(d-1)}(\check{\boldsymbol{A}})$ can be written as

$$\check{\boldsymbol{\Psi}}^{(d-1)} \left(\check{\boldsymbol{A}} \right) = (\boldsymbol{I}_M \otimes \boldsymbol{U}) \, \check{\boldsymbol{\Psi}}^{(d-1)} \left(\check{\boldsymbol{\Lambda}} \right) \left(\boldsymbol{I}_M \otimes \boldsymbol{U}^{\mathrm{H}} \right), \tag{3.100}$$

where $\check{\boldsymbol{\Lambda}} = \boldsymbol{I}_M \otimes \boldsymbol{\Lambda}$ is a diagonal positive real-valued $NM \times NM$ matrix. Plugging this result into Equation (3.98) and introducing the identity matrix $\check{\boldsymbol{\Lambda}}\check{\boldsymbol{\Lambda}}^{-1}$, yields

---

[23] Note that $\check{\boldsymbol{\Psi}}^{(d-1)}(\check{\boldsymbol{A}})$ is no polynomial in the sense of Equation (3.3) since the coefficients $\check{\boldsymbol{\Psi}}_\ell$ are $NM \times NM$ matrices and no scalars. Here, polynomials with matrix coefficients are represented by bold letters.

$$\left| \check{\boldsymbol{x}}_d = (\boldsymbol{I}_M \otimes \boldsymbol{U}) \, \check{\boldsymbol{\Psi}}^{(d-1)} \left(\check{\boldsymbol{\Lambda}} \right) \check{\boldsymbol{\Lambda}} \left(\boldsymbol{I}_M \otimes \boldsymbol{\Lambda}^{-1} \boldsymbol{U}^{\mathrm{H}} \right) \check{\boldsymbol{b}}. \right| \tag{3.101}$$

Here, the term

$$
\begin{aligned}
\check{\boldsymbol{\Psi}}^{(d-1)} \left(\check{\boldsymbol{\Lambda}} \right) \check{\boldsymbol{\Lambda}} &= \sum_{\ell=0}^{d-1} \left(\boldsymbol{\Psi}_\ell^{\mathrm{T}} \otimes \boldsymbol{\Lambda}^\ell \right) \left(\boldsymbol{I}_M \otimes \boldsymbol{\Lambda} \right) \\
&= \sum_{\ell=0}^{d-1} \left(\boldsymbol{\Psi}_\ell^{\mathrm{T}} \otimes \boldsymbol{\Lambda}^{\ell+1} \right) \in \mathbb{C}^{NM \times NM},
\end{aligned}
\tag{3.102}
$$

is a matrix with the diagonal matrices $\sum_{\ell=0}^{d-1} \psi_{\ell,j,i} \boldsymbol{\Lambda}^{\ell+1} \in \mathbb{C}^{N \times N}$, $i,j \in \{0,1,\dots,M-1\}$, at the $(i+1)$th block row and $(j+1)$th block column. The element in the $(i+1)$th row and $(j+1)$th column of $\boldsymbol{\Psi}_\ell \in \mathbb{C}^{M \times M}$ is denoted as $\psi_{\ell,i,j} \in \mathbb{C}$. Thus, if we compare Equation (3.95) with Equation (3.101), we see that $\boldsymbol{\Psi}^{(d-1)}(\boldsymbol{\Lambda})\boldsymbol{\Lambda}$ is not the desired filter factor matrix $\check{\boldsymbol{F}}_d$ because it is not diagonal. In other words, each column of $\boldsymbol{X}_d$ depends on all columns of the right-hand side $\boldsymbol{B}$ and not only on the corresponding column with the same index. If we transform the right part of Equation (3.101) to the right part of Equation (3.95) without changing the left-hand side, i. e., the approximate solution $\check{\boldsymbol{x}}_d$, we obtain the diagonal filter factor matrix of the BCG algorithm according to

$$\left| \begin{aligned} \check{\boldsymbol{F}}_d = \mathrm{diag} \Big\{ &\left(\check{\boldsymbol{\Psi}}^{(d-1)} \left(\check{\boldsymbol{\Lambda}} \right) \check{\boldsymbol{\Lambda}} \left(\boldsymbol{I}_M \otimes \boldsymbol{\Lambda}^{-1} \boldsymbol{U}^{\mathrm{H}} \right) \check{\boldsymbol{b}} \right) \\ &\oslash \left(\left(\boldsymbol{I}_M \otimes \boldsymbol{\Lambda}^{-1} \boldsymbol{U}^{\mathrm{H}} \right) \check{\boldsymbol{b}} \right) \Big\}, \end{aligned} \right| \tag{3.103}$$

where '$\oslash$' denotes elementwise division. Note that the diagonal entries of $\check{\boldsymbol{F}}_d$ are generally complex-valued since the non-diagonal matrix $\check{\boldsymbol{\Psi}}^{(d-1)}(\check{\boldsymbol{\Lambda}})\check{\boldsymbol{\Lambda}}$ has complex-valued elements due to the matrix $\boldsymbol{\Psi}_\ell \in \mathbb{C}^{M \times M}$ (cf. Equation 3.102), and because of the complex-valued vector $(\boldsymbol{I}_M \otimes \boldsymbol{\Lambda}^{-1}\boldsymbol{U}^{\mathrm{H}})\check{\boldsymbol{b}}$. However, the absolute values of the complex-valued filter factors still describe the attenuation of the small eigenvalues (see Appendix A.4 and Example 3.9) and can therefore be used to explain the regularizing effect of the BCG algorithm.

Finally, it remains to compute the polynomial $\check{\boldsymbol{\Psi}}^{(d-1)}(\check{\boldsymbol{\Lambda}})$ in order to calculate the filter factor matrix $\check{\boldsymbol{F}}_d$ based on Equation (3.103). To do so, we use the equations of the BCG algorithm derived in Subsection 3.4.1. Let $\ell \in \{0,1,\dots,d-1\}$. Replacing $\boldsymbol{R}_{\ell+1}$ in Equation (3.58) based on Equation (3.49), and using the result to replace $\boldsymbol{D}_\ell$ in Equation (3.30), yields

$$\boldsymbol{X}_{\ell+1} = \boldsymbol{X}_\ell + \boldsymbol{B}\boldsymbol{\Phi}_\ell - \boldsymbol{A}\boldsymbol{X}_\ell\boldsymbol{\Phi}_\ell + \boldsymbol{D}_{\ell-1}\boldsymbol{\Theta}_{\ell-1}\boldsymbol{\Phi}_\ell. \tag{3.104}$$

Then, if we solve Equation (3.30) with respect to $\boldsymbol{D}_\ell$ and use the result to replace $\boldsymbol{D}_{\ell-1}$ in Equation (3.104), we get finally

$$\boldsymbol{X}_{\ell+1} = \boldsymbol{X}_\ell - \boldsymbol{A}\boldsymbol{X}_\ell\boldsymbol{\Phi}_\ell + \left(\boldsymbol{X}_\ell - \boldsymbol{X}_{\ell-1} \right) \boldsymbol{\Phi}_{\ell-1}^{-1}\boldsymbol{\Theta}_{\ell-1}\boldsymbol{\Phi}_\ell + \boldsymbol{B}\boldsymbol{\Phi}_\ell. \tag{3.105}$$

Hence, we derived an iteration formula which computes the update $\boldsymbol{X}_{\ell+1}$ of the BCG algorithm based on the previous updates $\boldsymbol{X}_\ell$ and $\boldsymbol{X}_{\ell-1}$. Note that the iteration is started with the initial values $\boldsymbol{X}_0 = \mathbf{0}_{N \times M}$ and $\boldsymbol{X}_{-1} = \mathbf{0}_{N \times M}$. Besides, we see in Equation (3.105) that each iteration involves a multiplication of the matrix $\boldsymbol{A}$ to the previous update $\boldsymbol{X}_\ell$ which leads to the polynomial structure given in Equation (3.97).

With the definitions $\check{\boldsymbol{x}}_\ell = \mathrm{vec}\{\boldsymbol{X}_\ell\} \in \mathbb{C}^{NM}$, $\check{\boldsymbol{\Phi}}_\ell = \boldsymbol{\Phi}_\ell^{\mathrm{T}} \otimes \boldsymbol{I}_N \in \mathbb{C}^{NM \times NM}$, and $\check{\boldsymbol{\Theta}}_\ell = \boldsymbol{\Theta}_\ell^{\mathrm{T}} \otimes \boldsymbol{I}_N \in \mathbb{C}^{NM \times NM}$, and using Equations (A.32), (A.33), and (A.34), Equation (3.105) can be rewritten as

$$\check{\boldsymbol{x}}_{\ell+1} = \left(\boldsymbol{I}_{NM} - \check{\boldsymbol{\Phi}}_\ell \check{\boldsymbol{A}} + \check{\boldsymbol{\Phi}}_\ell \check{\boldsymbol{\Theta}}_{\ell-1} \check{\boldsymbol{\Phi}}_{\ell-1}^{-1}\right) \check{\boldsymbol{x}}_\ell - \check{\boldsymbol{\Phi}}_\ell \check{\boldsymbol{\Theta}}_{\ell-1} \check{\boldsymbol{\Phi}}_{\ell-1}^{-1} \check{\boldsymbol{x}}_{\ell-1} + \check{\boldsymbol{\Phi}}_\ell \check{\boldsymbol{b}}. \quad (3.106)$$

Replacing $\check{\boldsymbol{x}}_\ell$ by $\check{\boldsymbol{\Psi}}^{(\ell-1)}(\check{\boldsymbol{A}})\check{\boldsymbol{b}}$ (cf. Equation 3.98) and $\check{\boldsymbol{\Psi}}^{(\ell-1)}(\check{\boldsymbol{A}})$ based on Equation (3.100), yields

$$\begin{aligned}
\check{\boldsymbol{\Psi}}^{(\ell)}\left(\check{\boldsymbol{\Lambda}}\right) &= \left(\boldsymbol{I}_{NM} - \check{\boldsymbol{\Phi}}_\ell \check{\boldsymbol{\Lambda}} + \check{\boldsymbol{\Phi}}_\ell \check{\boldsymbol{\Theta}}_{\ell-1} \check{\boldsymbol{\Phi}}_{\ell-1}^{-1}\right) \check{\boldsymbol{\Psi}}^{(\ell-1)}\left(\check{\boldsymbol{\Lambda}}\right) \\
&\quad - \check{\boldsymbol{\Phi}}_\ell \check{\boldsymbol{\Theta}}_{\ell-1} \check{\boldsymbol{\Phi}}_{\ell-1}^{-1} \check{\boldsymbol{\Psi}}^{(\ell-2)}\left(\check{\boldsymbol{\Lambda}}\right) + \check{\boldsymbol{\Phi}}_\ell,
\end{aligned} \quad (3.107)$$

if we apply again Equation (A.32). Now, the iteration formula in Equation (3.107) can be used to compute iteratively the required polynomial $\check{\boldsymbol{\Psi}}^{(d-1)}\left(\check{\boldsymbol{\Lambda}}\right)$ starting with $\check{\boldsymbol{\Psi}}^{(\ell)}\left(\check{\boldsymbol{\Lambda}}\right) := \mathbf{0}_{NM \times NM}$ for $\ell \in \{-2, -1\}$.

Before presenting two examples revealing the significance of the filter factor matrix for the visualization of the regularizing effect of the BCG or CG procedure, respectively, we briefly derive the filter factor matrix of the CG method [110], i.e., for the special case where $M = 1$. With $D = d$, $\check{\boldsymbol{A}} = \boldsymbol{A}$, $\check{\boldsymbol{b}} = \boldsymbol{b}$, and $\boldsymbol{\Psi}_\ell = \psi_\ell$ for $M = 1$, and the eigen decomposition of $\boldsymbol{A}$ as given in Equation (3.74), the polynomial representation of the approximate CG solution $\boldsymbol{x}_D$ reads as (cf. Equation 3.98)

$$\boldsymbol{x}_D = \boldsymbol{\Psi}^{(D-1)}(\boldsymbol{A})\,\boldsymbol{b} = \sum_{\ell=0}^{D-1} \psi_\ell \boldsymbol{A}^\ell \boldsymbol{b} = \boldsymbol{U}\left(\sum_{\ell=0}^{D-1} \psi_\ell \boldsymbol{\Lambda}^{\ell+1}\right) \boldsymbol{\Lambda}^{-1} \boldsymbol{U}^{\mathrm{H}} \boldsymbol{b}, \quad (3.108)$$

if the CG algorithm is initialized with $\boldsymbol{x}_0 = \mathbf{0}_N$. Note that the polynomial coefficients $\psi_\ell \in \mathbb{R}$ since we assumed that $\boldsymbol{A}$ is Hermitian and positive definite (cf. Appendix B.2). If we compare Equation (3.108) with Equation (3.93), we see that the diagonal filter factor matrix $\boldsymbol{F}_D$ of the CG algorithm is given by

$$\boldsymbol{F}_D = \boldsymbol{\Psi}^{(D-1)}(\boldsymbol{\Lambda})\,\boldsymbol{\Lambda} = \sum_{\ell=0}^{D-1} \psi_\ell \boldsymbol{\Lambda}^{\ell+1} \in \mathbb{R}^{N \times N}. \quad (3.109)$$

Since $M = 1$, it holds that $\check{\boldsymbol{\Lambda}} = \boldsymbol{\Lambda}$, $\check{\boldsymbol{\Phi}}_\ell = \phi_\ell \in \mathbb{R}_+$, and $\check{\boldsymbol{\Theta}}_\ell = \theta_\ell \in \mathbb{R}_+$ (cf. Equations 3.65 and 3.66). Thus, according to Equation (3.107), the polynomial $\boldsymbol{\Psi}^{(D-1)}(\boldsymbol{\Lambda}) \in \mathbb{R}^{N \times N}$ required to calculate the filter factor matrix $\boldsymbol{F}_D$, can be computed iteratively based on the formula

$$\Psi^{(\ell)}(\Lambda) = \left(\left(\frac{\phi_\ell\theta_{\ell-1}}{\phi_{\ell-1}} + 1\right)I_N - \phi_\ell\Lambda\right)\Psi^{(\ell-1)}(\Lambda)$$
$$- \frac{\phi_\ell\theta_{\ell-1}}{\phi_{\ell-1}}\Psi^{(\ell-2)}(\Lambda) + \phi_\ell I_N, \quad \ell \in \{0,1,\ldots,D-1\}, \tag{3.110}$$

initialized with $\Psi^{(\ell)}(\Lambda) := \mathbf{0}_{N\times N}$ for $\ell \in \{-2,-1\}$.

Example 3.8 (Filter factors of the CG algorithm).
Consider the system $\boldsymbol{Ax} = \boldsymbol{b}$ of linear equations with $N = 40$ and where $\boldsymbol{A} = \boldsymbol{U\Lambda U}^{\mathrm{H}} \in \mathbb{C}^{40\times 40}$ is constructed based on an arbitrary unitary matrix $\boldsymbol{U} \in \mathbb{C}^{40\times 40}$ and the diagonal matrix $\boldsymbol{\Lambda} \in \mathbb{R}_{0,+}^{40\times 40}$ with the eigenvalues $\lambda_i = 1 - 19\,i/780$, $i \in \{0,1,\ldots,39\}$. Thus, the eigenvalues are linearly decreasing from $\lambda_0 = 1$ to $\lambda_{39} = 1/20$ and the condition number of the system matrix $\boldsymbol{A}$ is $\kappa = \lambda_0/\lambda_{39} = 20$ (cf. Equation A.13). Each element of the vector $\boldsymbol{b} \in \mathbb{C}^{40}$ is a realization of the complex normal distribution $\mathcal{N}_{\mathbb{C}}(0,1)$.[24] The filter factors $f_{D,i} = \boldsymbol{e}_{i+1}^{\mathrm{T}}\boldsymbol{F}_D\boldsymbol{e}_{i+1} \in \mathbb{R}_+$ of the CG algorithm solving the given system of linear equations, are depicted in Figure 3.4 for $D \in \{1,2,\ldots,10\}$. Here, the filter factor matrix $\boldsymbol{F}_D$ is computed according to Equations (3.109) and (3.110).

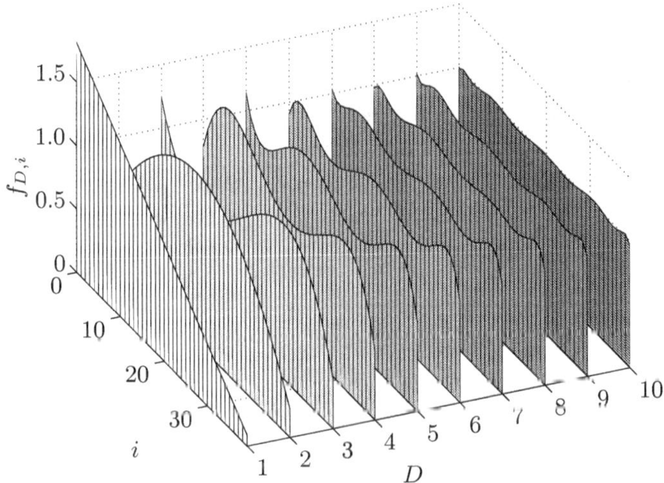

Fig. 3.4. Filter factors of the CG algorithm

In Figure 3.4, we observe the most important properties of the CG filter factor representation. First of all, we see that the influence of the small

---

[24] The probability density function of the complex normal distribution $\mathcal{N}_{\mathbb{C}}(m_x, c_x)$ with mean m_x and variance c_x reads as $\mathrm{p}_x(x) = \exp(-|x - m_x|^2/c_x)/(\pi c_x)$ (e.g., [183]).

eigenvalues (large index i) is attenuated by small values of $f_{D,i}$, if the CG algorithm is stopped before convergence ($D \ll N$). This attenuation explains the intrinsic regularizing effect of the CG algorithm because especially errors in the small eigenvalues have the highest influence on the quality of the solution due to their inversion in Equation (3.93). Besides, we see that all filter factors converge to one if D is chosen sufficiently large. Note that $D = 40$ which is not included in Figure 3.4, leads to $\boldsymbol{F}_D = \boldsymbol{I}_{40}$. Last but not least, due to the linear alignment of the eigenvalues, Figure 3.4 reveals the polynomial behavior of the filter factor matrix $\boldsymbol{F}_D$ as stated in Equation (3.109). Stopping the CG procedure after one iteration, the filter factor matrix reads as $\boldsymbol{F}_1 = \psi_0 \boldsymbol{\Lambda}$ resulting in the straight line at $D = 1$. Increasing the number of iterations D where the CG method is stopped, increases the order of the polynomial in Equation (3.109), i.e., at $D = 2$, the filter factors form a parabola, at $D = 3$, a cubic polynomial, etc. Note that these polynomials are very similar to the polynomials used to approximate the error norm of the CG method derived in the previous subsection (cf. Figure 3.3), which explains the fact that $f_{D,0} - 1$ is alternating between positive and negative values with increasing D.

Example 3.9 (Filter factors of the BCG algorithm).

Consider now the system $\boldsymbol{AX} = \boldsymbol{B}$ of linear equations with $M = 5$ right-hand sides and where $N = 40$. With $\boldsymbol{N} \in \mathbb{C}^{40 \times 40}$ being a matrix whose elements are realizations of the complex normal distribution $\mathcal{N}_{\mathbb{C}}(0, 1)$, the system matrix is defined as $\boldsymbol{A} = \boldsymbol{NN}^{\mathrm{H}} + \boldsymbol{I}_{40}/10 \in \mathbb{C}^{40 \times 40}$ and the matrix $\boldsymbol{B} = \boldsymbol{NS}_{(0,5,35)}^{\mathrm{T}} \in \mathbb{C}^{40 \times 5}$ with $\boldsymbol{S}_{(0,5,35)} = [\boldsymbol{I}_5, \boldsymbol{0}_{5 \times 35}] \in \{0, 1\}^{5 \times 40}$, comprises the first $M = 5$ columns of $\boldsymbol{N}$. The absolute values of the filter factors $\check{f}_{d,i} = \boldsymbol{e}_{i+1}^{\mathrm{T}} \check{\boldsymbol{F}}_d \boldsymbol{e}_{i+1} \in \mathbb{C}$ of the BCG procedure solving the given system of linear equations, are depicted in Figure 3.5 for $d \in \{1, 2, \ldots, 8\}$ and $i \in \{0, 1, \ldots, 39\}$, i.e., only for the filter factors corresponding to the pair $\boldsymbol{x}_{d,1}$ and $\boldsymbol{b}_1$ (cf. Equation 3.92). The filter factor matrix $\check{\boldsymbol{F}}_d$ of the BCG method is computed based on Equations (3.103) and (3.107).

We see that compared to Figure 3.4, the absolute values of the filter factors $\check{f}_{d,i}$ for a fixed number of iterations d where the BCG procedure is stopped, are no longer polynomials in i nor are they smooth. This is due to the facts that the eigenvalues are no longer aligned linearly and the filter factors of the BCG algorithm are a combination of all columns of the right-hand side $\boldsymbol{B}$ (see Equation 3.103). However, again, the filter factors corresponding to small eigenvalues (large index i) tend to zero for $d \ll L$ explaining the intrinsic regularization of the BCG algorithm, and the filter factor matrix converges to the identity matrix $\boldsymbol{I}_{40}$ if d goes to $L = N/M = 8$.

3.4.4 Dimension-Flexible Version

One problem of the original BCG procedure (cf. Algorithm 3.4.) is that the dimension D of the Krylov subspace $\mathcal{K}^{(D)}(\boldsymbol{A}, \boldsymbol{B} - \boldsymbol{AX}_0)$ in which the solution

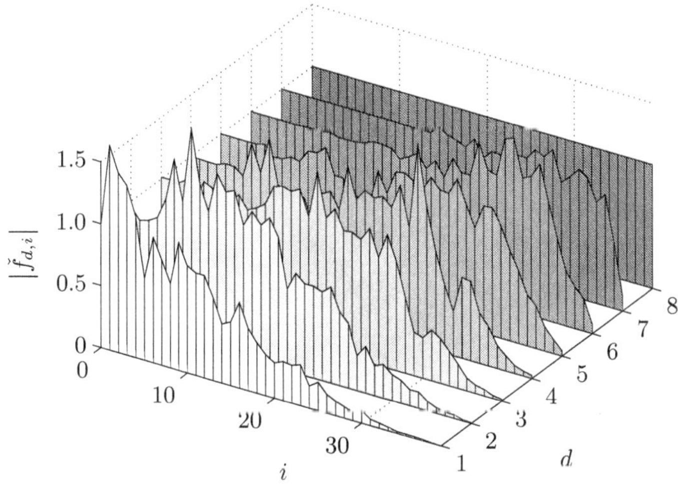

Fig. 3.5. Filter factors of the BCG algorithm ($j = 1$)

of the system $\boldsymbol{AX} = \boldsymbol{B}$ is approximated, has to be an integer multiple of the number M of right-hand sides. Similar to the Ruhe extension of the Lanczos algorithm described in Subsection 3.3.2, we derive in the following an efficient version of the BCG method which calculates an approximation of the exact solution $\boldsymbol{X}$ in the D-dimensional Krylov subspace $\mathcal{K}^{(D)}\left(\boldsymbol{A}, \boldsymbol{B} - \boldsymbol{AX}_0\right)$ where $D \in \{M, M-1, \ldots, N\}$ (cf. also [101]). Thus, the resulting algorithm allows a more flexible selection of the dimension D.

The fundamental idea of the dimension-flexible BCG version is as follows. The second last update $\boldsymbol{X}_{d-1}$, $d = \lceil D/M \rceil \in \{1, 2, \ldots, \lceil N/M \rceil\}$, is computed based on Algorithm 3.4.. Then, the final desired approximation $\boldsymbol{X}'_d \in \mathbb{C}^{N \times M}$ is calculated according to

$$\boxed{\boldsymbol{X}'_d = \boldsymbol{X}_{d-1} + \boldsymbol{D}'_{d-1}\boldsymbol{\Phi}'_{d-1},} \tag{3.111}$$

where the direction matrix $\boldsymbol{D}'_{d-1} \in \mathbb{C}^{N \times \mu}$ is based on $\mu = D - (d-1)M \in \{1, 2, \ldots, M\}$ instead of M search directions. As in the original BCG algorithm derived in Subsection 3.4.1, the step size matrix $\boldsymbol{\Phi}'_{d-1} \in \mathbb{C}^{\mu \times M}$ is the solution of the optimization

$$\boldsymbol{\Phi}'_{d-1} = \underset{\boldsymbol{\Phi} \in \mathbb{C}^{\mu \times M}}{\operatorname{argmin}} \, \varepsilon_{\boldsymbol{A}}\left(\boldsymbol{X}_{d-1} + \boldsymbol{D}'_{d-1}\boldsymbol{\Phi}\right), \tag{3.112}$$

and may be written as

$$\boldsymbol{\Phi}'_{d-1} = \left(\boldsymbol{D}'^{,\mathrm{H}}_{d-1}\boldsymbol{A}\boldsymbol{D}'_{d-1}\right)^{-1}\boldsymbol{D}'^{,\mathrm{H}}_{d-1}\boldsymbol{R}_{d-1}. \tag{3.113}$$

Again, the direction matrix $\boldsymbol{D}'_{d-1}$ is chosen to be $\boldsymbol{A}$-conjugate to the previous direction matrices, i. e., it holds for $i \in \{0, 1, \ldots, d-2\}$, $d > 1$:

$$D_i^{\mathrm{H}} A D_{d-1}' = \mathbf{0}_{M \times \mu}. \tag{3.114}$$

This property is obtained using the iteration formula (cf. Subsection 3.4.1)

$$\boxed{D_{d-1}' = R_{d-1}' + D_{d-2}\Theta_{d-2}',} \tag{3.115}$$

where $R_{d-1}' = [R_{d-1}]_{:,0:\mu-1} \in \mathbb{C}^{N \times \mu}$ with the notation as introduced in Subsection 2.2.1, comprises the first μ columns of the residuum matrix R_{d-1}, and the matrix $\Theta_{d-2}' \in \mathbb{C}^{M \times \mu}$ computes as (cf. Subsection 3.4.1)

$$\left| \Theta_{d-2}' = \left(R_{d-2}^{\mathrm{H}} R_{d-2}\right)^{-1} R_{d-1}^{\mathrm{H}} R_{d-1}'. \right. \tag{3.116}$$

Further, multiplying the Hermitian of Equation (3.115) by R_{d-1} on the right-hand side, yields $D_{d-1}'^{,\mathrm{H}} R_{d-1} = R_{d-1}'^{,\mathrm{H}} R_{d-1}$ if we apply Equation (3.45) of Proposition 3.5. Hence, Equation (3.113) can also be written as (cf. Equation 3.32)

$$\left| \Phi_{d-1}' = \left(D_{d-1}'^{,\mathrm{H}} A D_{d-1}'\right)^{-1} R_{d-1}'^{,\mathrm{H}} R_{d-1}. \right. \tag{3.117}$$

Note that the residuum matrix $R_{d-1} \in \mathbb{C}^{N \times M}$ with M columns is needed in Equation (3.117) to compute Φ_{d-1}' as well as in Equation (3.116) to get Θ_{d-2}'. Thus, it is not possible to further simplify the dimension-flexible BCG algorithm by computing only the first μ columns of R_{d-1}.

Algorithm 3.5. summarizes the dimension-flexible BCG procedure. For $\ell \in \{0, 1, \ldots, d-3\}$, $d > 2$, Algorithm 3.5. is identical to Algorithm 3.4.. However, at iteration step $\ell = d-2$, Equations (3.115) and (3.116) are taken into account to get an efficient implementation of computing the direction matrix D_{d-1}' with μ columns (cf. Lines 15 to 18 of Algorithm 3.5.). Besides, the last iteration based on Equations (3.111) and (3.117) is written outside the **for**-loop.

Algorithm 3.5. Dimension-flexible version of the BCG procedure

Choose an arbitrary matrix $\boldsymbol{X}_0 \in \mathbb{C}^{N \times M}$ (e.g., $\boldsymbol{X}_0 \leftarrow \boldsymbol{0}_{N \times M}$).

2: $\boldsymbol{R}_0 \leftarrow \boldsymbol{B} - \boldsymbol{A}\boldsymbol{X}_0$

$\quad \boldsymbol{G}_0 = \boldsymbol{G}_0' \leftarrow \boldsymbol{R}_0^{\mathrm{H}} \boldsymbol{R}_0$

4: $\boldsymbol{D}_0 = \boldsymbol{D}_0' \leftarrow \boldsymbol{R}_0$

$\quad$ **for** $\ell = 0, 1, \ldots, d-2 \wedge d > 1$ **do**

6: $\quad\quad \boldsymbol{V} \leftarrow \boldsymbol{A}\boldsymbol{D}_\ell$

$\quad\quad\quad \boldsymbol{\Phi}_\ell \leftarrow (\boldsymbol{D}_\ell^{\mathrm{H}} \boldsymbol{V})^{-1} \boldsymbol{G}_\ell$

8: $\quad\quad \boldsymbol{X}_{\ell+1} \leftarrow \boldsymbol{X}_\ell + \boldsymbol{D}_\ell \boldsymbol{\Phi}_\ell$

$\quad\quad\quad \boldsymbol{R}_{\ell+1} \leftarrow \boldsymbol{R}_\ell - \boldsymbol{V}\boldsymbol{\Phi}_\ell$

10: $\quad\quad$ **if** $\ell < d-2$ **then**

$\quad\quad\quad\quad \boldsymbol{G}_{\ell+1} \leftarrow \boldsymbol{R}_{\ell+1}^{\mathrm{H}} \boldsymbol{R}_{\ell+1}$

12: $\quad\quad\quad \boldsymbol{\Theta}_\ell \leftarrow \boldsymbol{G}_\ell^{-1} \boldsymbol{G}_{\ell+1}$

$\quad\quad\quad\quad \boldsymbol{D}_{\ell+1} \leftarrow \boldsymbol{R}_{\ell+1} + \boldsymbol{D}_\ell \boldsymbol{\Theta}_\ell$

14: $\quad\quad$ **else**

$\quad\quad\quad\quad \boldsymbol{R}_{d-1}' \leftarrow [\boldsymbol{R}_{d-1}]_{.,0.\mu-1}$

16: $\quad\quad\quad \boldsymbol{G}_{d-1}' \leftarrow \boldsymbol{R}_{d-1}'^{\mathrm{H}} \boldsymbol{R}_{d-1}'$

$\quad\quad\quad\quad \boldsymbol{\Theta}_{d-2}' \leftarrow \boldsymbol{G}_{d-2}^{-1} \boldsymbol{G}_{d-1}'$

18: $\quad\quad\quad \boldsymbol{D}_{d-1}' \leftarrow \boldsymbol{R}_{d-1}' + \boldsymbol{D}_{d-2} \boldsymbol{\Theta}_{d-2}'$

$\quad\quad$ **end if**

20: **end for**

$\quad \boldsymbol{\Phi}_{d-1}' \leftarrow (\boldsymbol{D}_{d-1}'^{,\mathrm{H}} \boldsymbol{A} \boldsymbol{D}_{d-1}')^{-1} \boldsymbol{G}_{d-1}'^{,\mathrm{H}}$

22: $\boldsymbol{X}_d' \leftarrow \boldsymbol{X}_{d-1} + \boldsymbol{D}_{d-1}' \boldsymbol{\Phi}_{d-1}'$

4

Reduced-Rank Matrix Wiener Filters in Krylov Subspaces

In 1976, Sidney P. Applebaum and Dean J. Chapman [3] presented an adaptive beamformer using a sidelobe cancellation technique which solves the *Minimum Variance Distortionless Response* (MVDR) constrained problem (see, e.g., [40, 25, 155, 244]). A few years later, this sidelobe cancellation structure has been extended by Lloyd J. Griffiths and Charles W. Jim [98] who denoted the result as the *Generalized Sidelobe Canceler* (GSC). The GSC plays a fundamental role in the derivation of the *MultiStage Wiener Filter* (MSWF) published by J. Scott Goldstein et al. [84, 90, 89, 83] in 1997. To be precise, the MSWF with two stages is equal to the GSC except from the fact that it solves the *Mean Square Error* (MSE) optimization (cf. Section 2.1) instead of the MVDR constrained problem. Besides, the MSWF with more than two stages exploits the GSC structure not only once but several times, leading to the characteristic multistage filter bank. The application of the MSWF to a multitude of signal processing scenarios revealed its excellent performance, especially compared to the eigensubspace methods which we introduced in Section 2.3, viz., the *Principal Component* (PC) and the *Cross-Spectral* (CS) method. Some of these applications are: space-time adaptive radar processing [272, 91, 102], space-time preprocessing for *Global Positioning Systems* (GPSs) [173, 174], adaptive interference suppression in *Direct Sequence Code Division Multiple-Access* (DS-CDMA) systems [120, 271, 33, 130, 121, 134] partially also including comparisons to the adaptive *Recursive Least Squares* (RLS) [187, 112] and *Least Mean Squares* (LMS) [257, 258, 112] *filter*, and low complexity receivers in space-time coded CDMA systems [229]. A very efficient implementation of the MSWF based on a lattice structure or *correlation-subtraction architecture* has been introduced by David C. Ricks and J. Scott Goldstein [199] in 2000. Note that this lattice structure reveals the fact that the MSWF can be implemented even without explicitly estimating the autocovariance matrix of the observation.

Michael L. Honig and Weimin Xiao proved in [128, 130] that the MSWF with a special choice of blocking matrices approximates the WF in the *Krylov subspace*. Precisely speaking, they showed that the reduced-rank MSWF can be seen as a truncated *polynomial expansion* [67, 169, 221, 222] which approximates the inverse of the auto-covariance matrix occurring when solving the *Wiener–Hopf equation*, by a weighted sum of its powers. By the way, the main contribution in [128, 130] is the analysis of the reduced-rank MSWF in a heavy loaded CDMA system exploiting results from *random matrix theory* (see, e. g., [171, 250]). Note that random matrix methods have also been used in [172, 161] to design reduced-rank WFs for *MultiUser* (MU) detection.

Revealing the connection between the MSWF and the Krylov subspace, Michael L. Honig and Weimin Xiao paved the way for a multitude of new MSWF implementations. For instance, Michael Joham et al. [141, 140] derived a MSWF algorithm based on the *Lanczos procedure* (cf. Section 3.3). In [56, 42, 57], it has been shown that this Lanczos based MSWF implementation is identical to the *Conjugate Gradient* (CG) *algorithm* (cf. Section 3.4). There followed many publications (see, e. g., [256, 23, 212]) discussing the resulting insights and consequences of the relationship between the MSWF and the CG algorithm.

Applications of the MSWF which arose from its relationship to Krylov subspace methods include: DS-CDMA interference suppression [264, 129, 125, 32, 265], long-term processing in DS-CDMA systems [48], linear equalization in *Global Systems for Mobile communications* (GSM) and its *Enhanced Data rates for GSM Evolution* (EDGE) extension [42, 56, 57], *Decision Feedback Equalization* (DFE) [230, 270, 51], adaptive decision feedback turbo equalization [124, 231, 49, 55], MVDR beamforming [209], and linear as well as non-linear transmit signal processing [47, 16]. Again, Krylov subspace methods have already been applied to signal processing problems before Michael L. Honig and Weimin Xiao revealed their link to the MSWF. Besides polynomial expansion as mentioned above, another example is the CG algorithm which has been applied amongst others to: adaptive spectral estimation [29], solving extreme eigenvalue problems [266], subspace tracking [73, 148], beamforming in CDMA systems [31], adaptive filtering [27], and *Least Squares* (LS) *estimation* in oceanography [215].

Before discussing the extension of the multistage decomposition to *Matrix Wiener Filters* (MWFs) as introduced in Section 2.1, it remains to mention the *Auxiliary Vector* (AV) *method* [178, 146, 179, 162, 181, 180, 194] whose derivation is also based on the GSC structure. In fact, some of the AV publications [146, 181] consider solely the implementation of the GSC method, thus, being also very related to the MSWF with two stages. Moreover, the formulation in [179] is similar to the one in [121, 141, 140] apart from the suboptimal combination of the transformed observation signal which forms the estimate of the desired signal. In [180], the authors corrected this deficiency by introducing non-orthogonal auxiliary vectors. Finally, it has been shown in [30] that the AV method is identical to the MSWF, therefore, being also related to other Krylov subspace methods. Due to this equivalence, the AV filter will be no longer considered in this book because its investigation is mainly covered by the detailed discussion of the MSWF.

The application of the multistage principle to MWFs has been firstly published by Goldstein et al. [88] who denoted the result as the *MultiStage Matrix Wiener Filter* (MSMWF). An application of the MSMWF for joint MU detection in an DS-CDMA system is given in [34]. Although the relationship of the MSMWF to *block Krylov methods* has firstly been studied in [44, 45], there exist previous publications where block Krylov methods have been applied to signal processing problems. One of them is the generation of high-resolution spectrograms based on the *Block Conjugate Gradient* (BCG) *algorithm* [269] (cf. Section 3.4). Again, there followed a couple of other publications about MSWF implementations based on block Krylov methods considering, e. g., reduced-rank equalization of a frequency-selective *Multiple-Input Multiple-Output* (MIMO) system by either receive [45, 46] or transmit signal processing [15], robust reduced-rank filtering in a MU *Single-Input Multiple-Output* (SIMO) system [58, 43], or adaptive iterative MU decision feedback detection [127, 232].

Note that in the following section, we present the derivation of a more general MSMWF than the one given in [88], and discuss the resulting properties. Besides, we reveal in this chapter the relationship of the MSMWF to the block Krylov methods as reviewed in Chapter 3. Finally, we use the obtained insights to present MSMWF algorithms based on block Krylov methods together with a detailed investigation of their computational complexity.

4.1 Multistage Matrix Wiener Filter

4.1.1 Multistage Decomposition

In this subsection, we derive the *multistage decomposition* of the MWF $W[n] \in \mathbb{C}^{N \times M}$, $N, M \in \mathbb{N}$, and $a[n] \in \mathbb{C}^M$ (cf. Section 2.1). Compared to the original *MultiStage Matrix Wiener Filter* (MSMWF) as introduced by J. Scott Goldstein et al. [88, 34], we present a generalized version thereof by removing some unnecessary assumptions and extending it for the application to random sequences which are restricted to be neither stationary nor zero-mean. Again, without loss of generality, let N be a multiple integer of M, i. e., $N = LM$, $L \in \mathbb{N}$.[1] Besides, recall from Section 2.1 (cf. Equation 2.9) that the matrix $W[n]$ of the general MWF, i. e., the combination of $W[n]$ and $a[n]$, can be interpreted as a MWF which computes the estimate $\hat{x}^\circ[n] = \hat{x}[n] - m_x[n] = W^{\mathrm{H}}[n]y^\circ[n] \in \mathbb{C}^M$ of the zero-mean version $x^\circ[n] = x[n] - m_x[n] \in \mathbb{C}^M$ of the unknown vector random sequence $x[n] \in \mathbb{C}^M$, based on the zero-mean version $y^\circ[n] = y[n] - m_y[n] \in \mathbb{C}^N$ of the observed vector random sequence $y[n] \in \mathbb{C}^N$. Remember that the estimation error $e^\circ[n] = x^\circ[n] - \hat{x}^\circ[n] = x[n] - \hat{x}[n] = e[n]$ is always zero-mean. The

---

[1] The observation vector $y[n] \in \mathbb{C}^N$ may always be filled with zeros such that this restriction is true.

following derivation of the MSMWF is based on the application of $\boldsymbol{W}[n]$ to the zero-mean versions of the vector random sequences.

In order to end up with the multistage decomposition, we firstly prefilter the zero-mean vector random sequence $\boldsymbol{y}^{\circ}[n]$ by the full-rank matrix $\bar{\boldsymbol{T}}[n] \in \mathbb{C}^{N \times N}$ and get the transformed zero-mean vector random sequence

$$\bar{\boldsymbol{y}}^{\circ}[n] = \bar{\boldsymbol{T}}^{\mathrm{H}}[n]\boldsymbol{y}^{\circ}[n] \in \mathbb{C}^{N}. \tag{4.1}$$

Note that $\bar{\boldsymbol{y}}^{\circ}[n] = \bar{\boldsymbol{y}}[n] - \boldsymbol{m}_{\bar{\boldsymbol{y}}}[n]$ is the zero-mean version of $\bar{\boldsymbol{y}}[n] \in \mathbb{C}^{N}$. Then, we estimate the unknown vector random sequence $\boldsymbol{x}^{\circ}[n]$ by applying the MWF (cf. Section 2.1)

$$\bar{\boldsymbol{G}}[n] = \boldsymbol{C}_{\bar{\boldsymbol{y}}}^{-1}[n]\boldsymbol{C}_{\bar{\boldsymbol{y}},\boldsymbol{x}} \in \mathbb{C}^{N \times M}, \tag{4.2}$$

to $\bar{\boldsymbol{y}}^{\circ}[n]$ such that the estimate computes as $\bar{\boldsymbol{G}}^{\mathrm{H}}[n]\bar{\boldsymbol{T}}^{\mathrm{H}}[n]\boldsymbol{y}^{\circ}[n]$. Here, $\boldsymbol{C}_{\bar{\boldsymbol{y}}}[n] = \bar{\boldsymbol{T}}^{\mathrm{H}}[n]\boldsymbol{C}_{\boldsymbol{y}}[n]\bar{\boldsymbol{T}}[n]$ denotes the auto-covariance matrix of $\bar{\boldsymbol{y}}^{\circ}[n]$ and $\boldsymbol{C}_{\bar{\boldsymbol{y}},\boldsymbol{x}}$ the cross-covariance matrix between $\bar{\boldsymbol{y}}^{\circ}[n]$ and $\boldsymbol{x}[n]$. Remember that the covariance matrices do not depend on the mean (see Equations 2.10 and 2.11), i.e., we can omit the circle at all random sequences occurring as their subscripts. Due to the fact that $\mathrm{rank}\{\bar{\boldsymbol{T}}[n]\} = N$, i.e., the prefilter matrix $\bar{\boldsymbol{T}}[n]$ is invertible, its combination with the MWF $\bar{\boldsymbol{G}}[n]$ can be written as (cf. also Proposition 2.1)

$$\begin{aligned}
\bar{\boldsymbol{T}}[n]\bar{\boldsymbol{G}}[n] &= \bar{\boldsymbol{T}}[n]\left(\bar{\boldsymbol{T}}^{\mathrm{H}}[n]\boldsymbol{C}_{\boldsymbol{y}}[n]\bar{\boldsymbol{T}}[n]\right)^{-1}\bar{\boldsymbol{T}}^{\mathrm{H}}[n]\boldsymbol{C}_{\boldsymbol{y},\boldsymbol{x}} \\
&= \boldsymbol{C}_{\boldsymbol{y}}^{-1}[n]\boldsymbol{C}_{\boldsymbol{y},\boldsymbol{x}} = \boldsymbol{W}[n].
\end{aligned} \tag{4.3}$$

In other words, the calculated estimate of $\boldsymbol{x}^{\circ}[n]$ is the same as in the case without prefiltering, i.e.,

$$\bar{\boldsymbol{G}}^{\mathrm{H}}[n]\bar{\boldsymbol{T}}^{\mathrm{H}}[n]\boldsymbol{y}^{\circ}[n] = \boldsymbol{W}^{\mathrm{H}}[n]\boldsymbol{y}^{\circ}[n] = \hat{\boldsymbol{x}}^{\circ}[n]. \tag{4.4}$$

Hence, Figure 4.1 depicts an exact decomposition of the MWF $\boldsymbol{W}[n]$ where we also included the transformation to the zero-mean random sequences and vice versa.

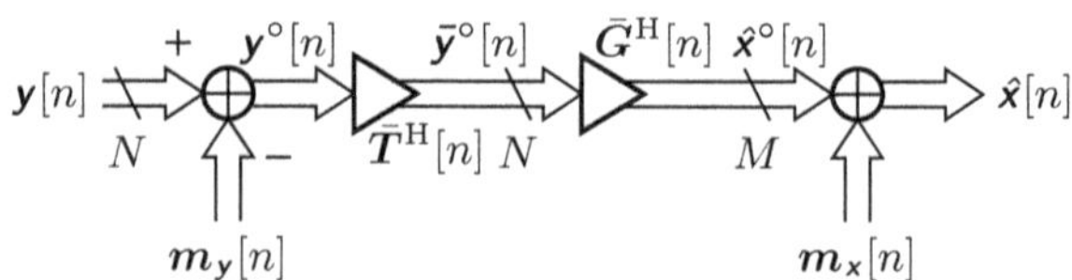

Fig. 4.1. Matrix Wiener Filter (MWF) with full-rank prefilter

Next, we rewrite the prefilter as

$$\bar{\boldsymbol{T}}[n] = \begin{bmatrix} \boldsymbol{M}_{0}[n] & \boldsymbol{B}_{0}[n] \end{bmatrix}, \tag{4.5}$$

where $M_0[n] \in \mathbb{C}^{N \times M}$ is an arbitrary $N \times M$ matrix with full column-rank, i. e., $\mathrm{rank}\{M_0[n]\} = M$, and $B_0[n] \in \mathbb{C}^{N \times (N-M)}$ is a matrix with full column-rank, i. e., $\mathrm{rank}\{B_0[n]\} = N - M$, which is orthogonal[2] to $M_0[n]$, i. e.,

$$\left| M_0^{\mathrm{H}}[n] B_0[n] = \mathbf{0}_{M \times (N-M)} \quad \Leftrightarrow \quad \mathrm{null}\left\{ M_0^{\mathrm{H}}[n]\right\} = \mathrm{range}\left\{B_0[n]\right\}. \right| \quad (4.6)$$

Note that $\mathrm{range}\{M_0[n]\} \oplus \mathrm{range}\{B_0[n]\} = \mathbb{C}^N$. With Equation (4.5) and the definitions $x_0^\circ[n] := M_0^{\mathrm{H}}[n] y^\circ[n] \in \mathbb{C}^M$ and $y_0^\circ[n] := B_0^{\mathrm{H}}[n] y^\circ[n] \in \mathbb{C}^{N-M}$, the transformed observation $\bar{y}^\circ[n]$ of Equation (4.1) may be written as

$$\bar{y}^\circ[n] = \begin{bmatrix} M_0^{\mathrm{H}}[n] \\ B_0^{\mathrm{H}}[n] \end{bmatrix} y^\circ[n] = \begin{bmatrix} x_0^\circ[n] \\ y_0^\circ[n] \end{bmatrix}, \quad (4.7)$$

and the auto-covariance matrix of $\bar{y}^\circ[n]$ can be partitioned according to

$$C_{\bar{y}}[n] = \begin{bmatrix} C_{x_0}[n] & C_{y_0,x_0}^{\mathrm{H}}[n] \\ C_{y_0,x_0}[n] & C_{y_0}[n] \end{bmatrix} \in \mathbb{C}^{N \times N}, \quad (4.8)$$

where (cf. Equation 4.7)

$$\left| \begin{aligned} C_{x_0}[n] &= M_0^{\mathrm{H}}[n] C_y[n] M_0[n] \in \mathbb{C}^{M \times M} \text{ and} \\ C_{y_0}[n] &= B_0^{\mathrm{H}}[n] C_y[n] B_0[n] \in \mathbb{C}^{(N-M) \times (N-M)}, \end{aligned} \right| \quad (4.9)$$

are the auto-covariance matrices of $x_0^\circ[n]$ and $y_0^\circ[n]$, respectively, and

$$\left| C_{y_0,x_0}[n] = B_0^{\mathrm{H}}[n] C_y[n] M_0[n] \in \mathbb{C}^{(N-M) \times M}, \right| \quad (4.10)$$

is the cross-covariance matrix between $y_0^\circ[n]$ and $x_0^\circ[n]$. Again, $x_0^\circ[n] = x_0[n] - m_{x_0}[n]$ and $y_0^\circ[n] = y_0[n] - m_{y_0}[n]$ are the zero-mean versions of $x_0[n] \in \mathbb{C}^M$ and $y_0[n] \in \mathbb{C}^{N-M}$, respectively, and the covariance matrices do not depend on the mean (cf. Equations 2.10 and 2.11). According to Appendix A.1 (cf. Equations A.2 and A.5), the inverse of the partitioned matrix $C_{\bar{y}}[n]$ computes as

$$C_{\bar{y}}^{-1}[n] = \begin{bmatrix} C_{\varepsilon_0}^{-1}[n] & -C_{\varepsilon_0}^{-1}[n] G_0^{\mathrm{H}}[n] \\ -G_0[n] C_{\varepsilon_0}^{-1}[n] & C_{y_0}^{-1}[n] + G_0[n] C_{\varepsilon_0}^{-1}[n] G_0^{\mathrm{H}}[n] \end{bmatrix}, \quad (4.11)$$

where the matrix

$$G_0[n] = C_{y_0}^{-1}[n] C_{y_0,x_0} \in \mathbb{C}^{(N-M) \times M}, \quad (4.12)$$

can be interpreted as a reduced-dimension MWF (cf. Equation 2.7) computing the estimate

---

[2] As introduced in Chapter 3, two matrices are denoted to be orthogonal if each column of the one matrix is orthogonal to all the columns of the second matrix.

$$\hat{x}_0^\circ[n] = G_0^{\mathrm{H}}[n]y_0^\circ[n] \in \mathbb{C}^M, \tag{4.13}$$

of the zero-mean vector random sequence $x_0^\circ[n]$ based on the observation $y_0^\circ[n]$, and the matrix

$$C_{\varepsilon_0}[n] = C_{x_0}[n] - C_{y_0,x_0}^{\mathrm{H}}[n]C_{y_0}^{-1}[n]C_{y_0,x_0}[n] \in \mathbb{C}^{M \times M}, \tag{4.14}$$

which can be seen as the auto-covariance matrix of the zero-mean estimation error $\varepsilon_0^\circ[n] = x_0^\circ[n] - \hat{x}_0^\circ[n] = x_0[n] - \hat{x}_0[n] = \varepsilon_0[n] \in \mathbb{C}^M$ when applying the MWF $G_0[n]$ (cf. Equation 2.8). Note that $\hat{x}_0^\circ[n] = \hat{x}_0[n] - m_{x_0} \in \mathbb{C}^M$ is the zero-mean version of $\hat{x}_0[n] \in \mathbb{C}^M$ and $C_{\varepsilon_0}[n]$ is the Schur complement of $C_{x_0}[n]$ (cf. Appendix A.1).

Further, with Equation (4.12) and the cross-covariance matrix (cf. Equations 4.5 and 4.7)

$$C_{\bar{y},x}[n] = \bar{T}^{\mathrm{H}}[n]C_{y,x}[n] = \begin{bmatrix} M_0^{\mathrm{H}}[n]C_{y,x}[n] \\ B_0^{\mathrm{H}}[n]C_{y,x}[n] \end{bmatrix} = \begin{bmatrix} C_{x_0,x}[n] \\ C_{y_0,x}[n] \end{bmatrix} \in \mathbb{C}^{N \times M}, \tag{4.15}$$

between $\bar{y}^\circ[n]$ and the unknown signal $x^\circ[n]$, we get the following decomposition of the MWF $\bar{G}[n]$:

$$\bar{G}[n] = \begin{bmatrix} \Delta_0[n] \\ -G_0[n]\Delta_0[n] + C_{y_0}^{-1}[n]C_{y_0,x}[n] \end{bmatrix}, \tag{4.16}$$

where

$$\Delta_0[n] = C_{\varepsilon_0}^{-1}[n]\left(C_{x_0,x}[n] - G_0^{\mathrm{H}}[n]C_{y_0,x}[n]\right) = C_{\varepsilon_0}^{-1}[n]C_{\varepsilon_0,x}[n], \tag{4.17}$$

is a quadratic MWF estimating the unknown signal $x^\circ[n]$ based on the estimation error $\varepsilon_0[n]$. With Equations (4.4), (4.5), and (4.16), the MWF estimate $\hat{x}^\circ[n]$ can finally be written as

$$\begin{aligned}
\hat{x}^\circ[n] = {}& \Delta_0^{\mathrm{H}}[n]\left(M_0^{\mathrm{H}}[n] - G_0^{\mathrm{H}}[n]B_0^{\mathrm{H}}[n]\right)y^\circ[n] \\
& + C_{y_0,x}^{\mathrm{H}}[n]C_{y_0}^{-1}[n]B_0^{\mathrm{H}}[n]y^\circ[n],
\end{aligned} \tag{4.18}$$

whose structure is depicted in Figure 4.2.

To sum up, the MWF $W[n]$ has been decomposed into the prefilters $M_0[n] \in \mathbb{C}^{N \times M}$ and $B_0[n] \in \mathbb{C}^{N \times (N-M)}$, and the reduced-dimension MWFs $G_0[n] \in \mathbb{C}^{(N-M) \times M}$ and $C_{y_0}^{-1}[n]C_{y_0,x}[n] \in \mathbb{C}^{(N-M) \times M}$, as well as the quadratic MWF $\Delta_0[n] \in \mathbb{C}^{M \times M}$. The question remains if there exists a more efficient decomposition of the MWF $W[n]$. Considering the reduced-dimension MWF $C_{y_0}^{-1}[n]C_{y_0,x}[n] = C_{y_0}^{-1}[n]B_0^{\mathrm{H}}[n]C_{y,x}[n]$, we see that it vanishes if the blocking matrix $B_0[n]$ is chosen to be orthogonal to the cross-covariance matrix $C_{y,x}[n]$. Roughly speaking, the blocking matrix must be chosen such that its output $y_0^\circ[n]$ is uncorrelated to the unknown signal $x^\circ[n]$. Only in this case, there is no need for the additional MWF $C_{y_0}^{-1}[n]C_{y_0,x}[n]$ which estimates $x^\circ[n]$ based on the component in $y_0^\circ[n]$ which is still correlated to $x^\circ[n]$. Since $B_0[n]$

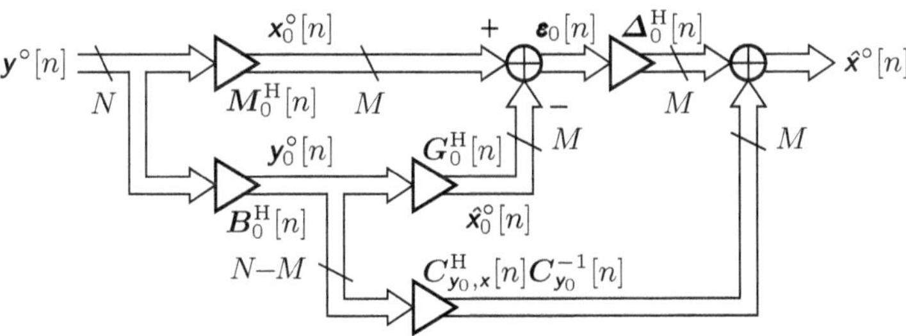

Fig. 4.2. Decomposition of the MWF $\boldsymbol{W}[n]$

is already orthogonal to the matrix $\boldsymbol{M}_0[n]$ (cf. Equation 4.6), choosing the range space of $\boldsymbol{B}_0[n]$ as the null space of $\boldsymbol{C}_{\boldsymbol{y},\boldsymbol{x}}^{\mathrm{H}}[n]$ is equal to selecting $\boldsymbol{M}_0[n]$ as an arbitrary basis of range$\{\boldsymbol{C}_{\boldsymbol{y},\boldsymbol{x}}[n]\}$, i. e.,

$$\left| \text{range}\left\{\boldsymbol{M}_0[n]\right\} = \text{range}\left\{\boldsymbol{C}_{\boldsymbol{y},\boldsymbol{x}}[n]\right\} = \mathcal{M}_0[n] \in G(N, M), \right| \qquad (4.19)$$

which is equal to the *correlated subspace*

$$\mathcal{M}_0[n] = \underset{\mathcal{M}' \in G(N,M)}{\mathrm{argmax}} \; \mathrm{tr}\left\{ \sqrt[\oplus]{\boldsymbol{C}_{\boldsymbol{y},\boldsymbol{x}}^{\mathrm{H}}[n]\boldsymbol{P}_{\mathcal{M}'}\boldsymbol{C}_{\boldsymbol{y},\boldsymbol{x}}[n]} \right\}, \qquad (4.20)$$

as introduced in Appendix B.5. Equation (4.19) implies that $\boldsymbol{C}_{\boldsymbol{y},\boldsymbol{x}}[n]$ must have full column-rank, i. e., rank$\{\boldsymbol{C}_{\boldsymbol{y},\boldsymbol{x}}[n]\} = M$. From now on, $\boldsymbol{M}_0[n]$ is chosen to fulfill Equation (4.19) which results in the simplified decomposition of the MWF $\boldsymbol{W}[n]$ as depicted in Figure 4.3. Note that besides the ambiguity of the blocking matrix $\boldsymbol{B}_0[n]$ which can be an arbitrary matrix as long as it is orthogonal to $\boldsymbol{M}_0[n]$, the choice of $\boldsymbol{M}_0[n]$ is also not unique because any basis of $\mathcal{M}_0[n]$ leads to the same estimate $\hat{\boldsymbol{x}}^\circ[n]$ as can be easily verified if we recall Equation (4.4). Since we do not restrict the matrix $\boldsymbol{M}_0[n]$ to be one particular basis of $\mathcal{M}_0[n]$, the presented MSMWF is a generalization of the original version introduced by J. Scott Goldstein et al. in [88] where $\boldsymbol{M}_0[n]$ has been chosen such that its columns are orthonormal. With Equations (4.19) and (4.6), $\boldsymbol{C}_{\boldsymbol{y}_0,\boldsymbol{x}}[n] = \boldsymbol{B}_0^{\mathrm{H}}[n]\boldsymbol{C}_{\boldsymbol{y},\boldsymbol{x}}[n] = \boldsymbol{0}_{(N-M)\times M}$ and the quadratic MWF $\boldsymbol{\Delta}_0[n]$ of Equation (4.17) reads as (cf. also Equation 4.15)

$$\left| \boldsymbol{\Delta}_0[n] = \boldsymbol{C}_{\boldsymbol{\varepsilon}_0}^{-1}[n]\boldsymbol{C}_{\boldsymbol{\varepsilon}_0,\boldsymbol{x}}[n] = \boldsymbol{C}_{\boldsymbol{\varepsilon}_0}^{-1}[n]\boldsymbol{C}_{\boldsymbol{x}_0,\boldsymbol{x}}[n] = \boldsymbol{C}_{\boldsymbol{\varepsilon}_0}^{-1}[n]\boldsymbol{M}_0^{\mathrm{H}}[n]\boldsymbol{C}_{\boldsymbol{y},\boldsymbol{x}}[n]. \right| \qquad (4.21)$$

Note that the filter block diagram of Figure 4.3 is strongly related to the *Generalized Sidelobe Canceler* (GSC) [3, 98], especially if we assume the vector case, i. e., $M = 1$. The output vector random sequence $\boldsymbol{x}_0^\circ[n]$ is still perturbed by interference because it is only correlated to the desired vector random sequence $\boldsymbol{x}^\circ[n]$. The MWF $\boldsymbol{G}_0[n]$ estimates the interference remained in $\boldsymbol{x}_0^\circ[n]$ from the transformed observation vector random sequence $\boldsymbol{y}_0^\circ[n]$ which is uncorrelated to $\boldsymbol{x}^\circ[n]$ but correlated to $\boldsymbol{x}_0^\circ[n]$. After removing

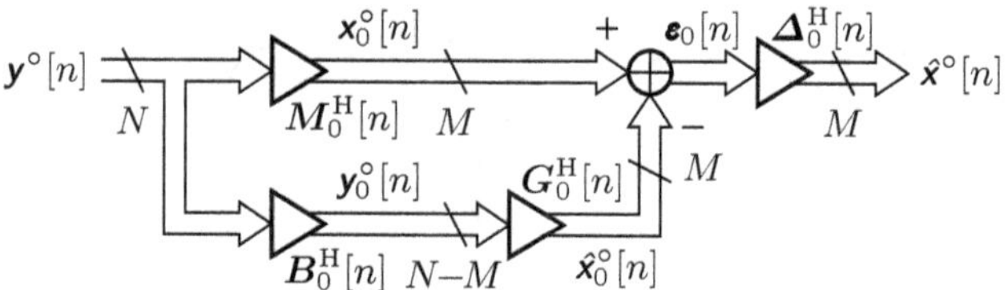

Fig. 4.3. Simplified decomposition of the MWF $\boldsymbol{W}[n]$

the interference $\hat{\boldsymbol{x}}_0^{\circ}[n]$ from $\boldsymbol{x}_0^{\circ}[n]$, the quadratic MWF $\boldsymbol{\Delta}_0[n]$ is needed to reconstruct the desired signal vector $\boldsymbol{x}^{\circ}[n]$ by minimizing the MSE. Nevertheless, the classical GSC has been designed to be the solution of the *Minimum Variance Distortionless Response* (MVDR) constrained optimization problem (see, e.g., [40, 25, 155, 244]) instead of the MMSE solution.

Up to now, without changing its performance (cf. Equation 4.4), the $N \times M$ MWF $\boldsymbol{W}[n]$ has been replaced by the prefilter matrices $\boldsymbol{M}_0[n]$ and $\boldsymbol{B}_0[n]$, the $(N - M) \times M$ MWF $\boldsymbol{G}_0[n]$, and the quadratic $M \times M$ MWF $\boldsymbol{\Delta}_0[n]$. The fundamental idea of the MSMWF is to repeatedly apply this GSC kind of decomposition to the reduced-dimension MWFs $\boldsymbol{G}_{\ell-1}[n] \in \mathbb{C}^{(N-\ell M) \times M}$, $\ell \in \{1, 2, \ldots, L-2\}$, $L > 1$, resulting in the structures shown in Figure 4.4.

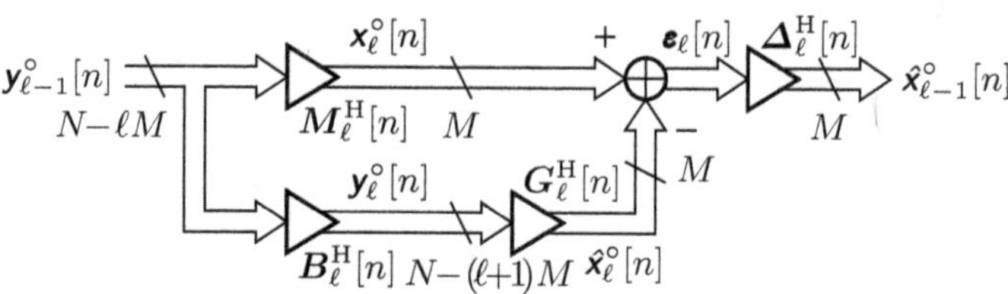

Fig. 4.4. Decomposition of the MWF $\boldsymbol{G}_{\ell-1}[n]$

This time, $\boldsymbol{M}_\ell[n] \in \mathbb{C}^{(N-\ell M) \times M}$ is an arbitrary matrix with full column-rank, i.e., $\mathrm{rank}\{\boldsymbol{M}_\ell[n]\} = M$, and whose columns span the correlated subspace

$$\mathcal{M}_\ell[n] = \underset{\mathcal{M}' \in G(N-\ell M, M)}{\mathrm{argmax}} \ \mathrm{tr}\left\{ \sqrt[\oplus]{\boldsymbol{C}_{\boldsymbol{y}_{\ell-1}, \boldsymbol{x}_{\ell-1}}^{\mathrm{H}}[n] \boldsymbol{P}_{\mathcal{M}'} \boldsymbol{C}_{\boldsymbol{y}_{\ell-1}, \boldsymbol{x}_{\ell-1}}[n]} \right\}, \qquad (4.22)$$

i.e., the subspace on which the projection of $\boldsymbol{y}_{\ell-1}^{\circ}[n] \in \mathbb{C}^{N-\ell M}$ still implies the signal which is maximal correlated to $\boldsymbol{x}_{\ell-1}^{\circ}[n] \in \mathbb{C}^M$. The output of $\boldsymbol{M}_\ell[n]$ is denoted as $\boldsymbol{x}_\ell^{\circ}[n] = \boldsymbol{M}_\ell^{\mathrm{H}}[n] \boldsymbol{y}_{\ell-1}^{\circ}[n] \in \mathbb{C}^M$. According to Appendix B.5, Equation (4.22) is solved by the subspace spanned by the columns of the cross-covariance matrix $\boldsymbol{C}_{\boldsymbol{y}_{\ell-1}, \boldsymbol{x}_{\ell-1}}[n] \in \mathbb{C}^{(N-\ell M) \times M}$ between $\boldsymbol{y}_{\ell-1}^{\circ}[n]$ and $\boldsymbol{x}_{\ell-1}^{\circ}[n]$, i.e.,

$$\left| \mathrm{range}\{\boldsymbol{M}_\ell[n]\} = \mathrm{range}\{\boldsymbol{C}_{\boldsymbol{y}_{\ell-1}, \boldsymbol{x}_{\ell-1}}[n]\} = \mathcal{M}_\ell[n]. \right| \qquad (4.23)$$

Further, the columns of the blocking matrix $\boldsymbol{B}_\ell[n] \in \mathbb{C}^{(N-\ell M)\times(N-(\ell+1)M)}$ with $\operatorname{rank}\{\boldsymbol{B}_\ell[n]\} = N - (\ell+1)M$, span the orthogonal complement of the range space of $\boldsymbol{M}_\ell[n]$ in $\mathbb{C}^{(N-\ell M)\times M}$, i.e.,

$$\boldsymbol{M}_\ell^{\mathrm{H}}[n]\boldsymbol{B}_\ell[n] = \boldsymbol{0}_{M\times(N-(\ell+1)M)} \Leftrightarrow \operatorname{null}\left\{\boldsymbol{M}_\ell^{\mathrm{H}}[n]\right\} = \operatorname{range}\left\{\boldsymbol{B}_\ell[n]\right\}, \tag{4.24}$$

and $\operatorname{range}\{\boldsymbol{M}_\ell[n]\} \oplus \operatorname{range}\{\boldsymbol{B}_\ell[n]\} = \mathbb{C}^{N-\ell M}$. Roughly speaking, $\boldsymbol{B}_\ell[n]$ is chosen such that the output vector random sequence $\boldsymbol{y}_\ell^\circ[n] = \boldsymbol{B}_\ell^{\mathrm{H}}[n]\boldsymbol{y}_{\ell-1}^\circ[n] \in \mathbb{C}^{N-(\ell+1)M}$ is uncorrelated to $\boldsymbol{x}_{\ell-1}^\circ[n]$. Next, the reduced-dimension MWF

$$\boldsymbol{G}_\ell[n] = \boldsymbol{C}_{\boldsymbol{y}_\ell}^{-1}[n]\boldsymbol{C}_{\boldsymbol{y}_\ell,\boldsymbol{x}_\ell}[n], \tag{4.25}$$

where

$$\boldsymbol{C}_{\boldsymbol{y}_\ell}[n] = \boldsymbol{B}_\ell^{\mathrm{H}}[n]\boldsymbol{C}_{\boldsymbol{y}_{\ell-1}}[n]\boldsymbol{B}_\ell[n] \in \mathbb{C}^{(N-(\ell+1)M)\times(N-(\ell+1)M)}, \tag{4.26}$$

denotes the auto-covariance matrix of $\boldsymbol{y}_\ell^\circ[n]$ and

$$\boldsymbol{C}_{\boldsymbol{y}_\ell,\boldsymbol{x}_\ell}[n] = \boldsymbol{B}_\ell^{\mathrm{H}}[n]\boldsymbol{C}_{\boldsymbol{y}_{\ell-1}}[n]\boldsymbol{M}_\ell[n] \in \mathbb{C}^{(N-(\ell+1)M)\times M}, \tag{4.27}$$

the cross-covariance matrix between $\boldsymbol{y}_\ell^\circ[n]$ and $\boldsymbol{x}_\ell^\circ[n]$, is designed to compute the estimate $\hat{\boldsymbol{x}}_\ell^\circ[n] = \boldsymbol{G}_\ell^{\mathrm{H}}[n]\boldsymbol{y}_\ell^\circ[n]$ of $\boldsymbol{x}_\ell^\circ[n]$ using $\boldsymbol{y}_\ell^\circ[n]$ as the observation. Based on the resulting estimation error $\boldsymbol{\varepsilon}_\ell[n] = \boldsymbol{x}_\ell^\circ[n] - \hat{\boldsymbol{x}}_\ell^\circ[n] \in \mathbb{C}^M$, the quadratic $M \times M$ MWF (cf. Equation 4.17)

$$\boldsymbol{\Delta}_\ell[n] = \boldsymbol{C}_{\boldsymbol{\varepsilon}_\ell}^{-1}[n]\boldsymbol{C}_{\boldsymbol{\varepsilon}_\ell,\boldsymbol{x}_{\ell-1}}[n] \in \mathbb{C}^{M\times M}, \tag{4.28}$$

with the output $\hat{\boldsymbol{x}}_{\ell-1}^\circ[n] = \boldsymbol{\Delta}_\ell^{\mathrm{H}}[n]\boldsymbol{\varepsilon}_\ell[n]$, estimates finally $\boldsymbol{x}_{\ell-1}^\circ[n]$. Here, the matrices

$$\boldsymbol{C}_{\boldsymbol{\varepsilon}_\ell}[n] = \boldsymbol{C}_{\boldsymbol{x}_\ell}[n] - \boldsymbol{C}_{\boldsymbol{y}_\ell,\boldsymbol{x}_\ell}^{\mathrm{H}}[n]\boldsymbol{C}_{\boldsymbol{y}_\ell}^{-1}[n]\boldsymbol{C}_{\boldsymbol{y}_\ell,\boldsymbol{x}_\ell}[n] \subset \mathbb{C}^{M\times M}, \tag{4.29}$$

and

$$\boldsymbol{C}_{\boldsymbol{x}_\ell}[n] = \boldsymbol{M}_\ell^{\mathrm{H}}[n]\boldsymbol{C}_{\boldsymbol{y}_{\ell-1}}[n]\boldsymbol{M}_\ell[n] \in \mathbb{C}^{M\times M}, \tag{4.30}$$

are the auto-covariance matrices of $\boldsymbol{\varepsilon}_\ell[n]$ and $\boldsymbol{x}_\ell^\circ[n]$, respectively, and

$$\boldsymbol{C}_{\boldsymbol{\varepsilon}_\ell,\boldsymbol{x}_{\ell-1}}[n] = \boldsymbol{C}_{\boldsymbol{x}_\ell,\boldsymbol{x}_{\ell-1}}[n] = \boldsymbol{M}_\ell^{\mathrm{H}}[n]\boldsymbol{C}_{\boldsymbol{y}_{\ell-1},\boldsymbol{x}_{\ell-1}}[n] \in \mathbb{C}^{M\times M}, \tag{4.31}$$

is the cross-covariance matrix between $\boldsymbol{\varepsilon}_\ell[n]$ and $\boldsymbol{x}_{\ell-1}^\circ[n]$. Note that $\boldsymbol{C}_{\boldsymbol{y}_\ell,\boldsymbol{x}_{\ell-1}}[n] = \boldsymbol{B}_\ell^{\mathrm{H}}[n]\boldsymbol{C}_{\boldsymbol{y}_{\ell-1},\boldsymbol{x}_{\ell-1}}[n] = \boldsymbol{0}_{(N-(\ell+1)M)\times M}$ due to Equations (4.23) and (4.24), thus, $\boldsymbol{C}_{\boldsymbol{\varepsilon}_\ell,\boldsymbol{x}_{\ell-1}}[n] = \boldsymbol{C}_{\boldsymbol{x}_\ell,\boldsymbol{x}_{\ell-1}}[n] - \boldsymbol{G}_\ell^{\mathrm{H}}[n]\boldsymbol{C}_{\boldsymbol{y}_\ell,\boldsymbol{x}_{\ell-1}}[n] = \boldsymbol{C}_{\boldsymbol{x}_\ell,\boldsymbol{x}_{\ell-1}}[n]$.

Finally, the decomposition of the last reduced-dimension MWF $\boldsymbol{G}_{L-2}[n] \in \mathbb{C}^{M\times M}$ is depicted in Figure 4.5. There, the matrix $\boldsymbol{M}_{L-1}[n] \in \mathbb{C}^{M\times M}$ spans the range space of $\boldsymbol{C}_{\boldsymbol{y}_{L-2},\boldsymbol{x}_{L-2}}[n] \in \mathbb{C}^{M\times M}$, i.e.,

$$\left| \text{range}\left\{\boldsymbol{M}_{L-1}[n]\right\} = \text{range}\left\{\boldsymbol{C}_{\boldsymbol{y}_{L-2},\boldsymbol{x}_{L-2}}[n]\right\} = \mathcal{M}_{L-1}[n], \right| \qquad (4.32)$$

obtained from the optimization problem defined in Equation (4.22) where $\ell = L - 1$. The output signal $\boldsymbol{x}_{L-1}^{\circ}[n] = \boldsymbol{M}_{L-1}^{\mathrm{H}}[n]\boldsymbol{y}_{L-2}^{\circ}[n] \in \mathbb{C}^M$ which can be defined as $\boldsymbol{\varepsilon}_{L-1}[n] := \boldsymbol{x}_{L-1}^{\circ}[n] \in \mathbb{C}^M$, is taken as the observation for the quadratic MWF

$$\left| \begin{aligned} \boldsymbol{\Delta}_{L-1}[n] &= \boldsymbol{C}_{\boldsymbol{\varepsilon}_{L-1}}^{-1}[n]\boldsymbol{C}_{\boldsymbol{\varepsilon}_{L-1},\boldsymbol{x}_{L-2}}[n] = \boldsymbol{C}_{\boldsymbol{x}_{L-1}}^{-1}[n]\boldsymbol{C}_{\boldsymbol{x}_{L-1},\boldsymbol{x}_{L-2}}[n] \\ &= \boldsymbol{C}_{\boldsymbol{x}_{L-1}}^{-1}[n]\boldsymbol{M}_{L-1}^{\mathrm{H}}[n]\boldsymbol{C}_{\boldsymbol{y}_{L-2},\boldsymbol{x}_{L-2}}[n], \end{aligned} \right| \qquad (4.33)$$

estimating $\boldsymbol{x}_{L-2}^{\circ}[n] \in \mathbb{C}^M$, i.e., the estimate computes as $\hat{\boldsymbol{x}}_{L-2}^{\circ}[n] = \boldsymbol{\Delta}_{L-1}^{\mathrm{H}}[n]\boldsymbol{x}_{L-1}^{\circ}[n] \in \mathbb{C}^M$.

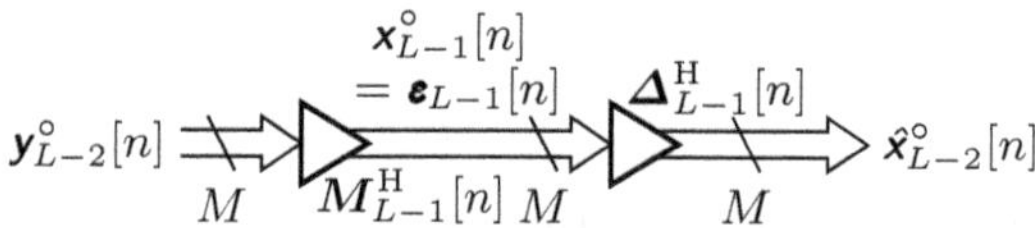

Fig. 4.5. Decomposition of the MWF $\boldsymbol{G}_{L-2}[n]$

Figure 4.6 presents the resulting full-rank MSMWF, i.e., the multistage decomposition of the MWF $\boldsymbol{W}[n]$ where we included again the zero-mean correction in $\boldsymbol{a}[n]$. Thus, all zero-mean versions of the random sequences have to be replaced by their originals, viz., $\boldsymbol{x}_\ell[n] = \boldsymbol{x}_\ell^{\circ}[n] + \boldsymbol{m}_{\boldsymbol{x}_\ell}[n] \in \mathbb{C}^M$, $\hat{\boldsymbol{x}}_\ell[n] = \hat{\boldsymbol{x}}_\ell^{\circ}[n] + \boldsymbol{m}_{\boldsymbol{x}_\ell}[n] \in \mathbb{C}^M$, and $\boldsymbol{y}_\ell[n] = \boldsymbol{y}_\ell^{\circ}[n] + \boldsymbol{m}_{\boldsymbol{y}_\ell}[n] \in \mathbb{C}^{N-(\ell+1)M}$, $\ell \in \{0, 1, \ldots, L - 1\}$. The most important property of the full-rank MSMWF is formulated in Proposition 4.1.

Proposition 4.1. *The full-rank MSMWF of Figure 4.6 is identical to the optimal MWF $\boldsymbol{W}[n]$ and $\boldsymbol{a}[n]$, i.e., they produce the same estimate $\hat{\boldsymbol{x}}[n]$. Thus, its performance is invariant to the choices of the filters $\boldsymbol{M}_\ell[n]$, $\ell \in \{0, 1, \ldots, L-1\}$, and $\boldsymbol{B}_\ell[n]$, $\ell \in \{0, 1, \ldots, L-2\}$, $L > 1$, as long as they fulfill Equations (4.6), (4.19), (4.23), (4.24), and (4.32).*

Proof. Proposition 4.1 holds per construction. Since all decompositions of the reduced-dimension MWFs $\boldsymbol{G}_{\ell-1}[n]$, $\ell \in \{1, 2, \ldots, L - 1\}$, as depicted in Figures 4.4 and 4.5, do not change the performance in analogy to the decomposition of the MWF $\boldsymbol{W}[n]$ as given in Figure 4.3 (cf. Equations 4.3 and 4.4), the MSMWF structure is equivalent to the optimal MWF $\boldsymbol{W}[n]$ and $\boldsymbol{a}[n]$ no matter which specific matrices $\boldsymbol{M}_\ell[n]$ and $\boldsymbol{B}_\ell[n]$ have been chosen. $\square$

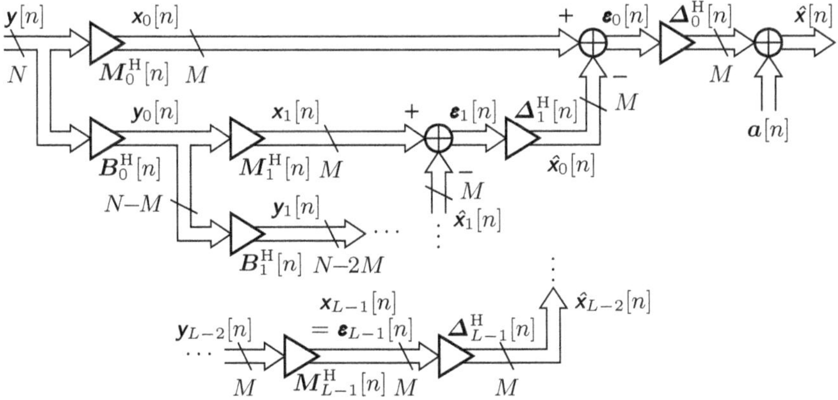

Fig. 4.6. Full-rank MultiStage Matrix Wiener Filter (MSMWF)

4.1.2 Reduced-Rank Multistage Matrix Wiener Filter

The reduced-rank approximation of the MWF $\boldsymbol{W}[n] \in \mathbb{C}^{N \times M}$ and $\boldsymbol{a}[n] \in \mathbb{C}^M$, the rank D MSMWF $\boldsymbol{W}_{\mathrm{MSMWF}}^{(D)}[n] \in \mathbb{C}^{N \times M}$ and $\boldsymbol{a}_{\mathrm{MSMWF}}^{(D)}[n] = \boldsymbol{m}_x[n] - \boldsymbol{W}_{\mathrm{MSMWF}}^{(D),\mathrm{H}}[n]\boldsymbol{m}_y[n] \in \mathbb{C}^M$, is obtained by neglecting the estimate $\hat{\boldsymbol{x}}_{d-1}[n] \in \mathbb{C}^M$ and approximating the error vector $\boldsymbol{\varepsilon}_{d-1}[n] \in \mathbb{C}^M$ by the signal vector $\boldsymbol{x}_{d-1}[n] \in \mathbb{C}^M$. Note that $D = dM$ with $d \in \{1, 2, \ldots, L\}$, i.e., D is restricted to be an integer multiple of M. Figure 4.7 depicts the resulting structure. We used the superscript '(D)' to emphasize that the corresponding signals and filters are different from those of the full-rank representation of the MWF as depicted in Figure 4.6.

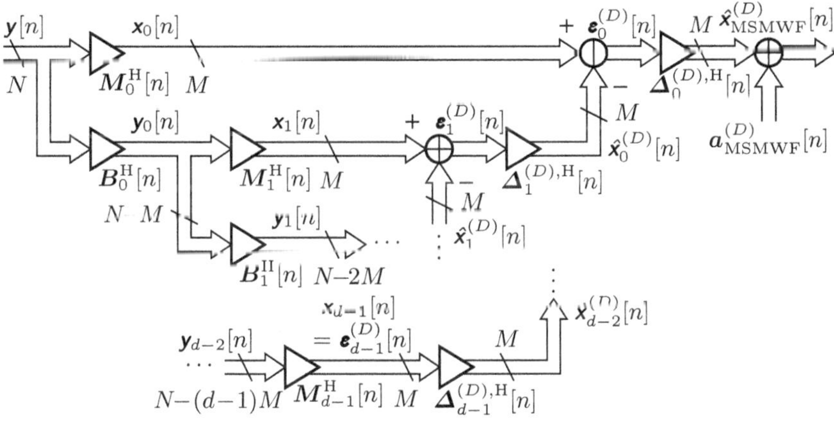

Fig. 4.7. Rank D MSMWF

Starting with $\boldsymbol{\varepsilon}_{d-1}^{(D)}[n] = \boldsymbol{x}_{d-1}[n]$, the error vector random sequence between $\boldsymbol{x}_\ell[n] \in \mathbb{C}^M$ and the output of the quadratic MWF (cf. Equation 4.28)

$$\left| \boldsymbol{\Delta}_{\ell+1}^{(D)}[n] = \boldsymbol{C}_{\boldsymbol{\varepsilon}_{\ell+1}^{(D)}}^{-1}[n] \boldsymbol{C}_{\boldsymbol{\varepsilon}_{\ell+1}^{(D)},\boldsymbol{x}_\ell}[n] \in \mathbb{C}^{M \times M}, \right| \tag{4.34}$$

with $\ell \in \{0,1,\ldots,d-2\}$, $d > 1$, reads as (cf. Figure 4.7)

$$\boldsymbol{\varepsilon}_\ell^{(D)}[n] = \boldsymbol{x}_\ell[n] - \hat{\boldsymbol{x}}_\ell^{(D)}[n] = \boldsymbol{x}_\ell[n] - \boldsymbol{\Delta}_{\ell+1}^{(D),\mathrm{H}}[n]\boldsymbol{\varepsilon}_{\ell+1}^{(D)}[n] \in \mathbb{C}^M. \tag{4.35}$$

Thus, the auto-covariance matrix $\boldsymbol{C}_{\boldsymbol{\varepsilon}_\ell^{(D)}}[n] \in \mathbb{C}^{M \times M}$ of $\boldsymbol{\varepsilon}_\ell^{(D)}[n]$ for $\ell \in \{0,1,\ldots,d-2\}$, $d > 1$, originally defined in Equation (4.29), can be rewritten as

$$\left| \boldsymbol{C}_{\boldsymbol{\varepsilon}_\ell^{(D)}}[n] = \boldsymbol{C}_{\boldsymbol{x}_\ell}[n] - \boldsymbol{C}_{\boldsymbol{\varepsilon}_{\ell+1}^{(D)},\boldsymbol{x}_\ell}^{\mathrm{H}}[n] \boldsymbol{C}_{\boldsymbol{\varepsilon}_{\ell+1}^{(D)}}^{-1}[n] \boldsymbol{C}_{\boldsymbol{\varepsilon}_{\ell+1}^{(D)},\boldsymbol{x}_\ell}[n], \right| \tag{4.36}$$

and for $\ell = d-1$, $\boldsymbol{C}_{\boldsymbol{\varepsilon}_{d-1}^{(D)}}[n] = \boldsymbol{C}_{\boldsymbol{x}_{d-1}}[n] \in \mathbb{C}^{M \times M}$.

Finally, the rank D MSMWF can be computed based on Algorithm 4.1. where all boxed equations of this and the previous subsection are summarized. The **for**-loop in Lines 3 to 9 of Algorithm 4.1., also denoted as the *forward recursion* (see, e.g., [89]), determines the prefilter matrices $\boldsymbol{M}_\ell[n]$ and $\boldsymbol{B}_\ell[n]$, $\ell \in \{0,1,\ldots,d-2\}$, $d > 1$, according to Equations (4.6), (4.19), (4.23), and (4.24), as well as the second order statistics as presented in Equations (4.9), (4.10), (4.26), (4.27), and (4.30) which we need for the *backward recursion* given in the second **for**-loop of Algorithm 4.1.. Lines 10 to 13 compute the prefilter $\boldsymbol{M}_{d-1}[n] \in \mathbb{C}^{(N-(d-1)M) \times M}$ and the quadratic MWF $\boldsymbol{\Delta}_{d-1}^{(D)}[n] \in \mathbb{C}^{M \times M}$ of the last stage (cf. Equations 4.32 and 4.33). Due to the fact that we neglected $\hat{\boldsymbol{x}}_{d-1}[n]$ in the computation of $\boldsymbol{\varepsilon}_{d-1}^{(D)}[n] = \boldsymbol{x}_{d-1}[n]$ and with Equation (4.30), the auto-covariance matrix of $\boldsymbol{\varepsilon}_{d-1}^{(D)}[n]$ computes as

$$\left| \boldsymbol{C}_{\boldsymbol{\varepsilon}_{d-1}^{(D)}}[n] = \boldsymbol{C}_{\boldsymbol{x}_{d-1}}[n] = \boldsymbol{M}_{d-1}^{\mathrm{H}}[n] \boldsymbol{C}_{\boldsymbol{y}_{d-2}}[n] \boldsymbol{M}_{d-1}[n]. \right| \tag{4.37}$$

Therefore, the last stage of the rank D MSMWF (cf. Figure 4.7) calculated in Line 14 of Algorithm 4.1. as

$$\left| \boldsymbol{G}_{d-2}^{(D)}[n] = \boldsymbol{M}_{d-1}[n] \boldsymbol{\Delta}_{d-1}^{(D)}[n] \in \mathbb{C}^{(N-(d-1)M) \times M}, \right| \tag{4.38}$$

is an approximation of the MWF $\boldsymbol{G}_{d-2}[n] \in \mathbb{C}^{(N-(d-1)M) \times M}$ (cf. Equation 4.25) in the subspace spanned by the columns of $\boldsymbol{M}_{d-1}[n]$. As already mentioned, the *backward recursion* (e.g., [89]) is summarized in the **for**-loop from Lines 15 to 20 of Algorithm 4.1. where we used Equations (4.36), (4.31), (4.28), and (4.21), as well as the decomposition of the MWF approximation

$$\left| \boldsymbol{G}_{\ell-1}^{(D)}[n] = \left(\boldsymbol{M}_\ell[n] - \boldsymbol{B}_\ell[n] \boldsymbol{G}_\ell^{(D)}[n] \right) \boldsymbol{\Delta}_\ell^{(D)}[n] \in \mathbb{C}^{(N-\ell M) \times M}, \right| \tag{4.39}$$

Algorithm 4.1. Rank D MultiStage Matrix Wiener Filter (MSMWF)

$\quad C_{\boldsymbol{y}_{-1},\boldsymbol{x}_{-1}}[n] \leftarrow C_{\boldsymbol{y},\boldsymbol{x}}[n]$

2: $\quad C_{\boldsymbol{y}_{-1}}[n] \leftarrow C_{\boldsymbol{y}}[n]$

$\quad$ **for** $\ell = 0,1,\ldots,d-2 \wedge d > 1$ **do**

4: $\quad\quad$ Choose $M_\ell[n]$ as an arbitrary basis of range$\{C_{\boldsymbol{y}_{\ell-1},\boldsymbol{x}_{\ell-1}}[n]\}$.

$\quad\quad$ Choose $B_\ell[n]$ as an arbitrary basis of null$\{M_\ell^{\mathrm{H}}[n]\}$.

6: $\quad\quad C_{\boldsymbol{x}_\ell}[n] \leftarrow M_\ell^{\mathrm{H}}[n]C_{\boldsymbol{y}_{\ell-1}}[n]M_\ell[n]$

$\quad\quad C_{\boldsymbol{y}_\ell}[n] \leftarrow B_\ell^{\mathrm{H}}[n]C_{\boldsymbol{y}_{\ell-1}}[n]B_\ell[n]$

8: $\quad\quad C_{\boldsymbol{y}_\ell,\boldsymbol{x}_\ell}[n] \leftarrow B_\ell^{\mathrm{H}}[n]C_{\boldsymbol{y}_{\ell-1}}[n]M_\ell[n]$

$\quad$ **end for**

10: Choose $M_{d-1}[n]$ as an arbitrary basis of range$\{C_{\boldsymbol{y}_{d-2},\boldsymbol{x}_{d-2}}[n]\}$.

$\quad C_{\boldsymbol{\varepsilon}_{d-1}^{(D)}}[n] \leftarrow M_{d-1}^{\mathrm{H}}[n]C_{\boldsymbol{y}_{d-2}}[n]M_{d-1}[n]$

12: $C_{\boldsymbol{\varepsilon}_{d-1}^{(D)},\boldsymbol{x}_{d-2}}[n] \leftarrow M_{d-1}^{\mathrm{H}}[n]C_{\boldsymbol{y}_{d-2},\boldsymbol{x}_{d-2}}[n]$

$\quad \boldsymbol{\Delta}_{d-1}^{(D)}[n] \leftarrow C_{\boldsymbol{\varepsilon}_{d-1}^{(D)}}^{-1}[n]C_{\boldsymbol{\varepsilon}_{d-1}^{(D)},\boldsymbol{x}_{d-2}}[n]$

14: $G_{d-2}^{(D)}[n] \leftarrow M_{d-1}[n]\boldsymbol{\Delta}_{d-1}^{(D)}[n]$

$\quad$ **for** $\ell = d-2,d-1,\ldots,0 \wedge d > 1$ **do**

16: $\quad\quad C_{\boldsymbol{\varepsilon}_\ell^{(D)}}[n] \leftarrow C_{\boldsymbol{x}_\ell}[n] - C_{\boldsymbol{\varepsilon}_{\ell+1}^{(D)},\boldsymbol{x}_\ell}^{\mathrm{H}}[n]C_{\boldsymbol{\varepsilon}_{\ell+1}^{(D)}}^{-1}[n]C_{\boldsymbol{\varepsilon}_{\ell+1}^{(D)},\boldsymbol{x}_\ell}[n]$

$\quad\quad C_{\boldsymbol{\varepsilon}_\ell^{(D)},\boldsymbol{x}_{\ell-1}}[n] \leftarrow M_\ell^{\mathrm{H}}[n]C_{\boldsymbol{y}_{\ell-1},\boldsymbol{x}_{\ell-1}}[n]$

18: $\quad\quad \boldsymbol{\Delta}_\ell^{(D)}[n] \leftarrow C_{\boldsymbol{\varepsilon}_\ell^{(D)}}^{-1}[n]C_{\boldsymbol{\varepsilon}_\ell^{(D)},\boldsymbol{x}_{\ell-1}}[n]$

$\quad\quad G_{\ell-1}^{(D)}[n] \leftarrow (M_\ell[n] - B_\ell[n]G_\ell^{(D)}[n])\boldsymbol{\Delta}_\ell^{(D)}[n]$

20: **end for**

$\quad W_{\mathrm{MSMWF}}^{(D)}[n] \leftarrow G_{-1}^{(D)}[n]$

22: $a_{\mathrm{MSMWF}}^{(D)}[n] \leftarrow m_{\boldsymbol{x}}[n] - G_{-1}^{(D),\mathrm{H}}[n]m_{\boldsymbol{y}}[n]$

in analogy to the one given in Figure 4.4. Finally, Lines 21 and 22 of Algorithm 4.1. define the filter coefficients of the rank D MSMWF.

Note that Algorithm 4.1. presents the reduced-rank MSMWF which is based on the fact that both the auto-covariance matrix $C_{\boldsymbol{y}}[n]$ as well as the cross-covariance matrix $C_{\boldsymbol{y},\boldsymbol{x}}[n]$, or estimations thereof, are available as in the application given in Part II of this book. There, the statistics is estimated based on a channel model and the *a priori* information obtained from the decoder. This type of estimation technique is necessary because the covariance matrices cannot be estimated with a correlation based method (*sample-mean estimation*) due to the non-stationarity of the random sequences. However, for applications where the covariance matrices in Lines 8, 11, 12, 16, and 17 of Algorithm 4.1. can be approximated by their sample-mean estimates, Lines 6 and 7 are no longer necessary. Consequently, the auto-covariance matrix $C_{\boldsymbol{y}}[n]$ needs not to be known at all.

Finally, it remains to mention that the MSMWF implementation given by Algorithm 4.1. can be easily extended to a rank-flexible version (cf. Subsection 3.4.4) if the rank D is no longer a multiple integer of M. With $d =$

$\lceil D/M \rceil \in \{1, 2, \ldots, \lceil N/M \rceil\}$ and $\mu = D - (d-1)M \in \{1, 2, \ldots M\}$, Line 10 of Algorithm 4.1. has to be changed such that $\boldsymbol{M}_{d-1}[n]$ is a $(N - (d-1)M) \times \mu$ matrix whose range space is the subspace spanned by the first μ columns of $\boldsymbol{C}_{\boldsymbol{y}_{d-2}, \boldsymbol{x}_{d-2}}[n]$.

In order to reveal the connection between the MSMWF and the general reduced-rank MWF as introduced in Section 2.3, we combine the matrices $\boldsymbol{M}_\ell[n]$, $\ell \in \{0, 1, \ldots, L - 1\}$, with the preceding blocking matrices $\boldsymbol{B}_i[n]$, $i \in \{0, 1, \ldots, \ell - 1\}$, $\ell \in \mathbb{N}$, to get the prefilter matrices

$$\boldsymbol{T}_\ell[n] = \left(\prod_{\substack{i=0 \\ \ell \in \mathbb{N}}}^{\ell-1} \boldsymbol{B}_i[n] \right) \boldsymbol{M}_\ell[n] \in \mathbb{C}^{N \times M}. \tag{4.40}$$

Hence, the reduced-rank MSMWF of Figure 4.7 may be redrawn as the filter bank plotted in Figure 4.8. In this case, we approximated the MWF $\boldsymbol{W}[n]$ and $\boldsymbol{a}[n]$ by d prefilter matrices $\boldsymbol{T}_\ell[n]$ and d quadratic MWFs $\boldsymbol{\Delta}_\ell^{(D)}[n]$, $\ell \in \{0, 1, \ldots, d-1\}$.

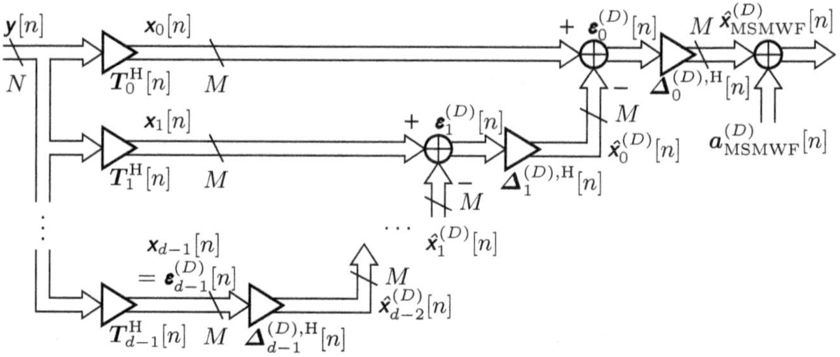

Fig. 4.8. Rank D MSMWF as a filter bank

In the sense of Section 2.3, this leads to the reduced-rank MWF $\boldsymbol{W}_{\mathrm{MSMWF}}^{(D)}[n]$ and $\boldsymbol{a}_{\mathrm{MSMWF}}^{(D)}[n]$ (cf. Equation 2.48) where the prefilter matrix is defined as

$$\boldsymbol{T}_{\mathrm{MSMWF}}^{(D)}[n] = [\boldsymbol{T}_0[n] \; \boldsymbol{T}_1[n] \cdots \boldsymbol{T}_{d-1}[n]] \in \mathbb{C}^{N \times D}. \tag{4.41}$$

If we compare Figure 2.3 with Figure 4.8, we see that the reduced-dimension MWF $\boldsymbol{G}_{\mathrm{MSMWF}}^{(D)}[n] \in \mathbb{C}^{D \times M}$ of the rank D MSMWF resolves in d quadratic MWFs ordered in the staircase structure as depicted in Figure 4.8. Finally, the reduced-rank MSMWF achieves the MMSE (cf. Equation 2.49)

$$\xi_{n,\mathrm{MSMWF}}^{(D)} = \xi_n\left(\boldsymbol{W}_{\mathrm{MSMWF}}^{(D)}[n], \boldsymbol{a}_{\mathrm{MSMWF}}^{(D)}[n]\right) \tag{4.42}$$

$$= \mathrm{tr}\left\{\boldsymbol{C}_{\boldsymbol{x}}[n] - \boldsymbol{C}_{\boldsymbol{y},\boldsymbol{x}}^{\mathrm{H}}[n]\boldsymbol{T}_{\mathrm{MSMWF}}^{(D)}[n]\boldsymbol{C}_{\boldsymbol{y}_{\mathrm{MSMWF}}^{(D)}}^{-1}[n]\boldsymbol{T}_{\mathrm{MSMWF}}^{(D),\mathrm{H}}[n]\boldsymbol{C}_{\boldsymbol{y},\boldsymbol{x}}[n]\right\},$$

where $\boldsymbol{C}_{\boldsymbol{y}_{\mathrm{MSMWF}}^{(D)}}[n] = \boldsymbol{T}_{\mathrm{MSMWF}}^{(D),\mathrm{H}}[n]\boldsymbol{C}_{\boldsymbol{y}}[n]\boldsymbol{T}_{\mathrm{MSMWF}}^{(D)}[n] \in \mathbb{C}^{D\times D}$ is the auto-covariance matrix of the transformed observation vector $\boldsymbol{y}_{\mathrm{MSMWF}}^{(D)}[n] \in \mathbb{C}^{D}$.

4.1.3 Fundamental Properties

This subsection reveals the fundamental properties of the reduced-rank MSMWF. For simplicity, we restrict ourselves to the case where the rank D is a multiple integer of M. However, the generalization to arbitrary $D \in \{M, M+1, \ldots N\}$ is straightforward if we recall the respective comments made in Subsection 4.1.2. Besides, although the matrices under consideration are still time-varying, we omit the time index n because the following characteristics of the rank D MSMWF are true for each n.

Proposition 4.2. *With the prefilter matrix $\boldsymbol{T}_{\mathrm{MSMWF}}^{(D)}$ of the reduced-rank MSMWF as given in Equation (4.41), the auto-covariance matrix $\boldsymbol{C}_{\boldsymbol{y}_{\mathrm{MSMWF}}^{(D)}} = \boldsymbol{T}_{\mathrm{MSMWF}}^{(D),\mathrm{H}}\boldsymbol{C}_{\boldsymbol{y}}\boldsymbol{T}_{\mathrm{MSMWF}}^{(D)} \in \mathbb{C}^{D\times D}$ of the transformed observation vector $\boldsymbol{y}_{\mathrm{MSMWF}}^{(D)} \in \mathbb{C}^{D}$ (cf. Section 2.3) is block tridiagonal[3].*

Proof. With Equations (4.40) and (4.41), the $M \times M$ submatrix $\boldsymbol{C}_{m,\ell}$, $m,\ell \in \{0,1,\ldots,d-1\}$, $d = D/M \in \mathbb{N}$, of $\boldsymbol{C}_{\boldsymbol{y}_{\mathrm{MSMWF}}^{(D)}}$ in the $(m+1)$th block row and $(\ell+1)$th block column reads as

$$\boldsymbol{C}_{m,\ell} = \boldsymbol{T}_m^{\mathrm{H}}\boldsymbol{C}_{\boldsymbol{y}}\boldsymbol{T}_\ell = \boldsymbol{M}_m^{\mathrm{H}}\left(\prod_{\substack{i=m-1\\ m\in\mathbb{N}}}^{0}\boldsymbol{B}_i^{\mathrm{H}}\right)\boldsymbol{C}_{\boldsymbol{y}}\left(\prod_{\substack{i=0\\ \ell\in\mathbb{N}}}^{\ell-1}\boldsymbol{B}_i\right)\boldsymbol{M}_\ell. \tag{4.43}$$

Besides, recall Equations (4.9), (4.10), (4.26), and (4.27), and use them to rewrite the cross-covariance matrix $\boldsymbol{C}_{\boldsymbol{y}_\ell,\boldsymbol{x}_\ell} \in \mathbb{C}^{(N-(\ell+1)M)\times M}$ according to

$$\boldsymbol{C}_{\boldsymbol{y}_\ell,\boldsymbol{x}_\ell} = \boldsymbol{B}_\ell^{\mathrm{H}}\boldsymbol{C}_{\boldsymbol{y}_{\ell-1}}\boldsymbol{M}_\ell = \left(\prod_{i=\ell}^{0}\boldsymbol{B}_i^{\mathrm{H}}\right)\boldsymbol{C}_{\boldsymbol{y}}\left(\prod_{\substack{i=0\\ \ell\in\mathbb{N}}}^{\ell-1}\boldsymbol{B}_i\right)\boldsymbol{M}_\ell, \tag{4.44}$$

where $\boldsymbol{C}_{\boldsymbol{y}_{-1}} := \boldsymbol{C}_{\boldsymbol{y}} \in \mathbb{C}^{N\times N}$. For the case $m > \ell+1$, plugging Equation (4.44) into Equation (4.43) yields

---

[3] A block tridiagonal matrix consists of block matrices in the main diagonal and the two adjacent subdiagonals.

$$C_{m,\ell} = M_m^{\mathrm{H}} \left(\prod_{i=m-1}^{\ell+1} B_i^{\mathrm{H}} \right) C_{\boldsymbol{y}_\ell, \boldsymbol{x}_\ell} = \mathbf{0}_{M \times M}, \qquad (4.45)$$

if we remember that $B_{\ell+1} \in \mathbb{C}^{(N-(\ell+1)M) \times (N-(\ell+2)M)}$ is chosen to be orthogonal to $M_{\ell+1} \in \mathbb{C}^{(N-(\ell+1)M) \times M}$ whose columns span the range space of $C_{\boldsymbol{y}_\ell, \boldsymbol{x}_\ell}$ (cf. Equations 4.24 and 4.23). For $m = \ell + 1$, the combination of Equations (4.43) and (4.44) results in (cf. Equation 4.31)

$$C_{m,\ell} = M_{\ell+1}^{\mathrm{H}} C_{\boldsymbol{y}_\ell, \boldsymbol{x}_\ell} = C_{\boldsymbol{x}_{\ell+1}, \boldsymbol{x}_\ell} = C_{\boldsymbol{\varepsilon}_{\ell+1}, \boldsymbol{x}_\ell}. \qquad (4.46)$$

Recall from Subsection 2.2.1 that we assume the auto-covariance matrix $C_{\boldsymbol{y}}$ to be positive definite. Thus, the matrices $C_{\ell,\ell} = T_\ell^{\mathrm{H}} C_{\boldsymbol{y}} T_\ell$ in the diagonal of $C_{\boldsymbol{y}_{\mathrm{MSMWF}}^{(D)}}$ are positive definite and therefore unequal to the zero matrix $\mathbf{0}_{M \times M}$. Finally, due to the Hermitian property of $C_{\boldsymbol{y}}$, the $M \times M$ blocks $C_{m,\ell}$ of $C_{\boldsymbol{y}_{\mathrm{MSMWF}}^{(D)}}$ can be written as

$$C_{m,\ell} = \begin{cases} C_{\boldsymbol{\varepsilon}_{\ell+1}, \boldsymbol{x}_\ell}^{\mathrm{H}}, & m = \ell - 1, \\ T_\ell^{\mathrm{H}} C_{\boldsymbol{y}} T_\ell, & m = \ell, \\ C_{\boldsymbol{\varepsilon}_{\ell+1}, \boldsymbol{x}_\ell}, & m = \ell + 1, \\ \mathbf{0}_{M \times M}, & \text{otherwise,} \end{cases} \qquad (4.47)$$

resulting in a block tridiagonal structure as depicted in Figure 4.9. Thus, Proposition 4.2 is proven. $\qquad \square$

As can be seen by Equation (4.45) of the proof of Proposition 4.2, the block tridiagonal structure of the auto-covariance matrix $C_{\boldsymbol{y}_{\mathrm{MSMWF}}^{(D)}}$ is mainly due to the fact that the blocking matrix $B_{\ell+1}$ is orthogonal to the cross-covariance matrix $C_{\boldsymbol{y}_\ell, \boldsymbol{x}_\ell}$. Remember from Subsection 4.1.1 that this condition was also necessary to get the simplified structure of the MSMWF (cf. Figure 4.3). Again, for this reason, we have chosen the matrix $M_{\ell+1}$ which is already orthogonal to $B_{\ell+1}$ (cf. Equation 4.24), such that its range space is the correlated subspace $\mathcal{M}_{\ell+1}$, i.e., the range space of $C_{\boldsymbol{y}_\ell, \boldsymbol{x}_\ell}$ (cf. Equation 4.23).

Proposition 4.3. *The prefilter matrices* T_ℓ, $\ell \in \{1, 2, \ldots, d-1\}$, *as defined in Equation (4.40), are all orthogonal to the first prefilter matrix* T_0, *i. e.,*

$$T_\ell^{\mathrm{H}} T_0 = \mathbf{0}_{M \times M}, \quad \ell \in \{1, 2, \ldots, d-1\}. \qquad (4.48)$$

However, they are generally not orthogonal to each other.

Proof. With Equation (4.40) and the fact that $B_0^{\mathrm{H}} M_0 = B_0^{\mathrm{H}} T_0 = \mathbf{0}_{(N-M) \times M}$ (cf. Equation 4.6), we get

$$T_\ell^{\mathrm{H}} T_0 = M_\ell^{\mathrm{H}} \left(\prod_{\substack{i=\ell-1 \\ \ell \in \mathbb{N}}}^{0} B_i^{\mathrm{H}} \right) M_0 = \mathbf{0}_{M \times M}, \qquad (4.49)$$

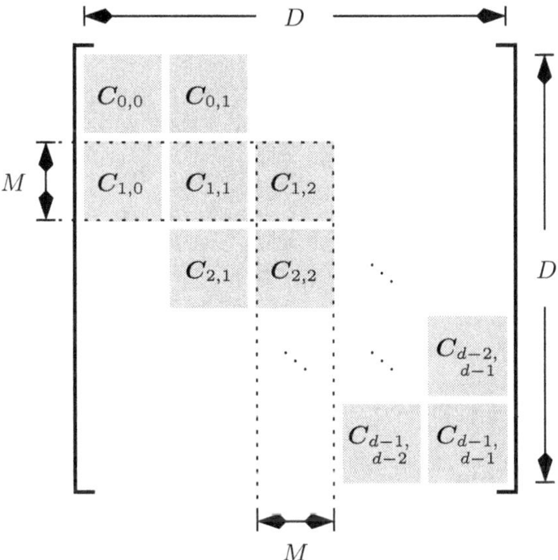

Fig. 4.9. Structure of a block tridiagonal $D \times D$ matrix with $M \times M$ blocks

which proves the first part of Proposition 4.3. The second part is proven if we find one couple of prefilter matrices which are not orthogonal to each other. Consider the two prefilter matrices $T_1 = B_0 M_1$ and $T_2 = B_0 B_1 M_2$ to get

$$T_1^{\mathrm{H}} T_2 = M_1^{\mathrm{H}} B_0^{\mathrm{H}} B_0 B_1 M_2. \tag{4.50}$$

Although $M_1^{\mathrm{H}} B_1 = \mathbf{0}_{M \times (N-2M)}$ due to Equation (4.24), B_0 under the condition that $B_0^{\mathrm{H}} M_0 = \mathbf{0}_{(N-M) \times M}$ can still be chosen such that the matrix $M_1^{\mathrm{H}} B_0^{\mathrm{H}} B_0 B_1$ is unequal to the zero matrix. In fact, this is the case if $B_0^{\mathrm{H}} B_0$ is unequal to a scaled identity matrix. Therefore, in general, $T_1^{\mathrm{H}} T_2 \neq \mathbf{0}_{M \times M}$ which completes the proof. $\square$

Note that the condition of the blocking matrices B_ℓ to be orthogonal to the basis matrices M_ℓ of the correlated subspaces $\mathcal{M}_\ell$ does not restrict the MSMWF to be an approximation of the MWF in a specific subspace. Since the subspace of a reduced-rank MWF is directly connected to its performance (see Proposition 2.1), reduced-rank MSMWFs with different blocking matrices can lead to different MSEs. An extreme case is formulated in the following proposition.

Proposition 4.4. *If the number of stages d is greater than one, the blocking matrices of the reduced-rank MSMWF can always be chosen such that its performance is equal to the one of the full-rank MWF.*

Proof. Remember from Subsections 4.1.1 and 4.1.2 (e. g., Proposition 4.1) that the multistage decomposition represents exactly the MWF as long as the reduced-dimension MWF (cf. Equation 4.25 and Figure 4.4)

$$G_{d-2} = C_{\mathbf{y}_{d-2}}^{-1} C_{\mathbf{y}_{d-2}, \mathbf{x}_{d-2}} \in \mathbb{C}^{(N-(d-1)M) \times M}, \tag{4.51}$$

is not replaced by its approximation in the range space of $M_{d-1} \in \mathbb{C}^{(N-(d-1)M) \times M}$, i.e.,

$$G_{d-2}^{(D)} = M_{d-1} \mathbf{\Delta}_{d-1}^{(D)} \in \mathbb{C}^{(N-(d-1)M) \times M}, \tag{4.52}$$

as appearing in Line 14 of Algorithm 4.1. (cf. also Equation 4.38). Thus, if we choose $B_{d-2} \in \mathbb{C}^{(N-(d-2)M) \times (N-(d-1)M)}$ such that it fulfills not only $M_{d-2}^{\mathrm{H}} B_{d-2} = \mathbf{0}_{M \times (N-(d-1)M)}$ (cf. Equation 4.24) but also

$$\operatorname{range}\{G_{d-2}\} = \operatorname{range}\{M_{d-1}\}, \tag{4.53}$$

the quadratic MWF $\mathbf{\Delta}_{d-1}^{(D)} \in \mathbb{C}^{M \times M}$ in Equation (4.52) ensures that $G_{d-2}^{(D)} = G_{d-2}$. In fact, this is the basis invariance of reduced-rank MWFs which we already presented in Proposition 2.1. Note that B_{d-2} directly influences G_{d-2} in Equation (4.51) via the covariance matrices $C_{\mathbf{y}_{d-2}} = B_{d-2}^{\mathrm{H}} C_{\mathbf{y}_{d-3}} B_{d-2}$ and $C_{\mathbf{y}_{d-2}, \mathbf{x}_{d-2}} = B_{d-2}^{\mathrm{H}} C_{\mathbf{y}_{d-3}} M_{d-2}$ (cf. Equations 4.26 and 4.27). Clearly, when $G_{d-2}^{(D)} = G_{d-2}$, the rank D MSMWF is equal to the MWF. From Equation (4.23), we know that the range space of M_{d-1} is equal to the range space of $C_{\mathbf{y}_{d-2}, \mathbf{x}_{d-2}}$. Hence, one solution to Equation (4.53) is to choose the blocking matrix B_{d-2} such that the inverse of the auto-covariance matrix $C_{\mathbf{y}_{d-2}} \in \mathbb{C}^{(N-(d-1)M) \times (N-(d-1)M)}$ in Equation (4.51) reduces to a scaled identity matrix, i.e.,

$$C_{\mathbf{y}_{d-2}} = B_{d-2}^{\mathrm{H}} C_{\mathbf{y}_{d-3}} B_{d-2} = \beta I_{N-(d-1)M}, \tag{4.54}$$

with $\beta \in \mathbb{R}_+$. It remains to find such a blocking matrix. With the factorization of $B_{d-2} = V B_{d-2}'$ where $V \in \mathbb{C}^{(N-(d-2)M) \times (N-(d-2)M)}$ has full-rank and $B_{d-2}' \in \mathbb{C}^{(N-(d-2)M) \times (N-(d-1)M)}$ has orthonormal rows, i.e., $B_{d-2}'^{\mathrm{,H}} B_{d-2}' = I_{N-(d-1)M}$, Equation (4.54) reads as

$$B_{d-2}'^{\mathrm{,H}} V^{\mathrm{H}} C_{\mathbf{y}_{d-3}} V B_{d-2}' = \beta I_{N-(d-1)M}. \tag{4.55}$$

Therefore, if the matrix V is chosen such that

$$V^{\mathrm{H}} C_{\mathbf{y}_{d-3}} V = \beta I_{N-(d-2)M}, \tag{4.56}$$

Equation (4.54) holds due to the fact that the columns of B_{d-2}' are orthonormal to each other. One solution to Equation (4.56) is given by

$$V = \sqrt{\beta} L^{-1,\mathrm{H}}, \tag{4.57}$$

where $L \in \mathbb{C}^{(N-(d-2)M) \times (N-(d-2)M)}$ is the lower triangular matrix of the *Cholesky factorization*

$\boldsymbol{C}_{\boldsymbol{y}_{d-3}} = \boldsymbol{L}\boldsymbol{L}^{\mathrm{H}} \in \mathbb{C}^{(N-(d-2)M)\times(N-(d-2)M)}$ which can be computed based on Algorithm 2.1.. Then, $\boldsymbol{B}'_{d-2}$ is chosen to fulfill $\boldsymbol{M}^{\mathrm{H}}_{d-2}\boldsymbol{B}_{d-2} = \boldsymbol{0}_{M\times(N-(d-1)M)}$, i. e.,

$$\boldsymbol{M}^{\mathrm{H}}_{d-2}\boldsymbol{V}\boldsymbol{B}'_{d-2} = \boldsymbol{0}_{M\times(N-(d-1)M)} \Leftrightarrow \mathrm{range}\left\{\boldsymbol{B}'_{d-2}\right\} = \mathrm{null}\left\{\boldsymbol{M}^{\mathrm{H}}_{d-2}\boldsymbol{V}\right\}. \quad (4.58)$$

Choosing the blocking matrices $\boldsymbol{B}_\ell$, $\ell \in \{0, 1, \ldots, d-3\}$, $d > 2$, arbitrarily but such that Equation (4.24) holds, and combining $\boldsymbol{V}$ of Equation (4.57) with $\boldsymbol{B}'_{d-2}$ of Equation (4.58) to the remaining blocking matrix $\boldsymbol{B}_{d-2} = \boldsymbol{V}\boldsymbol{B}'_{d-2}$ leads to a reduced-rank MSMWF which is equal to the MWF. $\qquad\square$

Note that the choice of the optimal blocking matrices as given in the proof of Proposition 4.4 has no practical relevance since their calculation is clearly too computationally intense, i. e., the resulting reduced-rank MSMWF is computationally more expensive than the Cholesky based implementation of the MWF as presented in Subsection 2.2.1.

The effect of the choice of the blocking matrices on the performance of the reduced-rank MSMWF is illustrated by the following example.

Example 4.5 (Influence of the blocking matrices on the MSMWF performance). In this example, we consider a scenario which is equal to the one given in Example 3.9 except from the fact that the cross-covariance matrix which represents the right-hand side matrix in Example 3.9, has only $M = 1$ column. This restriction is necessary because we want to make a performance comparison to the computationally efficient blocking matrices suggested by J. Scott Goldstein et al. in [89], which exist only for $M = 1$. With $N = 40$, the auto-covariance matrix is defined to be $\boldsymbol{C}_{\boldsymbol{y}} = \boldsymbol{N}\boldsymbol{N}^{\mathrm{H}} + \boldsymbol{I}_{40}/10 \in \mathbb{C}^{40\times40}$ where each element of $\boldsymbol{N} \in \mathbb{C}^{40\times40}$ is a realization of the complex normal distribution $\mathcal{N}_{\mathbb{C}}(0,1)$.[4] Since $M = 1$, the cross-covariance matrix $\boldsymbol{C}_{\boldsymbol{y},\mathrm{x}}$ reduces to the cross-covariance vector $\boldsymbol{c}_{\boldsymbol{y},\mathrm{x}}$ which we choose to be the first column of $\boldsymbol{N}$, i. e., $\boldsymbol{c}_{\boldsymbol{y},\mathrm{x}} = \boldsymbol{N}\boldsymbol{e}_1$. Finally, the auto-covariance matrix $\boldsymbol{C}_{\mathrm{x}} = c_{\mathrm{x}} = 1$. Figure 4.10 depicts the MSE $\xi^{(D)}_{n,\mathrm{MSMWF}}$ (cf. Equation 4.42) over the rank $D \in \{1, 2, \ldots, 40\}$ of the reduced-rank MSMWFs where we have averaged over several realizations $\boldsymbol{N}$. We compare the reduced-rank MSMWFs with

- optimal blocking matrices as derived in the proof of Proposition 4.4,
- blocking matrices which are scaled versions of matrices with orthonormal columns,
- arbitrary blocking matrices which have been computed from the orthonormal blocking matrices by multiplying them on the left-hand side by quadratic matrices where the entries are realizations of the complex normal distribution $\mathcal{N}_{\mathbb{C}}(0,1)$, and

---

[4] The probability density function of the complex normal distribution $\mathcal{N}_{\mathbb{C}}(m_{\mathrm{x}}, c_{\mathrm{x}})$ with mean m_{x} and variance c_{x} reads as $\mathrm{p}_{\mathrm{x}}(x) = \exp(-|x - m_{\mathrm{x}}|^2/c_{\mathrm{x}})/(\pi c_{\mathrm{x}})$ (e. g., [183]).

- the computationally efficient blocking matrices [89]

$$\boldsymbol{B}_\ell^{\mathrm{H}} = \left(\boldsymbol{S}_{(0,N-\ell-1,1)} - \boldsymbol{S}_{(1,N-\ell-1,1)}\right) \mathrm{diag}\left\{\boldsymbol{1}_{N-\ell} \oslash \boldsymbol{m}_\ell\right\},^5 \qquad (4.59)$$

where $\boldsymbol{S}_{(0,N-\ell-1,1)} = [\boldsymbol{I}_{N-\ell-1}, \boldsymbol{0}_{N-\ell-1}] \in \{0,1\}^{(N-\ell-1)\times(N-\ell)}$, $\boldsymbol{S}_{(1,N-\ell-1,1)} = [\boldsymbol{0}_{N-\ell-1}, \boldsymbol{I}_{N-\ell-1}] \in \{0,1\}^{(N-\ell-1)\times(N-\ell)}$, $\boldsymbol{1}_{N-\ell} \in \{1\}^{N-\ell}$ is the all-ones vector, '$\oslash$' denotes elementwise division, and $\boldsymbol{m}_\ell \in \mathbb{C}^{N-\ell}$ is the reduced version of the matrix $\boldsymbol{M}_\ell$ for $M = 1$.

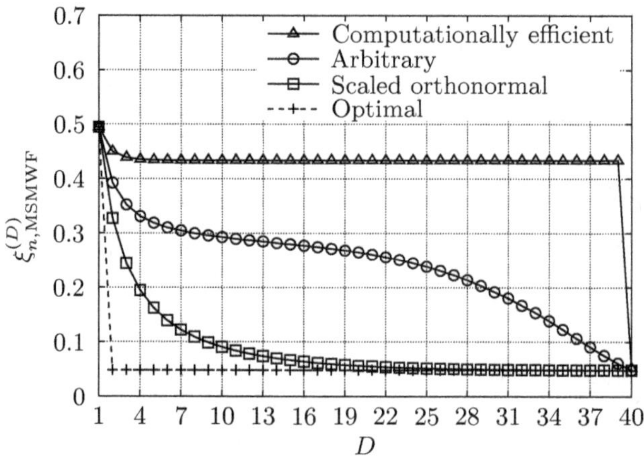

Fig. 4.10. MSMWF performance for different blocking matrices

It can be seen that the MSMWF with optimal blocking matrices achieves the MMSE of the full-rank MWF already after the second stage and is therefore the MSMWF with the best performance. However, it involves also the highest computational complexity. On the other hand, the computationally cheapest choice from J. Scott Goldstein et al. has the worst performance of all considered blocking matrices. If we compare the MSMWF with arbitrary blocking matrices and the MSMWF where the columns of the blocking matrices are orthonormalized, the latter one has a smaller MSE for all ranks D. Thus, it seems to be worth to introduce orthonormalization of the blocking matrix columns despite of the increase in computational complexity.

Whereas the blocking matrices have a strong influence on the performance of the reduced-rank MSMWF, the choice of the matrices $\boldsymbol{M}_\ell$, $\ell \in \{0, 1, \ldots, d-1\}$, as arbitrary bases of the correlated subspace $\mathcal{M}_\ell$, does not affect the MSE because they do not change the overall subspace spanned by the columns of $\boldsymbol{T}_{\mathrm{MSMWF}}^{(D)}$ (cf. Proposition 2.1).

---

^{5} The construction according to Equation (4.59) ensures the requirement $\boldsymbol{B}_\ell^{\mathrm{H}} \boldsymbol{m}_\ell = \boldsymbol{0}_{N-\ell-1}$ (cf. Equation 4.24).

4.2 Relationship Between Multistage Matrix Wiener Filter and Krylov Subspace Methods

4.2.1 Relationship to Krylov Subspace

Again, we omit the time index n due to simplicity although the variables under consideration are time-variant. From Example 4.5, we know that choosing the blocking matrices $\boldsymbol{B}_\ell \in \mathbb{C}^{(N-\ell M)\times(N-(\ell+1)M)}$, $\ell \in \{0, 1, \ldots, d-2\}$, $d > 1$, of the reduced-rank MSMWF as scaled versions of orthonormal matrices, i.e.,

$$\boxed{\boldsymbol{B}_\ell^{\mathrm{H}}\boldsymbol{B}_\ell = \beta_\ell \boldsymbol{I}_{N-(\ell+1)M},} \tag{4.60}$$

with $\beta_\ell \in \mathbb{R}_+$, leads to an acceptable performance-complexity trade-off. Note that restricting the blocking matrices to have orthonormal columns is widely used in publications (c. g., [130, 121]) although no specific motivation for this choice is given. Besides, note that the choice of $\boldsymbol{B}_\ell$ according to Equation (4.60) is equal to selecting $\boldsymbol{B}_\ell$ such that their non-zero singular values are identical and equal to $\sqrt{\beta_\ell}$. In the following, we present the consequences of this choice.

Proposition 4.6. *If the blocking matrices fulfill Equation (4.60), the prefilter matrices $\boldsymbol{T}_\ell$ as given by Equation (4.40) are orthogonal to each other, i.e.,*

$$\boldsymbol{T}_m^{\mathrm{H}}\boldsymbol{T}_\ell = \begin{cases} \breve{\beta}_{\ell-1}\boldsymbol{M}_\ell^{\mathrm{H}}\boldsymbol{M}_\ell, & m = \ell, \\ \boldsymbol{0}_{M\times M}, & m \neq \ell, \end{cases} \quad m, \ell \in \{0, 1, \ldots, d-1\}, \tag{4.61}$$

where $\breve{\beta}_{\ell-1} := \prod_{\substack{i=0 \\ \ell \in \mathbb{N}}}^{\ell-1} \beta_i \in \mathbb{R}_+$ and $\breve{\beta}_{-1} := 1$.

Proof. Clearly, if Equation (4.61) is true for $m \leq \ell$, it also holds for any combination of m and ℓ. Plugging Equation (4.40) into Equation (4.61) yields

$$\boldsymbol{T}_m^{\mathrm{H}}\boldsymbol{T}_\ell = \boldsymbol{M}_m^{\mathrm{H}}\left(\prod_{\substack{i=m-1 \\ m\in\mathbb{N}}}^{0} \boldsymbol{B}_i^{\mathrm{H}}\right)\left(\prod_{\substack{i=0 \\ \ell\in\mathbb{N}}}^{\ell-1} \boldsymbol{R}_i\right)\boldsymbol{M}_\ell = \left(\prod_{\substack{i=0 \\ m\in\mathbb{N}}}^{m-1} \beta_i\right)\boldsymbol{M}_m^{\mathrm{H}}\left(\prod_{\substack{i=m \\ \ell>m}}^{\ell-1} \boldsymbol{B}_i\right)\boldsymbol{M}_\ell$$

$$= \begin{cases} \breve{\beta}_{\ell-1}\boldsymbol{M}_\ell^{\mathrm{H}}\boldsymbol{M}_\ell, & m = \ell, \\ \boldsymbol{0}_{M\times M}, & m < \ell, \end{cases} \tag{4.62}$$

if we recall Equation (4.60) and remember from Equation (4.24) that $\boldsymbol{M}_m^{\mathrm{H}}\boldsymbol{B}_m = \boldsymbol{0}_{M\times(N-(m+1)M)}$. Thus, Proposition 4.6 is proven. $\qquad\square$

Before proving that the blocking matrices which fulfill Equation (4.60) lead to a reduced-rank MSMWF which is an approximation of the MWF in a Krylov subspace, we derive a recursion formula for the prefilter matrices $\boldsymbol{T}_\ell \in \mathbb{C}^{N\times M}$. Remember from Equations (4.19) and (4.23) that the range space of $M_\ell \in \mathbb{C}^{(N-\ell M) \times M}$, $\ell \in \{0, 1, \ldots, d-1\}$, is chosen to be the range space of $C_{y_{\ell-1}, x_{\ell-1}}$ where $C_{y_{-1}, x_{-1}} := C_{y,x} \in \mathbb{C}^{N \times M}$. Thus, we can write $M_\ell = C_{y_{\ell-1}, x_{\ell-1}} \Psi_\ell$ with the quadratic full-rank matrices $\Psi_\ell \in \mathbb{C}^{M \times M}$. With Equations (4.27), (4.26), and (4.9), it holds for $\ell > 0$:

$$M_\ell = \left(\prod_{i=\ell-1}^{0} B_i^{\mathrm{H}} \right) C_y \left(\prod_{\substack{i=0 \\ \ell>1}}^{\ell-2} B_i \right) M_{\ell-1} \Psi_\ell. \tag{4.63}$$

Finally, using the definition of the prefilter matrices T_ℓ from Equation (4.40) yields the recursion formula

$$T_\ell = \left(\prod_{i=0}^{\ell-1} B_i \right) \left(\prod_{i=\ell-1}^{0} B_i^{\mathrm{H}} \right) C_y T_{\ell-1} \Psi_\ell, \tag{4.64}$$

for $\ell \in \{1, 2, \ldots, d-1\}$, starting with $T_0 = M_0 = C_{y,x} \Psi_0$.

Proposition 4.7. *If the blocking matrices fulfill Equation (4.60), their combination in Equation (4.64) is a scaled projector onto the $(N-\ell M)$-dimensional subspace which is orthogonal to the range space of $[T_0, T_1, \ldots, T_{\ell-1}] \in \mathbb{C}^{N \times \ell M}$, and can be written as*

$$\left(\prod_{i=0}^{\ell-1} B_i \right) \left(\prod_{i=\ell-1}^{0} B_i^{\mathrm{H}} \right) = \breve{\beta}_{\ell-1} \prod_{i=0}^{\ell-1} P_{\perp T_i}, \quad \ell \in \{1, 2, \ldots, d-1\}. \tag{4.65}$$

Here, the matrix $P_{\perp T_i} \in \mathbb{C}^{N \times N}$ denotes the projector onto the $(N - M)$-dimensional subspace $\perp T_i \subset \mathbb{C}^N$ which is defined as the orthogonal complement to the range space of T_i, i. e., $(\perp T_i) \oplus \mathrm{range}\{T_i\} = \mathbb{C}^N$.

Proof. First, we define $\breve{B}_{\ell-1} := \prod_{i=0}^{\ell-1} B_i \in \mathbb{C}^{N \times (N-\ell M)}$ for $\ell \in \{1, 2, \ldots, d-1\}$. Then, it holds

$$\breve{B}_{\ell-1}^{\mathrm{H}} \breve{B}_{\ell-1} = \left(\prod_{i=\ell-1}^{0} B_i^{\mathrm{H}} \right) \left(\prod_{i=0}^{\ell-1} B_i \right) = \prod_{i=0}^{\ell-1} \beta_i I_{N-\ell M}$$
$$= \breve{\beta}_{\ell-1} I_{N-\ell M}, \tag{4.66}$$

if we recall Equation (4.60). Due to the fact that the matrix $\breve{B}_{\ell-1}$ is a scaled version of a matrix with orthonormal columns, i. e., its column-rank is $N-\ell M$, the product $\breve{B}_{\ell-1} \breve{B}_{\ell-1}^{\mathrm{H}}$ in Equation (4.65) is a scaled version of a projector with rank $N-\ell M$. Next, we show some important properties of this projector. Consider the product $T_m^{\mathrm{H}} \breve{B}_{\ell-1}$ for $m \in \{0, 1, \ldots, \ell-1\}$. With the above definition of $\breve{B}_{\ell-1}$ and Equation (4.40), we get

$$T_m^{\mathrm{H}} \breve{B}_{\ell-1} = M_m^{\mathrm{H}} \left(\prod_{\substack{i=m-1 \\ m\in\mathbb{N}}}^{0} B_i^{\mathrm{H}} \right) \left(\prod_{i=0}^{\ell-1} B_i \right) = \breve{\beta}_{m-1} M_m^{\mathrm{H}} \left(\prod_{i=m}^{\ell-1} B_i \right) \tag{4.67}$$

$$= \mathbf{0}_{M \times (N-\ell M)},$$

where the latter two equalities holds due to Equation (4.60) and (4.24), respectively. Hence, $\breve{B}_{\ell-1}$ is orthogonal to the matrix $[T_0, T_1, \ldots, T_{\ell-1}]$. Remember from Proposition 4.6 that the prefilter matrices T_m, $m \in \{0, 1, \ldots, \ell-1\}$, are orthogonal to each other. Besides, all prefilter matrices have full-column rank, i. e., their rank is M, because of Equation (4.61) and due to the fact that the matrices $M_m[n]$ have full-column rank since they are bases of M-dimensional subspaces (cf. Equations 4.19 and 4.23). Hence, the dimension of range$\{[T_0, T_1, \ldots, T_{\ell-1}]\}$ is ℓM which is N minus the dimension of the subspace described by the scaled projector $\breve{B}_{\ell-1}\breve{B}_{\ell-1}^{\mathrm{H}}$. Consequently, $\breve{B}_{\ell-1}\breve{B}_{\ell-1}^{\mathrm{H}}$ is a scaled projector onto the ℓM-dimensional subspace $\perp [T_0, T_1, \ldots, T_{\ell-1}]$. With Equation (3.14) from the previous chapter, it can be easily shown that the projector onto $\perp [T_0, T_1, \ldots, T_{\ell-1}]$ is equal to the product of the projectors $P_{\perp T_i}$, $i \in \{0, 1, \ldots, \ell-1\}$, if the prefilter matrices are orthogonal to each other. To sum up, it holds

$$\breve{B}_{\ell-1}\breve{B}_{\ell-1}^{\mathrm{H}} = \breve{\beta}_{\ell-1} \prod_{i=0}^{\ell-1} P_{\perp T_i}, \tag{4.68}$$

which completes the proof. $\qquad\square$

With Proposition 4.7, the recursion formula of Equation (4.64) can be rewritten as

$$\left| T_\ell = \breve{\beta}_{\ell-1} \left(\prod_{i=0}^{\ell-1} P_{\perp T_i} \right) C_y T_{\ell-1} \Psi_\ell, \right| \tag{4.69}$$

which we use to prove Proposition 4.8. Note that Michael L. Honig et al. [130, 121] derived already the same recursion formula for the special case where $M = 1$ and used the result to prove that the MSWF is an approximation of the WF in a Krylov subspace (cf. Proposition 4.8). Also Dimitris A. Pados et al. (e. g., [178, 162]) ended up with a similar expression as given in Equation (4.69) for the AV method but did not consider the orthogonality of the prefilter matrices which lead to a more complicated formula. The missing connection to the Arnoldi or Lanczos algorithm will be presented in Subsection 4.2.2.

Proposition 4.8. *If the blocking matrices fulfill Equation (4.60), the rank D MSMWF is an approximation of the MWF in the D-dimensional Krylov subspace $\mathcal{K}^{(D)}(C_y, M_0) = \mathcal{K}^{(D)}(C_y, C_{y,x})$ (cf. Equation 3.1), thus, independent of the choices made for the matrices M_ℓ, $\ell \in \{0, 1, \ldots, d-1\}$, as defined in Equations (4.19) and (4.23).*

Proof. In order to show that the rank D MSMWF is an approximation of the MWF in a Krylov subspace, we must prove that the columns of the prefilter matrices $\boldsymbol{T}_\ell$, $\ell \in \{0, 1, \ldots, d-1\}$, defined by the recursion formula in Equation (4.69), fulfill

$$\text{range}\left\{\begin{bmatrix} \boldsymbol{T}_0 \ \boldsymbol{T}_1 \ \cdots \ \boldsymbol{T}_{d-1} \end{bmatrix}\right\} = \mathcal{K}^{(D)}(\boldsymbol{C}_{\boldsymbol{y}}, \boldsymbol{M}_0) = \mathcal{K}^{(D)}(\boldsymbol{C}_{\boldsymbol{y}}, \boldsymbol{C}_{\boldsymbol{y},\boldsymbol{x}}). \tag{4.70}$$

We prove by induction on $d = D/M \in \mathbb{N}$. Consider the case $d = 1$ where $\boldsymbol{T}_0 = \boldsymbol{M}_0 = \boldsymbol{C}_{\boldsymbol{y},\boldsymbol{x}}\boldsymbol{\Psi}_0$ is the only available prefilter matrix. Remember that $\boldsymbol{\Psi}_0 \in \mathbb{C}^{M \times M}$ defines an arbitrary full-rank matrix which does not change the range space of $\boldsymbol{C}_{\boldsymbol{y},\boldsymbol{x}}$ when applying it to its right-hand side, i. e., range$\{\boldsymbol{T}_0\} = $ range$\{\boldsymbol{M}_0\} = $ range$\{\boldsymbol{C}_{\boldsymbol{y},\boldsymbol{x}}\}$. With Equation (3.1) from the previous chapter, the M-dimensional subspace spanned by the columns of the prefilter matrix is clearly range$\{\boldsymbol{T}_0\} = \mathcal{K}^{(M)}(\boldsymbol{C}_{\boldsymbol{y}}, \boldsymbol{M}_0) = \mathcal{K}^{(M)}(\boldsymbol{C}_{\boldsymbol{y}}, \boldsymbol{C}_{\boldsymbol{y},\boldsymbol{x}})$. Now, assume that Equation (4.70) holds for $d = \ell \in \{1, 2, \ldots, L-1\}$, i. e., the range space of $\boldsymbol{T}_{\ell-1}$ is already a subset of $\mathcal{K}^{(\ell M)}(\boldsymbol{C}_{\boldsymbol{y}}, \boldsymbol{M}_0) = \mathcal{K}^{(\ell M)}(\boldsymbol{C}_{\boldsymbol{y}}, \boldsymbol{C}_{\boldsymbol{y},\boldsymbol{x}})$. In the polynomial representation similar to the one given in Equation (3.2), this fact reads as

$$\boldsymbol{T}_{\ell-1} = \sum_{j=0}^{\ell-1} \boldsymbol{C}_{\boldsymbol{y}}^j \boldsymbol{M}_0 \boldsymbol{\Psi}_{\boldsymbol{M}_0,j,\ell-1} = \sum_{j=0}^{\ell-1} \boldsymbol{C}_{\boldsymbol{y}}^j \boldsymbol{C}_{\boldsymbol{y},\boldsymbol{x}} \boldsymbol{\Psi}_{\boldsymbol{C}_{\boldsymbol{y},\boldsymbol{x}},j,\ell-1}, \tag{4.71}$$

where $\boldsymbol{\Psi}_{\boldsymbol{C}_{\boldsymbol{y},\boldsymbol{x}},j,\ell-1} = \boldsymbol{\Psi}_0 \boldsymbol{\Psi}_{\boldsymbol{M}_0,j,\ell-1}$ due to the fact that $\boldsymbol{M}_0 = \boldsymbol{C}_{\boldsymbol{y},\boldsymbol{x}}\boldsymbol{\Psi}_0$. Next, we have to prove that Equation (4.70) is also true for $d = \ell + 1$. Replace the prefilter matrix $\boldsymbol{T}_{\ell-1}$ in Equation (4.69) by the polynomial representation as given in Equation (4.71). Then, the term $\check{\boldsymbol{T}}_\ell := \boldsymbol{C}_{\boldsymbol{y}}\boldsymbol{T}_{\ell-1}\boldsymbol{\Psi}_\ell \in \mathbb{C}^{N \times M}$ in Equation (4.69) can be written as

$$\check{\boldsymbol{T}}_\ell = \sum_{j=0}^{\ell-1} \boldsymbol{C}_{\boldsymbol{y}}^{j+1} \boldsymbol{M}_0 \boldsymbol{\Psi}_{\boldsymbol{M}_0,j,\ell-1}\boldsymbol{\Psi}_\ell = \sum_{j=0}^{\ell-1} \boldsymbol{C}_{\boldsymbol{y}}^{j+1} \boldsymbol{C}_{\boldsymbol{y},\boldsymbol{x}} \boldsymbol{\Psi}_{\boldsymbol{C}_{\boldsymbol{y},\boldsymbol{x}},j,\ell-1}\boldsymbol{\Psi}_\ell, \tag{4.72}$$

which proves that range$\{\check{\boldsymbol{T}}_\ell\} \subset \mathcal{K}^{((\ell+1)M)}(\boldsymbol{C}_{\boldsymbol{y}}, \boldsymbol{M}_0) = \mathcal{K}^{((\ell+1)M)}(\boldsymbol{C}_{\boldsymbol{y}}, \boldsymbol{C}_{\boldsymbol{y},\boldsymbol{x}})$ because the power of the auto-covariance matrix $\boldsymbol{C}_{\boldsymbol{y}}$ has been increased by one. Finally, the multiplication of the projectors $\boldsymbol{P}_{\perp \boldsymbol{T}_i}$, $i \in \{0, 1, \ldots, \ell-1\}$, to $\check{\boldsymbol{T}}_\ell$ in Equation (4.69) represents the orthogonalization of $\check{\boldsymbol{T}}_\ell$ to the already computed prefilter matrices $\boldsymbol{T}_0$, $\boldsymbol{T}_1$, $\ldots$, and $\boldsymbol{T}_{\ell-1}$. Since this procedure does not change the subspace spanned by the columns of all prefilter matrices, it holds

$$\begin{aligned} \text{range}\left\{\begin{bmatrix} \boldsymbol{T}_0 \ \boldsymbol{T}_1 \ \cdots \ \boldsymbol{T}_\ell \end{bmatrix}\right\} &= \mathcal{K}^{((\ell+1)M)}(\boldsymbol{C}_{\boldsymbol{y}}, \boldsymbol{M}_0) \\ &= \mathcal{K}^{((\ell+1)M)}(\boldsymbol{C}_{\boldsymbol{y}}, \boldsymbol{C}_{\boldsymbol{y},\boldsymbol{x}}), \end{aligned} \tag{4.73}$$

and Equation (4.70) is also true for $d = \ell + 1$ which completes the proof. $\qquad\square$

Note that Proposition 4.8 includes also that all possible blocking matrices which fulfill Equation (4.60) lead to the same reduced-rank MWF performance (see also [30]) if we remember from Proposition 2.1 of Section 2.3 that the performance of a reduced-rank MWF depends solely on the subspace and not on the specific basis thereof. In the following, we restrict ourselves to the special case of blocking matrices which fulfill Equation (4.60).

Before revealing the connection of the reduced-rank MSMWF to block Krylov methods, we present an alternative way of computing the prefilter matrices T_ℓ when the blocking matrices have been chosen to be scaled matrices with orthonormal columns. Due to Proposition 4.3 and the fact that the matrices M_ℓ in Equation (4.40) are bases of the correlated subspaces as defined in Equations (4.20) and (4.22), the prefilter matrices T_ℓ, $\ell \in \{1, 2, \ldots, d-1\}$, can be chosen as arbitrary bases of the following correlated subspaces

$$\mathcal{T}_\ell = \operatorname*{argmax}_{\mathcal{T}' \in G(N,M)} \operatorname{tr}\left\{ \sqrt[\oplus]{C_{y,x_{\ell-1}}^{\mathrm{H}} P_{\mathcal{T}'} C_{y,x_{\ell-1}}} \right\} \quad \text{s.t.} \quad \mathcal{T}' \perp \bigoplus_{i=0}^{\ell-1} \mathcal{T}_i, \qquad (4.74)$$

initialized with $\mathcal{T}_0 = \mathcal{M}_0$. Note that T_0 is an arbitrary basis of $\mathcal{T}_0$. The solution to Equation (4.74) computes to $\mathcal{T}_\ell = \operatorname{range}\{T_\ell\}$ with T_ℓ as given in Equation (4.69). Thus, the full-rank matrix Ψ_ℓ in Equation (4.69) represents the variety of the number of bases which can be chosen to get a specific implementation of the reduced-rank MSMWF.

4.2.2 Block Arnoldi–Lanczos Connection

We have shown with Proposition 4.8 that the rank D MSMWF with blocking matrices according to Equation (4.60) operates in a D-dimensional Krylov subspace. It remains to find the detailed connection between the MSMWF and each of the block Krylov methods presented in Chapter 3. Remember from Section 3.2 that the fundamental recursion formula of the block Arnoldi algorithm reads as (cf. Equation 3.15)

$$Q_\ell H_{\ell,\ell-1} = \left(\prod_{i=0}^{\ell-1} P_{\perp Q_i} \right) A Q_{\ell-1} \in \mathbb{C}^{N \times M}, \quad \ell \in \{1, 2, \ldots, d-1\}, \qquad (4.75)$$

where the left-hand side of Equation (4.75) is the QR factorization of its right-hand side, i.e., the matrices $Q_\ell \in \mathbb{C}^{N \times M}$ have orthonormal columns and $H_{\ell,\ell-1} \in \mathbb{C}^{M \times M}$ are upper triangular matrices. Recall that the matrix $[Q_0, Q_1, \ldots, Q_{d-1}] \in \mathbb{C}^{N \times D}$, $D = dM$, is constructed to form an orthonormal basis of $\mathcal{K}^{(D)}(A, Q_0)$.

If we transform Equation (4.60) to

$$\breve{\beta}_{\ell-1}^{-1} T_\ell \Psi_\ell^{-1} = \left(\prod_{i=0}^{\ell-1} P_{\perp T_i} \right) C_y T_{\ell-1}, \qquad (4.76)$$

and compare the result to the recursion formula of the block Arnoldi algorithm, i.e., Equation (4.75), we see that both are equal when the left-hand side of Equation (4.76) is again a QR factorization of its right-hand side, i. e., the columns of the matrix T_ℓ has to be mutually orthonormal and such that $\breve{\beta}_{\ell-1}^{-1}\boldsymbol{\Psi}_\ell^{-1}$ is an upper triangular matrix. Besides, it must hold $A = C_y$ and $(Q_0, H_{0,-1}) = \mathrm{QR}(M_0)$. Note that $\mathcal{K}^{(M)}(C_y, M_0) = \mathcal{K}^{(M)}(C_y, T_0)$ due to $T_0 = M_0$.

Now, the following question remains: Which special choice of matrices $M_\ell \in \mathbb{C}^{(N-\ell M)\times M}$, $\ell \in \{0, 1, \ldots, d-1\}$, of the reduced-rank MSMWF with blocking matrices according to Equation (4.60) result in prefilter matrices T_ℓ which are equal to the matrices Q_ℓ of the block Arnoldi algorithm? To this end, the matrices M_ℓ must orthonormalize the columns of the prefilter matrix T_ℓ from Equation (4.40), i. e.,

$$T_\ell^{\mathrm{H}} T_\ell = M_\ell^{\mathrm{H}} \left(\prod_{\substack{i=\ell-1 \\ \ell\in\mathbb{N}}}^{0} B_i^{\mathrm{H}}\right) \left(\prod_{\substack{i=0 \\ \ell\in\mathbb{N}}}^{\ell-1} B_i\right) M_\ell \overset{!}{=} I_M. \tag{4.77}$$

Due to Equation (4.66), the condition

$$M_\ell^{\mathrm{H}} M_\ell = \breve{\beta}_{\ell-1}^{-1} I_M, \tag{4.78}$$

fulfills Equation (4.77). In other words, the columns of the matrix $\breve{\beta}_{\ell-1}^{1/2} M_\ell$ have to be mutually orthonormal. Since $\breve{\beta}_{\ell-1}^{-1}\boldsymbol{\Psi}_\ell^{-1}$ has to be additionally upper triangular, we get the solution for $M_\ell = C_{y_{\ell-1}, x_{\ell-1}}\boldsymbol{\Psi}_\ell$ via the QR factorization

$$\left|\left(\sqrt{\breve{\beta}_{\ell-1}} M_\ell, \breve{\beta}_{\ell-1}^{-1}\boldsymbol{\Psi}_\ell^{-1}\right) = \mathrm{QR}\left(\sqrt{\breve{\beta}_{\ell-1}^{-1}} C_{y_{\ell-1}, x_{\ell-1}}\right),\right| \tag{4.79}$$

where the columns of $\breve{\beta}_{\ell-1}^{1/2} M_\ell$ are mutually orthonormal and $\breve{\beta}_{\ell-1}^{-1}\boldsymbol{\Psi}_\ell^{-1}$ is an upper triangular matrix.

Again, if we choose the matrices M_ℓ according to Equation (4.79) and the blocking matrices according to Equation (4.60), the prefilter matrices of the reduced-rank MSMWF are implicitly computed by the block Arnoldi algorithm. Thus, the reduced-rank MSMWF for this case is computationally very inefficient because it does not exploit the Hermitian structure of the auto-covariance matrix C_y. Nevertheless, since we revealed the connection between the MSMWF and the block Arnoldi algorithm, we can use the computationally much cheaper block Lanczos algorithm for the calculation of the prefilter matrices. However, also the block Lanczos algorithm can be only applied to the special version of the MSMWF where the bases of the correlated subspaces as well as blocking matrices are chosen adequately. To sum up, the prefilter matrix of this special MSMWF version reads as

$$\left|T_{\mathrm{MSMWF}}^{(D)}\right|_{\substack{\mathrm{Eq.\ (4.60)} \\ \mathrm{Eq.\ (4.79)}}} = \begin{bmatrix} Q_0\ Q_1\ \cdots\ Q_{d-1} \end{bmatrix}. \tag{4.80}$$

4.2.3 Block Conjugate Gradient Connection

Recall from Section 3.4 that the BCG algorithm at the $(\ell + 1)$th iteration minimizes the cost function (cf. Equation 3.29)

$$\varepsilon_A \left(X' \right) = \| E' \|_A^2 = \| X - X' \|_A^2$$
$$= \mathrm{tr} \left\{ X'^{,\mathrm{H}} A X' - X'^{,\mathrm{H}} B - B^{\mathrm{H}} X' + X^{\mathrm{H}} B \right\}, \tag{4.81}$$

for $X' = X_\ell + D_\ell \Phi' \in \mathbb{C}^{N \times M}$ according to the step size matrix $\Phi' \in \mathbb{C}^{M \times M}$. Further, we have shown with Proposition 3.4 that this M-dimensional search in the range space of the direction matrix $D_\ell \in \mathbb{C}^{N \times M}$ is equal to the $(\ell + 1)M$-dimensional search in the range space of $[D_0, D_1, \dots, D_\ell] \in \mathbb{C}^{N \times (\ell+1)M}$ which is further equal to the $(\ell + 1)M$-dimensional Krylov subspace $\mathcal{K}^{((\ell+1)M)}(A, D_0) = \mathcal{K}^{((\ell+1)M)}(A, R_0)$ due to Proposition 3.6. In essence, the approximation X_d of X at the dth BCG iteration, minimizes the cost function $\varepsilon_A \left(X' \right)$ in the D-dimensional Krylov subspace $\mathcal{K}^{(D)}(A, B)$, $D = dM$. Here, we assume that the BCG algorithm have been initialized with $X_0 = 0_{N \times M}$. Thus, $R_0 = B - A X_0 = B \in \mathbb{C}^{N \times M}$.

On the other hand, the rank D MSMWF $W_{\mathrm{MSMWF}}^{(D)}$ and $a_{\mathrm{MSMWF}}^{(D)}$ with blocking matrices according to Equation (4.60) is an approximation of the MWF W and a in the D-dimensional Krylov subspace $\mathcal{K}^{(D)}(C_y, C_{y,x})$ which we have proven in Proposition 4.8 of Subsection 4.2.1. Besides, the rank D MSMWF as a reduced-rank MWF (cf. Equations 2.45 and 2.47 of Section 2.3) minimizes the MSE $\xi(W', a')$ defined in Equation (2.4) over all linear matrix filters $W' = T_{\mathrm{MSMWF}}^{(D)} G'$ and $a' = m_x - W'^{,\mathrm{H}} m_y$. Here, the matrix $T_{\mathrm{MSMWF}}^{(D)}$ is the prefilter of the reduced-rank MSMWF as defined in Equation (4.41) but where the blocking matrices fulfill Equation (4.60). With a' as defined above, the MSE of Equation (2.4) can be rewritten as

$$\xi \left(W', m_x - W'^{,\mathrm{H}} m_y \right) = \mathrm{E} \left\{ \| x^\circ - W'^{,\mathrm{H}} y^\circ \|_2^2 \right\}$$
$$= \mathrm{tr} \left\{ C_x - W'^{,\mathrm{H}} C_{y,x} - C_{y,x}^{\mathrm{H}} W' + W'^{,\mathrm{H}} C_y W' \right\}. \tag{4.82}$$

If we compare the cost function of the BCG algorithm in Equation (4.81) with the cost function of the MSMWF in Equation (4.82), we see that except from the constants $\mathrm{tr}\{C_x\}$ and $\mathrm{tr}\{X^{\mathrm{H}} B\}$, respectively, both are equal if we make the substitutions $A = C_y$, $B = C_{y,x}$, and $X = W$. Thus, when applying the BCG algorithm to solve the Wiener–Hopf equation $C_y W = C_{y,x}$ as the system of linear equations $A X = B$, the approximated solution at the dth iteration is equal to the rank D MSMWF, i. e.,

$$\boxed{W_{\mathrm{MSMWF}}^{(dM)}\Big|_{\mathrm{Eq.\ (4.60)}} = X_d,} \tag{4.83}$$

because they are both approximations of $W = X$ in the same subspace, i. e., the Krylov subspace $\mathcal{K}^{(D)}(C_y, C_{y,x})$, and minimize the same cost function up to a constant which does not change the optimum.

This equality between the special version of the reduced-rank MSMWF where the scaled blocking matrices have orthonormal columns and the aborted BCG algorithm allows to make the following conclusions:

- The convergence analysis of the BCG algorithm in Subsection 3.4.2 can be directly assigned to the reduced-rank MSMWF.
- The methods presented in Subsection 3.4.3 to understand the regularizing effect of the BCG algorithm can be also used to describe the inherent regularization of the reduced-rank MSMWF as has been reported in literature (see, e. g., [119, 43]).
- The investigation of the MSMWF performance for a multitude of communication systems have shown that the reduced-rank MSMWF is a very good approximation of the full-rank MWF. Thus, the BCG algorithm can also yield very good approximations in these practical cases although the involved system matrices are far from the main application area of Krylov subspace methods, i. e., large sparse systems of linear equations (cf., e. g., [204]).

4.3 Krylov Subspace Based Multistage Matrix Wiener Filter Implementations

4.3.1 Block Lanczos–Ruhe Implementation

Again, due to simplicity, we omit the time index n throughout this chapter although the matrices under consideration are still time-varying. In Subsection 4.2.2, we have shown that the prefilter matrices of the reduced-rank MSMWF can be computed based on the block Lanczos algorithm if the blocking matrices and the bases of the correlated subspaces are chosen properly, and if the rank D is an integer multiple of the dimension M of the desired signal vector. Proposition 4.8 revealed that these prefilter matrices span the D-dimensional Krylov subspace $\mathcal{K}^{(D)}(C_y, C_{y,x})$. Thus, if D is an arbitrary integer with $M \leq D \leq N$, the *Block Lanczos–Ruhe* (BLR) algorithm which has been introduced in Section 3.3.2 to generate an orthonormal basis of the D-dimensional Krylov subspace $\mathcal{K}^{(D)}(C_y, C_{y,x})$ where $D \in \{M, M+1, \ldots, N\}$, can be used to compute the prefilter matrix

$$T_{\mathrm{BLR}}^{(D)} = \begin{bmatrix} q_0 & q_1 & \cdots & q_{D-1} \end{bmatrix} \in \mathbb{C}^{N \times D}, \tag{4.84}$$

of the rank-flexible MSMWF version. Note that the following derivation leads to an extension of the MSWF implementation presented in [141, 140] for the matrix case, i. e., where the cross-covariance matrix $C_{y,x}$ may have more than one column (see also [45]).

To end up with an algorithm which computes the reduced-rank MSMWF $\boldsymbol{W}_{\mathrm{BLR}}^{(D)} \in \mathbb{C}^{N \times M}$ and $\boldsymbol{a}_{\mathrm{BLR}}^{(D)} = \boldsymbol{m}_{\mathsf{x}} - \boldsymbol{W}_{\mathrm{BLR}}^{(D),\mathrm{H}} \boldsymbol{m}_{\mathsf{y}} \in \mathbb{C}^{M}$ based on the BLR algorithm, it remains to compute the reduced-dimension MWF $\boldsymbol{G}_{\mathrm{BLR}}^{(D)} \in \mathbb{C}^{D \times M}$ if we recall Equations (2.46) and (2.48), i. e.,

$$\boldsymbol{W}_{\mathrm{BLR}}^{(D)} = \boldsymbol{T}_{\mathrm{BLR}}^{(D)} \boldsymbol{G}_{\mathrm{BLR}}^{(D)} = \boldsymbol{T}_{\mathrm{BLR}}^{(D)} \boldsymbol{C}_{\mathsf{y}_{\mathrm{BLR}}^{(D)}}^{-1} \boldsymbol{C}_{\mathsf{y}_{\mathrm{BLR}}^{(D)},\mathsf{x}}. \tag{4.85}$$

Due to the orthogonality of the columns of $\boldsymbol{T}_{\mathrm{BLR}}^{(D)}$ and the fact that the first M columns $\boldsymbol{Q}_0 = [\boldsymbol{q}_0, \boldsymbol{q}_1, \ldots, \boldsymbol{q}_{M-1}] \in \mathbb{C}^{N \times M}$ thereof are obtained by a QR factorization, i. e., (cf. Subsection 3.3.2 and Appendix B.3)

$$\boxed{(\boldsymbol{Q}_0, \boldsymbol{C}_{0,-1}) = \mathrm{QR}\,(\boldsymbol{C}_{\mathsf{y},\mathsf{x}})\,,} \tag{4.86}$$

where $\boldsymbol{C}_{0,-1} \in \mathbb{C}^{M \times M}$ is an upper triangular matrix, the cross-covariance matrix of the transformed observation vector $\boldsymbol{y}_{\mathrm{BLR}}^{(D)} = \boldsymbol{T}_{\mathrm{BLR}}^{(D),\mathrm{H}} \boldsymbol{y}$ which has been filtered by the prefilter matrix $\boldsymbol{T}_{\mathrm{BLR}}^{(D)}$ reads as

$$\boldsymbol{C}_{\mathsf{y}_{\mathrm{BLR}}^{(D)},\mathsf{x}} = \boldsymbol{T}_{\mathrm{BLR}}^{(D),\mathrm{H}} \boldsymbol{C}_{\mathsf{y},\mathsf{x}} = \begin{bmatrix} \boldsymbol{C}_{0,-1} \\ \boldsymbol{0}_{(D-M) \times M} \end{bmatrix} \in \mathbb{C}^{D \times M}. \tag{4.87}$$

Thus, only the first M columns of the inverse of $\boldsymbol{C}_{\mathsf{y}_{\mathrm{BLR}}^{(D)}} = \boldsymbol{T}_{\mathrm{BLR}}^{(D),\mathrm{H}} \boldsymbol{C}_{\mathsf{y}} \boldsymbol{T}_{\mathrm{BLR}}^{(D)} \in \mathbb{C}^{D \times D}$ has to be computed (cf. Equation 4.85). We derive in the following a recursive method which performs this task.

For simplicity, let us define

$$\boldsymbol{D}^{(j)} := \boldsymbol{C}_{\mathsf{y}_{\mathrm{BLR}}^{(j)}}^{-1} \in \mathbb{C}^{j \times j}, \quad j \in \{M, M+1, \ldots, D\}. \tag{4.88}$$

Starting with $j = M$, the inverse of the corresponding auto-covariance matrix is

$$\boxed{\boldsymbol{D}^{(M)} = (\boldsymbol{Q}_0^{\mathrm{H}} \boldsymbol{C}_{\mathsf{y}} \boldsymbol{Q}_0)^{-1} \in \mathbb{C}^{M \times M},} \tag{4.89}$$

with the matrix $\boldsymbol{Q}_0$ from the QR factorization introduced above. Next, we have to find a computationally efficient way to compute the inverse $\boldsymbol{D}^{(j+1)}$ based on the inverse $\boldsymbol{D}^{(j)}$ where $j \in \{M, M+1, \ldots, D-1\}$. It has been shown in Propositions 3.1 and 3.2 that the auto-covariance matrix $\boldsymbol{C}_{\mathsf{y}_{\mathrm{BLR}}^{(j+1)}} = \boldsymbol{T}_{\mathrm{BLR}}^{(j+1),\mathrm{H}} \boldsymbol{C}_{\mathsf{y}} \boldsymbol{T}_{\mathrm{BLR}}^{(j+1)}$ of the transformed observation vector is a Hermitian band Hessenberg matrix with bandwidth $2M+1$ (cf. Figure 3.2). Hence, it may be written as

$$\boldsymbol{C}_{\mathsf{y}_{\mathrm{BLR}}^{(j+1)}} = \left[\begin{array}{c|c} \boldsymbol{C}_{\mathsf{y}_{\mathrm{BLR}}^{(j)}} & \boldsymbol{0}_{j-M} \\ & \boldsymbol{c}_j \\ \hline \boldsymbol{0}_{j-M}^{\mathrm{T}} \quad \boldsymbol{c}_j^{\mathrm{H}} & c_{j,j} \end{array} \right] \in \mathbb{C}^{(j+1) \times (j+1)}, \tag{4.90}$$

with the entries

$$\boxed{c_j := \begin{bmatrix} c_{j-M,j} \\ c_{j-M+1,j} \\ \vdots \\ c_{j-1,j} \end{bmatrix} = \begin{bmatrix} q_{j-M}^{\mathrm{H}} \\ q_{j-M+1}^{\mathrm{H}} \\ \vdots \\ q_{j-1}^{\mathrm{H}} \end{bmatrix} C_y q_j \in \mathbb{C}^M,} \tag{4.91}$$

and

$$\boxed{c_{j,j} := q_j^{\mathrm{H}} C_y q_j \in \mathbb{R}_+.} \tag{4.92}$$

Note that $C_{y_{\mathrm{BLR}}^{(j)}} = T_{\mathrm{BLR}}^{(j),\mathrm{H}} C_y T_{\mathrm{BLR}}^{(j)} \in \mathbb{C}^{j \times j}$ which depends on the prefilter vectors calculated in all previous steps, has also a Hermitian band Hessenberg structure. Generally, the element in the $(i+1)$th row and $(m+1)$th column of the auto-covariance matrix $C_{y_{\mathrm{BLR}}^{(j+1)}}$, $i, m \in \{0, 1, \ldots, j\}$, is denoted as

$$c_{i,m} = \left[C_{y_{\mathrm{BLR}}^{(j+1)}} \right]_{i,m} = q_i^{\mathrm{H}} C_y q_m, \tag{4.93}$$

where $q_i \in \mathbb{C}^N$ is the $(i+1)$th column of $T_{\mathrm{BLR}}^{(j+1)} \in \mathbb{C}^{N \times (j+1)}$. Recall the notation for submatrices defined in Subsection 2.2.1 or Section 1.2, respectively.

Since $C_{y_{\mathrm{BLR}}^{(j+1)}}$ is a partitioned matrix (cf. Equation 4.90), its inverse can be computed by using Equation (A.8) of Appendix A.1. It follows:

$$D^{(j+1)} = \begin{bmatrix} D^{(j)} & 0_j \\ 0_j^{\mathrm{T}} & 0 \end{bmatrix} + \delta_{j+1}^{-1} d_{j+1} d_{j+1}^{\mathrm{H}}, \tag{4.94}$$

with the vector

$$d_{j+1} = \begin{bmatrix} D^{(j)} \begin{bmatrix} 0_{j-M} \\ c_j \\ -1 \end{bmatrix} \end{bmatrix} = \begin{bmatrix} D_{\to M}^{(j)} c_j \\ -1 \end{bmatrix} = \begin{bmatrix} d^{(j)} \\ -1 \end{bmatrix} \in \mathbb{C}^{j+1}, \tag{4.95}$$

and the Schur complement

$$\begin{aligned} \delta_{j+1} &= c_{j,j} - \begin{bmatrix} 0_{j-M}^{\mathrm{T}} & c_j^{\mathrm{H}} \end{bmatrix} D^{(j)} \begin{bmatrix} 0_{j-M} \\ c_j \end{bmatrix} \\ &= c_{j,j} - c_j^{\mathrm{H}} d_{\downarrow M}^{(j)} \in \mathbb{R}_+, \end{aligned} \tag{4.96}$$

of $c_{j,j}$.[6] In Equations (4.95) and (4.96), we used the abbreviations

$$D_{\to M}^{(j)} = \left[D^{(j)} \right]_{:, j-M:j-1} \in \mathbb{C}^{j \times M}, \tag{4.97}$$

$$d_{\downarrow M}^{(j)} = \left[d^{(j)} \right]_{j-M:j-1} \in \mathbb{C}^M, \tag{4.98}$$

---

[6] Note that the Schur complements of a positive definite matrix are again positive definite (see, e. g., [131]).

for the last M columns of $\boldsymbol{D}^{(j)}$ (cf. Figure 4.11(a)) and the last M elements of the vector

$$\boxed{\boldsymbol{d}^{(j)} = \boldsymbol{D}^{(j)}_{\to M}\boldsymbol{c}_j \in \mathbb{C}^j,}$$
(4.99)

respectively. Note that $[\boldsymbol{a}]_{n:m}$ denotes the $(n-m+1)$-dimensional subvector of $\boldsymbol{a}$ from its $(n+1)$th to its $(m+1)$th element.

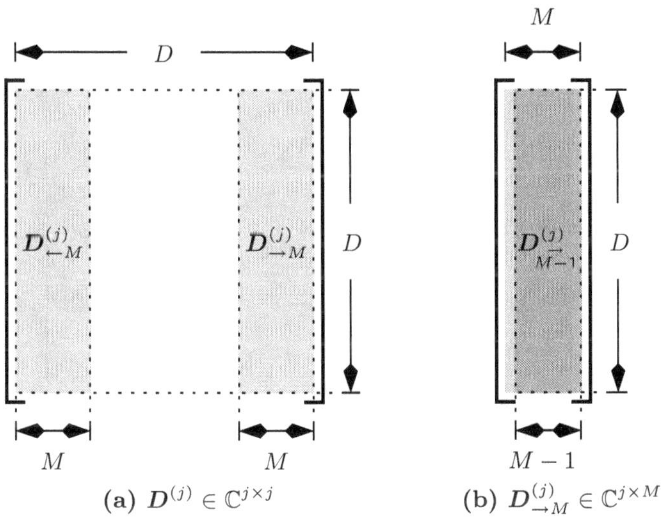

(a) $\boldsymbol{D}^{(j)} \in \mathbb{C}^{j \times j}$ (b) $\boldsymbol{D}^{(j)}_{\to M} \in \mathbb{C}^{j \times M}$

Fig. 4.11. Submatrices $\boldsymbol{D}^{(j)}_{\leftarrow M} \in \mathbb{C}^{j \times M}$, $\boldsymbol{D}^{(j)}_{\to M} \in \mathbb{C}^{j \times M}$, and $\boldsymbol{D}^{(j)}_{\to M-1} \in \mathbb{C}^{j \times (M-1)}$ of the matrices $\boldsymbol{D}^{(j)}$ and $\boldsymbol{D}^{(j)}_{\to M}$, respectively, $j \in \{M, M+1, \ldots, D\}$

Remember from Equations (4.85) and (4.87) that we are only interested in the first M columns of the inverse $\boldsymbol{D}^{(D)}$. At the last recursion step where $j = D-1$, these columns are the first M columns of the inverse $\boldsymbol{D}^{(j+1)}$. Using Equation (4.94), these first M columns may be written as

$$
\begin{aligned}
\boldsymbol{D}^{(j+1)}_{\leftarrow M} &= \begin{bmatrix} \boldsymbol{D}^{(j)}_{\leftarrow M} \\ \boldsymbol{0}^{\mathrm{T}}_M \end{bmatrix} + \delta^{-1}_{j+1} \begin{bmatrix} \boldsymbol{d}^{(j)} \\ -1 \end{bmatrix} \boldsymbol{d}^{(j),\mathrm{H}}_{\uparrow M} \\
&= \begin{bmatrix} \boldsymbol{D}^{(j)}_{\leftarrow M} + \delta^{-1}_{j+1} \boldsymbol{d}^{(j)} \boldsymbol{d}^{(j),\mathrm{H}}_{\uparrow M} \\ -\delta^{-1}_{j+1} \boldsymbol{d}^{(j),\mathrm{H}}_{\uparrow M} \end{bmatrix} \in \mathbb{C}^{(j+1) \times M}.
\end{aligned}
$$
(4.100)

Here,

$$\boldsymbol{D}^{(j)}_{\leftarrow M} = \left[\boldsymbol{D}^{(j)} \right]_{:,0:M-1} \in \mathbb{C}^{j \times M},$$
(4.101)

$$\boldsymbol{d}^{(j)}_{\uparrow M} = \left[\boldsymbol{d}^{(j)} \right]_{0:M-1} \in \mathbb{C}^M,$$
(4.102)

denote the first M columns of $\boldsymbol{D}^{(j)}$ (cf. Figure 4.11(a)) and the first M elements of $\boldsymbol{d}^{(j)}$, respectively. It can be seen that the first M columns of $\boldsymbol{D}^{(j+1)}$ only depend on the first and last M columns of $\boldsymbol{D}^{(j)}$ as well as the entries c_j and $c_{j,j}$ of the auto-covariance matrix as defined in Equation (4.90). Again with Equation (4.94), we get the last M columns of $\boldsymbol{D}^{(j+1)}$ as follows:

$$
\begin{aligned}
\boldsymbol{D}_{\to M}^{(j+1)} &= \begin{bmatrix} \boldsymbol{D}_{\to M-1}^{(j)} & \boldsymbol{0}_j \\ \boldsymbol{0}_{M-1}^{\mathrm{T}} & 0 \end{bmatrix} + \delta_{j+1}^{-1} \begin{bmatrix} \boldsymbol{d}^{(j)} \\ -1 \end{bmatrix} \begin{bmatrix} \boldsymbol{d}_{\downarrow M-1}^{(j),\mathrm{H}} & -1 \end{bmatrix} \\
&= \begin{bmatrix} \boldsymbol{D}_{\to M-1}^{(j)} + \delta_{j+1}^{-1} \boldsymbol{d}^{(j)} \boldsymbol{d}_{\downarrow M-1}^{(j),\mathrm{H}} & -\delta_{j+1}^{-1} \boldsymbol{d}^{(j)} \\ -\delta_{j+1}^{-1} \boldsymbol{d}_{\downarrow M-1}^{(j),\mathrm{H}} & \delta_{j+1}^{-1} \end{bmatrix} \in \mathbb{C}^{(j+1)\times M}.
\end{aligned}
\tag{4.103}
$$

The variables

$$
\boldsymbol{D}_{\to M-1}^{(j)} = \left[\boldsymbol{D}^{(j)} \right]_{:,\, j-M+1:j-1} \in \mathbb{C}^{j\times(M-1)},
\tag{4.104}
$$

$$
\boldsymbol{d}_{\downarrow M-1}^{(j)} = \left[\boldsymbol{d}^{(j)} \right]_{j-M+1:j-1} \in \mathbb{C}^{M-1},
\tag{4.105}
$$

denote the last $M-1$ columns of $\boldsymbol{D}^{(j)}$ (cf. Figure 4.11(b)) and the last $M-1$ elements of $\boldsymbol{d}^{(j)}$, respectively. Again, the last M columns of $\boldsymbol{D}^{(j+1)}$ only depend on the last M columns of $\boldsymbol{D}^{(j)}$ and the entries c_j and $c_{j,j}$. Thus, at each recursion step, it is sufficient to update the first M as well as the last M columns of $\boldsymbol{D}^{(j+1)}$ because we need only the first M columns $\boldsymbol{D}_{\leftarrow M}^{(D)}$ of the auto-covariance matrix inverse $\boldsymbol{D}^{(D)}$ to compute the reduced-rank MSMWF according to

$$
\boxed{\boldsymbol{W}_{\mathrm{BLR}}^{(D)} = \boldsymbol{T}_{\mathrm{BLR}}^{(D)} \boldsymbol{D}_{\leftarrow M}^{(D)} \boldsymbol{C}_{0,-1},}
\tag{4.106}
$$

if we recall Equations (4.85) and (4.87).

Finally, Algorithm 4.2. summarizes the BLR implementation of the reduced-rank MSMWF. The numbers in angle brackets are explained in Section 4.4. Lines 1 to 4 of Algorithm 4.2. initialize the recursion according to Equations (4.86) and (4.89). Lines 6 to 16 produce the columns of the prefilter matrix $\boldsymbol{T}_{\mathrm{BLR}}^{(D)}$ according to the BLR method, i.e., Algorithm 3.3., which we explained in great detail in Subsection 3.3.2. However, compared to Algorithm 3.3., we replace j by $j-1$, h by c, and $\boldsymbol{A}$ by $\boldsymbol{C_y}$, and move the computation of the vector $\boldsymbol{C_y}\boldsymbol{q}_j$ to Line 16, i.e., after the inner **for**-loop, because it is already needed in the second part of the outer **for**-loop of Algorithm 4.2. where the submatrices of the inverse $\boldsymbol{D}^{(j+1)}$ are calculated. Since we need previous versions of this vector afterwards for the computation of a new basis vector according to the BLR method, it is stored in different variables $\boldsymbol{v}_j$. Remember that the vector $\boldsymbol{C_y}\boldsymbol{q}_j$ in Algorithm 3.3. has been computed and stored in the single variable $\boldsymbol{v}$ before its inner **for**-loop. Besides, since the entries $\boldsymbol{c}_j = [c_{m,j}, c_{m+1,j}, \ldots, c_{m+M-1,j}]^{\mathrm{T}} \in \mathbb{C}^M$ and $c_{j,j}$ of the auto-covariance matrix $\boldsymbol{C}_{\boldsymbol{y}_{\mathrm{BLR}}^{(j+1)}}$ (cf. Equation 4.90) are already computed in Lines 17 and 18 according to Equations (4.91) and (4.92), respectively, they need no longer be calculated in the **if**-statement from Lines 8 to 12 as done in Algorithm 3.3.. By the way, as in the algorithm of Subsection 3.3.2, the **if**-statement is used because only the elements $c_{i,m}$ on the main diagonal and the upper M sub-diagonals, i.e., for $i \leq m$, of the auto-covariance matrix $\boldsymbol{C}_{\boldsymbol{y}_{\mathrm{BLR}}^{(j+1)}}$ are directly available. The elements on the lower M subdiagonals are obtained exploiting the Hermitian structure of the auto-covariance matrix, i.e., $c_{i,m} = c_{m,i}^*$ for $i > m$. Finally, the remaining lines of the outer **for**-loop of Algorithm 4.2., viz., Lines 19 to 25, compute the first and last M columns of the inverse $\boldsymbol{D}^{(j+1)}$ as described in Equations (4.96), (4.99), (4.100), and (4.103), and in Lines 28 to 30, the reduced-rank MSMWF is constructed according to Equations (4.84) and (4.106) (see also Equation 2.48).

4.3.2 Block Conjugate Gradient Implementation

Again, in this subsection, D is not restricted to be an integer multiple of M, i.e., $D \in \{M, M + 1, \ldots, N\}$. In Subsection 4.2.3, we revealed the relationship between the BCG algorithm and the reduced-rank MSMWF with rank $D = dM$ and blocking matrices $\boldsymbol{B}_\ell$ such that $\boldsymbol{B}_\ell^{\mathrm{H}} \boldsymbol{B}_\ell = \beta_\ell \boldsymbol{I}_{N-(\ell+1)M}$ for all $\ell \in \{0, 1, \ldots, d - 2\}$, $d > 1$ (cf. Equation 4.60). If we extend this connection to the case where $D \in \{M, M + 1, \ldots, N\}$, the dimension-flexible BCG algorithm of Subsection 3.4.4, i.e., Algorithm 3.5., can be used to get an efficient implementation of the reduced-rank MSMWF. With the replacements $\boldsymbol{A} = \boldsymbol{C}_{\boldsymbol{y}}$, $\boldsymbol{B} = \boldsymbol{C}_{\boldsymbol{y},\boldsymbol{x}}$, and $\boldsymbol{X} = \boldsymbol{W}$ as already made in Subsection 4.2.3, the resulting BCG based MSMWF implementation $\boldsymbol{W}_{\mathrm{BCG}}^{(D)}$ and $\boldsymbol{a}_{\mathrm{BCG}}^{(D)}$ can be formulated as Algorithm 4.3.. Again, the number in angle brackets are explained in Section 4.4.

One advantage of Algorithm 4.3. compared to the MSMWF implementation as given in Algorithm 4.1. is the possibility to be initialized with $\boldsymbol{W}_0 \neq \boldsymbol{0}_{N \times M}$. This can improve convergence if we exploit available *a priori* knowledge for the choice of $\boldsymbol{W}_0$ like, e.g., the MWF approximation from the previous communication block (see, e.g., [32]).

4.4 Computational Complexity Considerations

This section is to investigate the computational efficiency of the different reduced-rank MSMWF implementations compared to the full-rank MWF which we investigated in Section 2.2. As in Chapter 2, we use the number of FLOPs as the complexity measure. Again, an addition, subtraction, multiplication, division, or root extraction is counted as one FLOP, no matter if operating on real or complex values. The following list summarizes repeatedly occurring operations together with the required number of FLOPs:

- The QR factorization of an $n \times m$ matrix requires $\zeta_{\mathrm{QR}}(n, m) = (2m + 1)mn - (m - 1)m/2$ FLOPs (cf. Equation B.16 of Appendix B.3).

Algorithm 4.2. Block Lanczos–Ruhe (BLR) implementation of the MSMWF

$$(\boldsymbol{Q}_0 = [\boldsymbol{q}_0, \boldsymbol{q}_1, \ldots \boldsymbol{q}_{M-1}], \boldsymbol{C}_{0,-1}) \leftarrow \mathrm{QR}(\boldsymbol{C}_{y,x}) \qquad \langle \zeta_{\mathrm{QR}}(N, M) \rangle$$

2: $\boldsymbol{V} = [\boldsymbol{v}_0, \boldsymbol{v}_1, \ldots, \boldsymbol{v}_{M-1}] \leftarrow \boldsymbol{C}_y \boldsymbol{Q}_0 \qquad \langle MN(2N-1) \rangle$

$\boldsymbol{D}_{\leftarrow M}^{(M)} = \boldsymbol{D}_{\rightarrow M}^{(M)} \leftarrow (\boldsymbol{Q}_0^{\mathrm{H}} \boldsymbol{V})^{-1} \qquad \langle \zeta_{\mathrm{HI}}(M) + M(M+1)(2N-1)/2 \rangle$

4: $\boldsymbol{D}_{\rightarrow M-1}^{(M)} \leftarrow [\boldsymbol{D}_{\rightarrow M}^{(M)}]_{:,1:M-1}$

    for $j = M, M+1, \ldots, D-1 \wedge D > M$ **do**

6:     $m \leftarrow j - M$

        for $i = j - 2M, j - 2M + 1, \ldots, j - 1 \wedge i \in \mathbb{N}_0$ **do**

8:         **if** $i \leq m$ **then**

$$\boldsymbol{v}_m \leftarrow \boldsymbol{v}_m - \boldsymbol{q}_i c_{i,m} \qquad \langle 2N \rangle$$

10:         **else**

$$\boldsymbol{v}_m \leftarrow \boldsymbol{v}_m - \boldsymbol{q}_i c_{m,i}^* \qquad \langle 2N \rangle$$

12:         **end if**

        end for

14:     $c_{j,m} = c_{m,j} \leftarrow \|\boldsymbol{v}_m\|_2 \qquad \langle 2N \rangle$

    $\boldsymbol{q}_j \leftarrow \boldsymbol{v}_m / c_{j,m} \qquad \langle N \rangle$

16:     $\boldsymbol{v}_j \leftarrow \boldsymbol{C}_y \boldsymbol{q}_j \qquad \langle N(2N-1) \rangle$

$$\boldsymbol{c}_j \leftarrow \begin{bmatrix} c_{m,j} \\ [\boldsymbol{q}_{m+1}, \boldsymbol{q}_{m+2}, \ldots, \boldsymbol{q}_{m+M-1}]^{\mathrm{H}} \boldsymbol{v}_j \end{bmatrix} \qquad \langle (M-1)(2N-1) \rangle$$

18:     $c_{j,j} \leftarrow \boldsymbol{q}_j^{\mathrm{H}} \boldsymbol{v}_j \qquad \langle 2N - 1 \rangle$

    $\boldsymbol{d}^{(j)} \leftarrow \boldsymbol{D}_{\rightarrow M}^{(j)} \boldsymbol{c}_j \qquad \langle j(2M-1) \rangle$

20:     $\delta_{j+1}^{-1} \leftarrow \left(c_{j,j} - \boldsymbol{c}_j^{\mathrm{H}} \boldsymbol{d}_{\downarrow M}^{(j)} \right)^{-1} \qquad \langle 2M + 1 \rangle$

    $\bar{\boldsymbol{d}}^{(j)} \leftarrow \delta_{j+1}^{-1} \boldsymbol{d}^{(j)} \qquad \langle j \rangle$

22:     $\boldsymbol{D}_{\leftarrow M}^{(j+1)} \leftarrow \begin{bmatrix} \boldsymbol{D}_{\leftarrow M}^{(j)} + \boldsymbol{d}^{(j)} \bar{\boldsymbol{d}}_{\uparrow M}^{(j),\mathrm{H}} \\ -\bar{\boldsymbol{d}}_{\uparrow M}^{(j),\mathrm{H}} \end{bmatrix} \qquad \langle 2jM + M(1-M) \rangle$

    if $j < D - 1$ **then**

24:         $\boldsymbol{D}_{\rightarrow M}^{(j+1)} \leftarrow \begin{bmatrix} \boldsymbol{D}_{\rightarrow M-1}^{(j)} + \boldsymbol{d}^{(j)} \bar{\boldsymbol{d}}_{\downarrow M-1}^{(j),\mathrm{H}} & -\bar{\boldsymbol{d}}^{(j)} \\ -\bar{\boldsymbol{d}}_{\downarrow M-1}^{(j),\mathrm{H}} & \delta_{j+1}^{-1} \end{bmatrix} \qquad \langle (2j - M + 2)(M-1) \rangle$

        $\boldsymbol{D}_{\rightarrow M-1}^{(j+1)} \leftarrow [\boldsymbol{D}_{\rightarrow M}^{(j+1)}]_{:,j-M+1:j-1}$

26:     **end if**

    end for

28: $\boldsymbol{T}_{\mathrm{BLR}}^{(D)} \leftarrow [\boldsymbol{q}_0, \boldsymbol{q}_1, \ldots, \boldsymbol{q}_{D-1}]$

$\boldsymbol{W}_{\mathrm{BLR}}^{(D)} \leftarrow \boldsymbol{T}_{\mathrm{BLR}}^{(D)} \boldsymbol{D}_{\leftarrow M}^{(D)} \boldsymbol{C}_{0,-1} \qquad \langle DM(2M-1) + NM(2D-1) \rangle$

30: $\boldsymbol{a}_{\mathrm{BLR}}^{(D)} \leftarrow \boldsymbol{m}_x - \boldsymbol{W}_{\mathrm{BLR}}^{(D),\mathrm{H}} \boldsymbol{m}_y$

- The inversion of a Hermitian and positive definite $n \times n$ matrix requires $\zeta_{\mathrm{HI}}(n) = n^3 + n^2 + n$ FLOPs (cf. Equation B.7 of Appendix B.1).
- The Hermitian product of an $n \times m$ and $m \times n$ matrix requires $(2m-1)n(n+1)/2$ FLOPs since only its lower triangular part has to be computed.
- The product of an $n \times m$ matrix by an $m \times p$ matrix requires $np(2m-1)$ FLOPs and reduces to $(np+p(1-p)/2)(2m-1)$ FLOPs if a $p \times p$ submatrix of the result is Hermitian. Without loss of generality, we assumed for the latter case that $p < n$.

Algorithm 4.3. BCG implementation of the MSMWF

$$\boldsymbol{W}_0 \leftarrow \boldsymbol{0}_{N\times M}$$

2:   $\boldsymbol{R}_0 \leftarrow \boldsymbol{C}_{\boldsymbol{y},\boldsymbol{x}}$

    $\boldsymbol{G}_0 = \boldsymbol{G}'_0 \leftarrow \boldsymbol{R}_0^{\mathrm{H}} \boldsymbol{R}_0$ $\langle M(M+1)(2N-1)/2\rangle$

4:   $\boldsymbol{D}_0 = \boldsymbol{D}'_0 \leftarrow \boldsymbol{R}_0$

    for $\ell = 0, 1, \ldots, d-2 \wedge d > 1$ **do**

6:       $\boldsymbol{V} \leftarrow \boldsymbol{C}_{\boldsymbol{y}} \boldsymbol{D}_\ell$ $\langle MN(2N-1)\rangle$

      $\boldsymbol{\Phi}_\ell \leftarrow (\boldsymbol{D}_\ell^{\mathrm{H}} \boldsymbol{V})^{-1} \boldsymbol{G}_\ell$ $\langle \zeta_{\mathrm{MWF}}(M, M) + M(M+1)(2N-1)/2\rangle$

8:       $\boldsymbol{W}_{\ell+1} \leftarrow \boldsymbol{W}_\ell + \boldsymbol{D}_\ell \boldsymbol{\Phi}_\ell$ $\langle 2M^2 N\rangle$

      $\boldsymbol{R}_{\ell+1} \leftarrow \boldsymbol{R}_\ell - \boldsymbol{V} \boldsymbol{\Phi}_\ell$ $\langle 2M^2 N\rangle$

10:      **if** $\ell < d-2$ **then**

        $\boldsymbol{G}_{\ell+1} \leftarrow \boldsymbol{R}_{\ell+1}^{\mathrm{H}} \boldsymbol{R}_{\ell+1}$ $\langle M(M+1)(2N-1)/2\rangle$

12:         $\boldsymbol{\Theta}_\ell \leftarrow \boldsymbol{G}_\ell^{-1} \boldsymbol{G}_{\ell+1}$ $\langle \zeta_{\mathrm{MWF}}(M, M)\rangle$

        $\boldsymbol{D}_{\ell+1} \leftarrow \boldsymbol{R}_{\ell+1} + \boldsymbol{D}_\ell \boldsymbol{\Theta}_\ell$ $\langle 2M^2 N\rangle$

14:      **else**

        $\boldsymbol{R}'_{d-1} \leftarrow [\boldsymbol{R}_{d-1}]_{\cdot, 0:\mu-1}$

16:         $\boldsymbol{G}'_{d-1} \leftarrow \boldsymbol{R}_{d-1}^{\prime\,\mathrm{H}} \boldsymbol{R}'_{d-1}$ $\langle (\mu M + \mu(1-\mu)/2)(2N-1)\rangle$

        $\boldsymbol{\Theta}'_{d-2} \leftarrow \boldsymbol{G}_{d-2}^{\prime\,-1} \boldsymbol{G}'_{d-1}$ $\langle \zeta_{\mathrm{MWF}}(M, \mu)\rangle$

18:         $\boldsymbol{D}'_{d-1} \leftarrow \boldsymbol{R}'_{d-1} + \boldsymbol{D}_{d-2} \boldsymbol{\Theta}'_{d-2}$ $\langle 2\mu M N\rangle$

     end if

20: **end for**

    $\boldsymbol{\Phi}'_{d-1} \leftarrow (\boldsymbol{D}_{d-1}^{\prime\,\mathrm{H}} \boldsymbol{C}_{\boldsymbol{y}} \boldsymbol{D}'_{d-1})^{-1} \boldsymbol{G}_{d-1}^{\prime\,\mathrm{H}}$ $\langle \zeta_{\mathrm{MWF}}(\mu, M) + (\mu N + \mu(\mu+1)/2)(2N-1)\rangle$

22: $\boldsymbol{W}_{\mathrm{BCG}}^{(D)} \leftarrow \boldsymbol{W}_{d-1} + \boldsymbol{D}'_{d-1} \boldsymbol{\Phi}'_{d-1}$ $\langle 2\mu M N\rangle$

    $\boldsymbol{a}_{\mathrm{BCG}}^{(D)} \leftarrow \boldsymbol{m}_{\boldsymbol{x}} - \boldsymbol{W}_{\mathrm{BCG}}^{(D),\mathrm{H}} \boldsymbol{m}_{\boldsymbol{y}}$

- The addition of two Hermitian $n \times n$ matrices requires $n(n+1)/2$ FLOPs if we exploit again the fact that only the lower triangular part of the result has to be computed.

- The addition of two $n \times m$ matrices requires nm FLOPs and reduces to $nm + m(1-m)/2$ FLOPs if the result contains an $m \times m$ block which is Hermitian. Here, without loss of generality, $n > m$.

- Solving a system of linear equations with an $n \times n$ system matrix and m right-hand sides requires $\zeta_{\mathrm{MWF}}(n, m) = n^3/3 + (2m + 1/2)n^2 + n/6$ FLOPs (cf. Equation 2.28) if we use the Cholesky based approach from Subsection 2.2.1. Remember from Appendix B.1 that this method is cheaper than directly inverting the system matrix, i.e., $\zeta_{\mathrm{MWF}}(n, m) < \zeta_{\mathrm{HI}}(n) + mn(2n-1)$, if $m < 2n^2/3 + n/2 + 5/6$ which is implicitly fulfilled for $m \leq n$.

First, consider the BLR implementation of the MSMWF in Algorithm 4.2.. As in Section 2.2, the number in angle brackets are the FLOPs required to perform the corresponding line of Algorithm 4.2.. Note that in Lines 22 and 24, we exploit the Hermitian structure of the $M \times M$ submatrix in the first M rows of the matrix $\boldsymbol{D}_{\leftarrow M}^{(j)} + \boldsymbol{d}^{(j)} \bar{\boldsymbol{d}}_{\uparrow M}^{(j),\mathrm{H}}$, and the $(M-1) \times (M-1)$ submatrix in the last $M - 1$ rows of the matrix $\boldsymbol{D}_{\rightarrow M-1}^{(j)} + \boldsymbol{d}^{(j)} \bar{\boldsymbol{d}}_{\downarrow M-1}^{(j),\mathrm{H}}$, respectively. After summing up, the total number of FLOPs involved when applying Algorithm 4.2. computes as

$$\left| \begin{aligned} \zeta_{\mathrm{BLR}}(N, M, D) &= 2DN^2 + 3MD^2 - D^2 - MD + 2D \\ &\quad + \begin{cases} \left(4MD + D^2 + D - M\right) N, & D \le 2M, \\ \left(8MD - 4M^2 + 2D - 3M\right) N, & D > 2M. \end{cases} \end{aligned} \right| \quad (4.107)$$

The different cases occur due to the inner **for**-loop of Algorithm 4.2. which is performed j times for $j \le 2M$ and $2M$ times for $j > 2M$. Thus, we get a different dependency on N for $D > 2M$ compared to the case where $D \le 2M$.

Next, recall the BCG implementation of the MSMWF in Algorithm 4.3.. Again, each number in angle brackets denotes the FLOPs required to perform the corresponding line of the algorithm. In Lines 7, 12, and 17 of Algorithm 4.3., it is computationally cheaper to solve the corresponding system of linear equations based on the Cholesky method of Subsection 2.2.1 instead of explicitly inverting the system matrix because the size of the system matrix, i.e., M, is equal to or greater than the number of right-hand sides, i.e., either μ or M. Contrary to that, it would be computationally less expensive to perform Line 21 based on the HI method of Appendix B.1 in the special cases where $M > 2\mu^2/3 + \mu/2 + 5/6$. However, we apply the Cholesky based method for every choice of D because the reduction in computational complexity is negligible when using the HI approach in the seldom cases where the above inequality is fulfilled. Finally, the BCG implementation of the MSMWF requires $\zeta_{\mathrm{BCG}}(N, M, D)$ FLOPs where

$$\left| \begin{aligned} \zeta_{\mathrm{BCG}}(N, M, D) &= 2DN^2 + \left(8MD - 2M^2 - 2\mu M + D\right) N \\ &\quad + \frac{14}{3}M^2 D - 2M^3 - \frac{8}{3}\mu M^2 - \frac{2}{3}D + 2\mu^2 M \\ &\quad - \mu M + \frac{1}{3}\mu^3 + \frac{1}{2}\mu^2 - \frac{1}{6}\mu, \\ \mu &= D - (d-1)M, \quad d = \lceil D/M \rceil. \end{aligned} \right| \quad (4.108)$$

Table 4.1 summarizes the computational complexity of the presented time-variant MWF implementations per time index n where only the terms with the highest order with respect to N are given. Besides the number of FLOPs of the suboptimal approximations (cf. Equations 4.107 and 4.108), we added the computational complexity of the optimal Cholesky based implementation given in Equation (2.28) as well as the one of the RC implementation which we presented in Equation (2.41) of Section 2.2. Remember that S is the length of the signal to be estimated.

For the same reasons as already mentioned in Section 2.3, i.e., the poor performance in the application of Part II, we exclude a detailed investigation of the computational complexity when approximating the MWF in eigensubspaces. Nevertheless, remember that the PC based rank D MWF has a

Table 4.1. Computational complexity of different time-variant MWF implementations per time index n

	Implementation	Number of FLOPs (highest order term)
Optimal	MWF	$N^3/3$
	RCMWF	$6MN^2 + (N^3 + N^2)/S$
Suboptimal	BLR	$2DN^2$
for $D < N$	BCG	$2DN^2$

computational complexity of $O(DN^2)$, i. e., the same order as the one of the reduced-rank BLR or BCG implementation of the MSMWF, whereas the CS based reduced-rank MWF has the same order as the optimal Cholesky based approach. Further, the computational complexity of the original MSMWF implementation as given in Algorithm 4.1. is also not investigated here due to its strong relationship to the block Arnoldi algorithm (cf. Subsection 4.2.2). Recall that the block Arnoldi algorithm is computationally very inefficient because it does not exploit the Hermitian structure of the auto-covariance matrix.

In Table 4.1, we see that the BLR and BCG implementation of the MSMWF decreases the computational complexity by one order compared to the optimal MWF implementations if $N \to \infty$. However, for realistic values of N, the computational complexity is only reduced if the rank D is chosen sufficiently small as can be seen by the following investigations of the exact number of FLOPs. Note that we use here the same parameters as for the simulation of the multiuser system in Chapter 6.

Figure 4.12 depicts the N-dependency of the exact number of FLOPs required to perform the different MWF implementations for $M = 4$ right-hand sides and a block length of $S = 512$ symbols (cf. Chapter 5). We see that although the RCMWF implementation reduces the computational complexity from cubic to squared order for $N \to \infty$ by exploiting the structured time-dependency of the auto-covariance matrix, the exact number of FLOPs is only smaller than the one of the MWF implementation based on the Cholesky factorization for $N > 42$. Besides, we can observe that the BLR and the BCG method, which perform almost equally at the given set of parameters, beat the MWF already for $N > 15$ and $N > 14$, respectively, if the rank can be reduced to $D = 4$. For $D = 8$, N must be greater than 38 to achieve a decrease in computational complexity.

The dependency of the computational complexity on D is plotted in Figure 4.13 for $N = 50$ and $S = 512$. Here, the markers denote the values of D which can be accessed by the classical BCG and block Lanczos algorithm. However, due to the dimension-flexible extensions of the algorithms, they can be applied to any $D \in \{M, M+1, \ldots, N\}$. The resulting computational

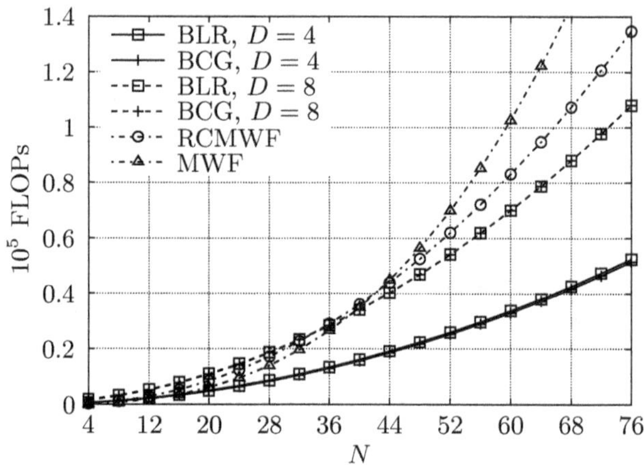

Fig. 4.12. Computational complexity versus N of different MWF implementations ($M = 4$, $S = 512$)

complexities are plotted as lines without markers. Again, we observe that the rank D has to be drastically reduced to achieve a reduction in computational complexity. For $M = 8$, D must be smaller than 12 when using the BLR algorithm or even smaller than 11 when using the BCG method. For $M = 4$, it must hold $D < 9$ to get a computationally efficient alternative to the MWF implementation. Note that the RCMWF is computationally cheaper than the MWF for $M = 4$ whereas the latter one is cheaper for $M = 8$. Again, this is due to the fact that the highest order term of the RCMWF's computational complexity in Table 4.1 has a higher prefactor. Besides, the RCMWF requires the explicit inversion of a matrix at the beginning of each block. It remains to explain why the increase in the number of FLOPs of the BCG based MSMWF implementation is especially high when increasing the rank D from an integer multiple of M by one. The reason is mainly Line 9 of Algorithm 4.3. which is performed $d - 1 = \lceil D/M \rceil - 1$ times. Thus, if d increases by one, a new residuum matrix has to be computed which is valid as long as d does not change.

It seems that the reduction in computational complexity depends strongly on the number M of right-hand sides, i. e., the number of columns of the cross-covariance matrix. Figure 4.14 depicts the computational complexity versus M for $N = 50$ and $S = 512$. First, we see that the BCG algorithm outperforms the BLR implementation of the MSMWF in terms of computational efficiency for higher M. We also observe that the Cholesky based MWF is the computationally cheapest solution for $M > 13$ or for $M > 5$ if we compare it with the block Krylov methods of rank $D = M$ and $D = 2M$, respectively. Note that the RCMWF implementation is never the computational cheapest solution for the given set of parameters.

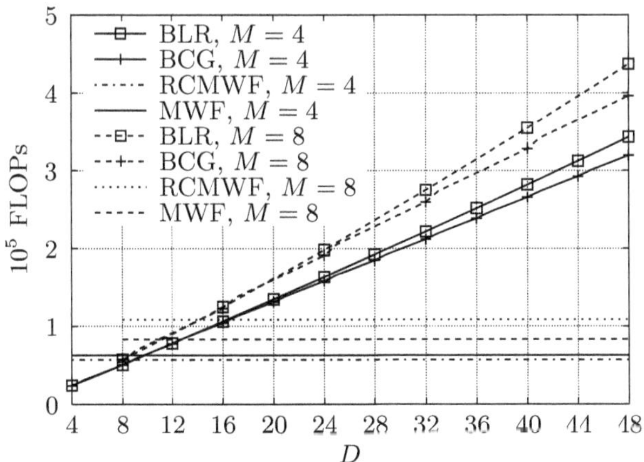

Fig. 4.13. Computational complexity versus D of different MWF implementations ($N = 50$, $S = 512$)

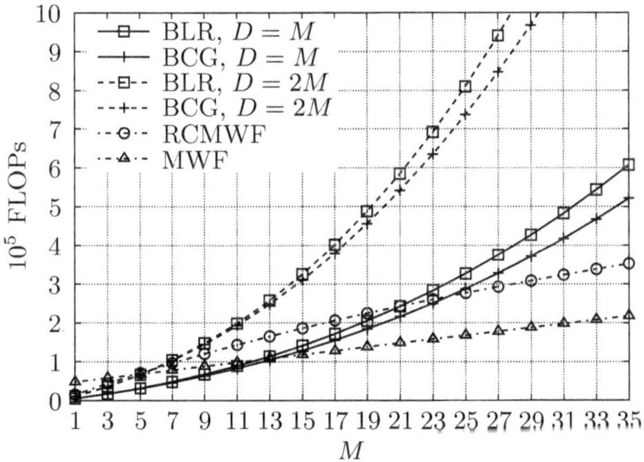

Fig. 4.14. Computational complexity versus M of different MWF implementations ($N = 50$, $S = 512$)

Up to now, we analyzed the dependency of the exact number of FLOPs on the given parameters and observed that the rank D of the MWF approximations in the Krylov subspace has to be drastically reduced in order to achieve a smaller computational complexity than the one of the optimal MWF implementations. However, to decide which algorithm is preferable compared to others, we must also investigate their performance which is done in Chapter 6 for the communication system introduced in Chapter 5.

Part II

Application: Iterative Multiuser Detection

5
System Model for Iterative Multiuser Detection

Although some fundamental ideas of *spread spectrum systems* arose earlier, *Direct Sequence* (DS) spread spectrum communications has been introduced in the 1950's (see, e. g., [216, 186]). Initially applied to military communications in order to preserve immunity and suppress jammers, DS spread spectrum techniques are currently used for multiplexing users in mobile communication systems, also known as *Code Division Multiple-Access* (CDMA). In [81], it was shown that CDMA based terrestrial cellular systems can achieve a higher capacity compared to competing digital techniques. The popularity of CDMA becomes apparent by its usage in many cellular standards like, e. g., *IS-95* and *cdma2000* in North America, or *Wideband CDMA* (WCDMA) in Europe and Japan (see, e. g., [191]). Note that a detailed historical overview over spread spectrum communications can be found in [216] and a tutorial of its theory is given in [186].

The first receiver dealing with multipath channels in a CDMA system, leading to both *Multiple-Access Interference* (MAI) and *InterSymbol Interference* (ISI), was the computationally inexpensive *Rake receiver* [192] which is based on a bank of single-user matched filters. However, Sergio Verdú showed in [246] that joint multiuser *Maximum Likelihood* (ML) *detection* can achieve much better results than the conventional Rake receiver. Unfortunately, its computational complexity increases exponentially in the number of users. A compromise between computational complexity and performance can be found by replacing the non linear ML approach with a linear *Minimum Mean Square Error* (MMSE) *receiver* or *Matrix Wiener Filter* (MWF) as reviewed in Section 2.1 (see, e. g., [247, 125]).

An even better system performance can be achieved by considering CDMA combined with channel coding [79, 80]. Instead of equalizing and decoding the received signal once in a row, the optimal receiver performs jointly equalization and decoding. However, this optimal receiver is computationally not feasible. More than 10 years ago, Catherine Douillard et al. [61] proposed a receiver structure performing equalization and decoding alternating in an iterative process in order to remove ISI caused by frequency-selective channels.

This so-called *Turbo equalizer* consists of an optimal *Maximum A Posteriori* (MAP) equalizer and a MAP decoder exchanging iteratively soft information about the coded data bits. Simulation results have shown that this procedure performs very close to the optimal receiver since it eliminates ISI after several iteration steps such that the *Bit Error Rate* (BER) of coded transmission over the corresponding *Additive White Gaussian Noise* (AWGN) channel can be achieved. Note that the proposed Turbo system can be interpreted as an iterative decoding scheme for serial concatenated codes where the inner code is the channel and the inner decoder is the equalizer (see, e. g., [106]). Claude Berrou et al. [12] developed this iterative decoding procedure originally for parallel concatenated codes. The first applications of the Turbo equalizer to coded CDMA systems can be found in [105, 166, 196].

Although the computational complexity of the iterative receiver is tremendously smaller than the one of the optimal receiver, the complexity is still very high for practical implementations. Again, researchers approximated the optimal non-linear MAP equalizer using an MMSE filter. Whereas Alain Glavieux et al. [82, 159] firstly used this idea to equalize a frequency-selective channel, Xiadong Wang and H. Vincent Poor [254, 198] applied it to suppress MAI and ISI in a coded CDMA system. In fact, they proposed *interference cancellation* [78, 170] for iterative processing of soft information and *parallel decoding* to detect jointly the data bits of all users. In the literature, this receiver is sometimes also denoted as the parallel *Decision Feedback Detector* (DFD) (see, e. g., [127]). Contrary to parallel decoding where all users are decoded before using the outputs as *a priori* information for the equalization at the next Turbo iteration, there exists the method of *successive decoding* [245] where each user is decoded separately and the output is immediately fed back to the equalizer in order to perform the next Turbo step. Thus, this so-called successive DFD needs several iterations before all users are decoded for the first time. However, its performance is better than the one of the parallel DFD if the order in which the users are decoded is chosen correctly (see, e. g., [263]). Applications of the DFD to coded CDMA systems can be found in [262, 126, 261, 123, 263, 122]. Note that early DFD publications [245, 262, 263] considered neither interleavers nor processing of soft information which are two essential prerequisites to achieve the characteristic Turbo gain.

The Turbo receiver which is presented in this chapter is neither based on interference cancellation nor does it involve a DFD. Instead, we follow the approach of Michael Tüchler et al. [240, 239], i. e., we use the available *a priori* information from the decoder to adopt the statistics needed to compute the linear MWF (cf. Section 2.1) which reequalizes the received signal sequence at the next Turbo iteration. Besides, we reveal the relationship between the resulting Turbo structure and the DFD or the iterative interference cancellation and decoding scheme (see also [50, 54, 55]). Finally, we apply the reduced-rank MWF approximations from Section 2.3 and Chapter 4 to the Turbo equalizer in order to further decrease its computational complexity (see also [49, 55, 14]). Note that the results are very related to the reduced-rank approximations presented by Michael L. Honig in [127] or to the more recently published ones in [63] despite of the fact that our Turbo receiver is based on the approach introduced by Michael Tüchler et al. [240, 239]. Besides reduced-rank algorithms as presented in this book, there exist several alternative methods which reduce the computational complexity of Turbo equalizers like, e. g., *channel-shortening* [8], equalizer design based on the *Fast Fourier Transform* (FFT) [158], or *sequential Monte Carlo sampling techniques* [60].

5.1 Transmitter Structure

Figure 5.1 depicts the transmitter structure of user $k \in \{0, 1, \ldots, K - 1\}$ for a coded *Direct Sequence Code Division Multiple-Access* (DS-CDMA) system with $K \in \mathbb{N}$ users. Note that the transmitter model is based on the receiver's point of view, i. e., all signals are assumed to be random since they are obviously not known at the receiver.

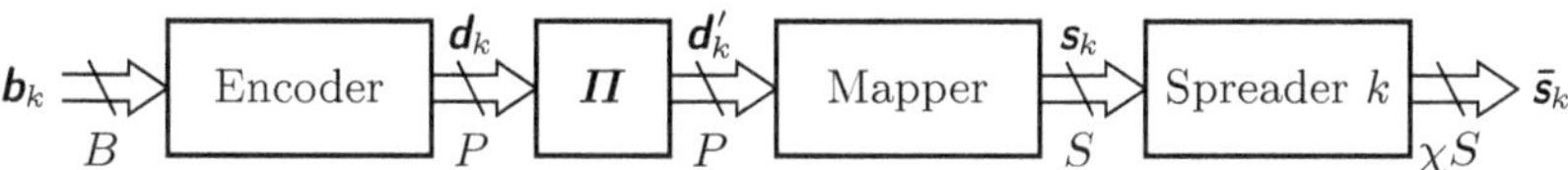

Fig. 5.1. Transmitter structure of user k for a coded DS-CDMA system

The binary data block $b_k \in \{0, 1\}^B$ composed of $B \in \mathbb{N}$ information bits of user k is encoded with code rate $r = B/P \in (0, 1]$ based on a *Forward Error Correction* (FEC) *code* (e. g., [163]). Afterwards, the block $d_k \in \{0, 1\}^P$, $P \in \mathbb{N}$, of coded data bits is interleaved using the permutation matrix $\Pi \in \{0, 1\}^{P \times P}$ and mapped to the symbol block $s_k \in \mathbb{M}^S$ of length $S \in \mathbb{N}$ using the complex symbol alphabet $\mathbb{M} \subset \mathbb{C}$ whose cardinality is 2^Q. Note that $Q \subset \mathbb{N}$ is the number of bits used to represent one symbol. Thus, P interleaved and coded data bits per block result in $S = P/Q = B/(rQ)$ symbols per block. Finally, the symbol block is spread with a code of length $\chi \in \mathbb{N}$ to get the chip block $\bar{s}_k \in \bar{\mathbb{M}}^{\chi S}$ of length χS which is transmitted over the channel as defined in Section 5.2. Note that the cardinality of the chip alphabet $\bar{\mathbb{M}}$ and the symbol alphabet $\mathbb{M}$ is the same. The following subsections describe each block of the diagram in Figure 5.1 in more detail.

5.1.1 Channel Coding

The *channel encoder* adds redundant information to the data bits in order to decrease transmission errors caused by the channel. We consider systems with FEC where, contrary to *Automatic Repeat reQuest* (ARQ) error control systems (see, e. g., [163]), errors which have been detected at the receiver can be automatically corrected without any retransmission of the information block. One possible way of constructing FEC codes is via *convolutional encoders* as introduced by Peter Elias [65] in 1955. Here, we restrict ourselves to *non-systematic*[1] *feedforward (non-recursive) convolutional encoders* (cf., e. g., [163]) with code rate

$$r = \frac{B}{P} = \frac{1}{p} \in (0, 1], \tag{5.1}$$

i. e., each data bit produces $p \in \mathbb{N}$ code bits.

In order to formulate the feedforward convolutional encoder based on convolution over the *binary Galois field $GF(2)$* as introduced in Appendix A.7, we represent the binary tuples $\boldsymbol{b}_k \in \{0, 1\}^B$ and $\boldsymbol{d}_k \in \{0, 1\}^P$, i. e., the input and output blocks of the channel encoder of user $k \in \{0, 1, \ldots, K - 1\}$, respectively, by random sequences over $GF(2)$,[2] i. e.,

$$b_k[n] = \boxplus_{i=0}^{B-1} b_{k,i} \boxdot \delta[n - i] \in \{0, 1\}, \tag{5.2}$$

$$d_k[n] = \boxplus_{i=0}^{P-1} d_{k,i} \boxdot \delta[n - i] \in \{0, 1\}, \quad n \in \mathbb{Z}, \tag{5.3}$$

where $b_{k,i} = \boldsymbol{e}_{i+1}^{\mathrm{T}} \boldsymbol{b}_k$ and $d_{k,i} = \boldsymbol{e}_{i+1}^{\mathrm{T}} \boldsymbol{d}_k$ are the $(i + 1)$th element of $\boldsymbol{b}_k$ and $\boldsymbol{d}_k$, respectively, and $\delta[n] = \delta_{n,n}$ is the *unit impulse*. Then, the feedforward convolutional encoder as depicted in Figure 5.2 is described by its *impulse response*, i. e., the vector-valued *generator sequence $\boldsymbol{p}[n] \in \{0, 1\}^p$* of length $m+1$. Note that m denotes the *order* or *memory* of the convolutional encoder. Besides, the generator sequence $\boldsymbol{p}[n]$ has no index k since we assume that the channel encoders are identical for all K users.

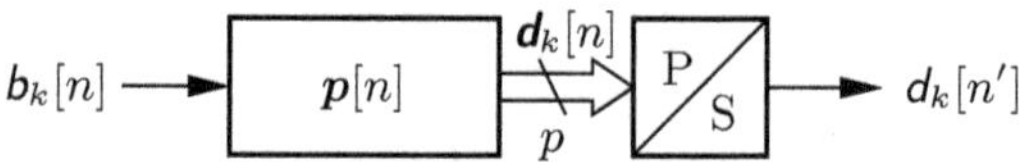

Fig. 5.2. Rate $r = 1/p$ feedforward convolutional encoder with generator sequence $\boldsymbol{p}[n]$

With the definition of the p-dimensional vector random sequence

---

[1] An encoder is denoted as *systematic* if all input sequences occur unchanged at the output of the encoder besides the additional redundant information. All encoders without this characteristic are denoted as *non-systematic*.

[2] Note that the convolution over $GF(2)$ can be also formulated if the binary tuples are represented by *algebraic polynomials*. Then, the rules known from *analytical polynomials* can be used to simplify calculations.

$$\boldsymbol{d}_k[n] = \begin{bmatrix} d_k[pn] \\ d_k[pn+1] \\ \vdots \\ d_k[p(n+1)-1] \end{bmatrix} \in \{0,1\}^p, \tag{5.4}$$

the output of the feedforward convolutional encoder in Figure 5.2 before *parallel-to-serial conversion* reads as

$$\boldsymbol{d}_k[n] = \underset{i=0}{\overset{m}{\boxplus}} \boldsymbol{p}[i] \boxdot b_k[n-i], \tag{5.5}$$

which is the convolution of the data bit sequence $b_k[n]$ by the generator sequence $\boldsymbol{p}[n]$ over $GF(2)$. Finally, the coded data bit sequence $d_k[n']$ is obtained by parallel-to-serial conversion of $\boldsymbol{d}_k[n]$, i.e.,

$$d_k[n'] = \boldsymbol{e}^{\mathrm{T}}_{n'-\lfloor \frac{n'}{n} \rfloor n+1} \boldsymbol{d}_k[n]. \tag{5.6}$$

If we define the *generator matrix*

$$\boldsymbol{P} = \begin{bmatrix} \boldsymbol{p}[0] \ \boldsymbol{p}[1] \ \cdots \ \boldsymbol{p}[m] \end{bmatrix} \in \{0,1\}^{p \times (m+1)}, \tag{5.7}$$

and the *state*

$$\boldsymbol{b}_k[n] = \begin{bmatrix} b_k[n] \\ b_k[n-1] \\ \vdots \\ b_k[n-m] \end{bmatrix} \in \{0,1\}^{m+1}, \tag{5.8}$$

of the convolutional encoder, Equation (5.5) can be also written in matrix-vector notation as

$$\boldsymbol{d}_k[n] = \boldsymbol{P} \boxdot \boldsymbol{b}_k[n]. \tag{5.9}$$

In the sequel of this subsection, we present an example of a non-systematic feedforward convolutional encoder which we will use in the simulations of Chapter 6.

Example 5.1 ((7,5)-convolutional encoder with rate $r = 1/2$ and memory $m = 2$). The (7,5) convolutional encoder with memory $m = 2$ and rate $r = 1/2$, i.e., each data bit generates $p = 2$ code bits, is defined via its generator sequence

$$\boldsymbol{p}[n] = \begin{bmatrix} 1 \\ 1 \end{bmatrix} \boxdot \delta[n] \boxplus \begin{bmatrix} 1 \\ 0 \end{bmatrix} \boxdot \delta[n-1] \boxplus \begin{bmatrix} 1 \\ 1 \end{bmatrix} \boxdot \delta[n-2] \in \{0,1\}^2, \tag{5.10}$$

or its generator matrix

$$\boldsymbol{P} = \begin{bmatrix} 1 & 1 & 1 \\ 1 & 0 & 1 \end{bmatrix} \in \{0,1\}^{2 \times 3}, \tag{5.11}$$

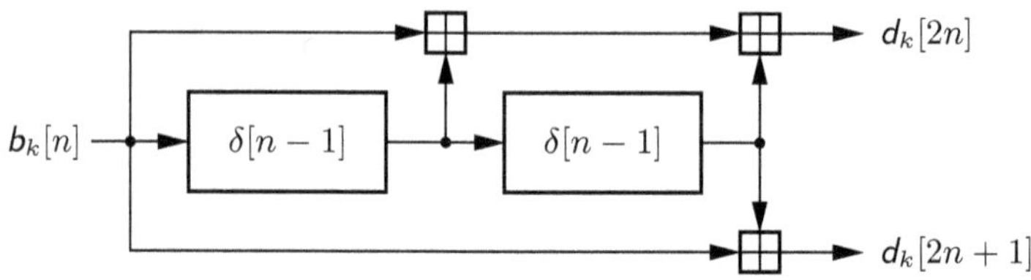

Fig. 5.3. $(7,5)$-convolutional encoder with rate $r = 1/2$ and memory $m = 2$

respectively. The resulting *Finite Impulse Response* (FIR) structure before parallel-to-serial conversion is shown in Figure 5.3.

Note that the term '$(7,5)$' denotes where the taps in the FIR structure of Figure 5.3 are placed (cf., e. g., [163]). More precisely speaking, the number '7' is the octal representation of the binary word interpretation of the first row of the generator matrix $\boldsymbol{P}$, i. e., $111_2 = 7_8$, which denotes that $d_k[2n]$ is the modulo 2 sum of $b_k[n]$, $b_k[n-1]$, and $b_k[n-2]$, and the number '5' the octal representation of the binary word in the second row of $\boldsymbol{P}$, i. e., $101_2 = 5_8$, which denotes that $d_k[2n+1]$ is the modulo 2 sum of $b_k[n]$ and $b_k[n-2]$.

It remains to mention that the *Bahl–Cocke–Jelinek–Raviv* (BCJR) *algorithm*[3] [7] which we use to decode the convolutional coded data bits in the iterative receiver of Section 5.3 (cf. especially Subsection 5.3.3), requires defined initial and ending states, e. g., the zero vector. Unfortunately, this termination procedure reduces the number of processed data bits for each block resulting in a rate loss of the system as described in, e. g., [163].

5.1.2 Interleaving

The main task of the *interleaver* is to resolve, at least locally, statistical dependencies between coded data bits. Thus, interleaving is essential for the performance of the Turbo principle (e. g., [106]). Additionally, the interleaver reduces error bursts caused by bad channel realizations. Note that interleaving can be also seen as a way of constructing a very powerful encoder if we consider the whole transmission chain, viz., encoder, interleaver, mapper, spreader, and channel, as one channel encoder for an *Additive White Gaussian Noise* (AWGN) channel. Here, we discuss only *bitwise interleaving* although early publications (e. g., [82, 159]) of Turbo equalizers with modulation alphabets of a higher order than the one of *Binary Phase Shift Keying* (BPSK) have been based on *symbolwise interleaving*. Comparisons between both interleaving techniques can be found in [41].

The interleaver can be described by an invertible *permutation matrix* $\boldsymbol{\Pi} \in \{0, 1\}^{P \times P}$ with one '1' in each row and each column. Applied to the coded

---

[3] The BCJR algorithm is named after its inventors, viz., Lalit R. Bahl, John Cocke, Frederick Jelinek, and Josef Raviv.

data block $\boldsymbol{d}_k \in \{0,1\}^P$ of user k, the output of the interleaver reads as

$$\boxed{\boldsymbol{d}'_k = \boldsymbol{\Pi}\boldsymbol{d}_k \in \{0,1\}^P.} \tag{5.12}$$

Finding the optimal interleaver for a system with an iterative receiver (cf. Section 5.3) is nearly impossible because it depends on too many parameters. However, simulations have shown (see, e.g., [113]) that suboptimal solutions achieve good performances in terms of *Bit Error Rate* (BER). One *ad hoc* approach is the so-called *block interleaver* with a permutation matrix where each '1' has the same distance to the next one. In other words, after interleaving, adjacent coded data bits have a constant distance to each other where the distance between the ith and jth element of a vector is defined as $|i - j|$. Whereas this kind of interleaving may be adequate for Turbo decoding, the BER performance of a system with Turbo equalization does not improve over several iteration steps.

A better choice for Turbo equalization is the *random interleaver* where the permutation matrix is randomly occupied. Despite of its simplicity, this interleaving scheme yields good results as shown in [113]. Unfortunately, the random interleaver has the drawback that symbols can stay close to each other which breaks the statistical independence between the interleaved and coded data bits. Therefore, researchers suggested the *S-random interleaver* ensuring a minimal distance between the symbols which is equivalent to a permutation matrix with a minimal distance between the '1'-entries. The S-random interleaver can be seen as a random interleaver with the constraint that the minimal distance between adjacent coded data bits is guaranteed. A more detailed explanation of all mentioned interleavers can be found in [113].

5.1.3 Symbol Mapping

This subsection is to describe the mapping of the interleaved and coded data block $\boldsymbol{d}'_k \in \{0,1\}^P$ of user $k \in \{0,1,\ldots,K-1\}$ to its symbol block $\boldsymbol{s}_k \in \mathbb{M}^S$ (cf. Figure 5.1). Again, $\mathbb{M} \subset \mathbb{C}$ is the modulation alphabet with cardinality 2^Q where Q is the number of bits used to represent one symbol. Thus, the length of each symbol block is $S = P/Q$. Since $S \in \mathbb{N}$, we must restrict P to be a multiple integer of Q. Further, we assume that the average of all elements in $\mathbb{M}$ is zero, i.e.,

$$\frac{1}{2^Q} \sum_{s \in \mathbb{M}} s = 0, \tag{5.13}$$

and that the average of the squared elements is ϱ_s, i.e.,

$$\frac{1}{2^Q} \sum_{s \in \mathbb{M}} |s|^2 = \varrho_s. \tag{5.14}$$

Then, the *symbol mapper M* is defined via the bijective function

$$\left| M : \{0,1\}^Q \rightarrow \mathbb{M}, \; \boldsymbol{d}'_{k,n} \mapsto s_{k,n} = M\left(\boldsymbol{d}'_{k,n}\right), \right| \qquad (5.15)$$

mapping the Q bits $\boldsymbol{d}'_{k,n} = \boldsymbol{S}_{(nQ,Q,P-Q)}\boldsymbol{d}'_k \in \{0,1\}^Q$, $n \in \{0,1,\dots,S-1\}$, i.e., the Q adjacent bits in $\boldsymbol{d}'_k$ beginning from its $(nQ+1)$th element, to the $(n+1)$th symbol $s_{k,n} = \boldsymbol{e}^{\mathrm{T}}_{n+1}\boldsymbol{s}_k \in \mathbb{M}$ of $\boldsymbol{s}_k$. Remember that $\boldsymbol{S}_{(nQ,Q,P-Q)} = [\boldsymbol{0}_{Q \times nQ}, \boldsymbol{I}_Q, \boldsymbol{0}_{Q \times (P-(n+1)Q)}] \in \{0,1\}^{Q \times P}$ is the selection matrix as defined in Section 1.2. In Example 5.2, we present the symbol mapper for *QuadriPhase Shift Keying* (QPSK) *modulation* based on the *Gray code* and with $\varrho_s = 1$ (see, e.g., [193]) which we will use for modulation in the sequel of this book. If the reader is interested in iterative receivers (cf. Section 5.3) for other modulation alphabets like, e.g., BPSK where $Q = 1$, *M-ary Quadrature Amplitude Modulation* (*M*-QAM), or *M-ary Phase Shift Keying* (*M*-PSK), we refer to [61, 82, 159, 240, 41].

Example 5.2 (QPSK symbol mapper based on the Gray code and with $\varrho_s = 1$).

For QPSK symbol mapping where $Q = 2$, the modulation alphabet with cardinality 4 reads as (cf. also Figure 5.4)

$$\mathbb{M} = \left\{ \frac{1}{\sqrt{2}} + \frac{j}{\sqrt{2}}, \frac{1}{\sqrt{2}} - \frac{j}{\sqrt{2}}, -\frac{1}{\sqrt{2}} + \frac{j}{\sqrt{2}}, -\frac{1}{\sqrt{2}} - \frac{j}{\sqrt{2}} \right\}, \qquad (5.16)$$

if we assume $\varrho_s = 1$ (cf. Equation 5.14).

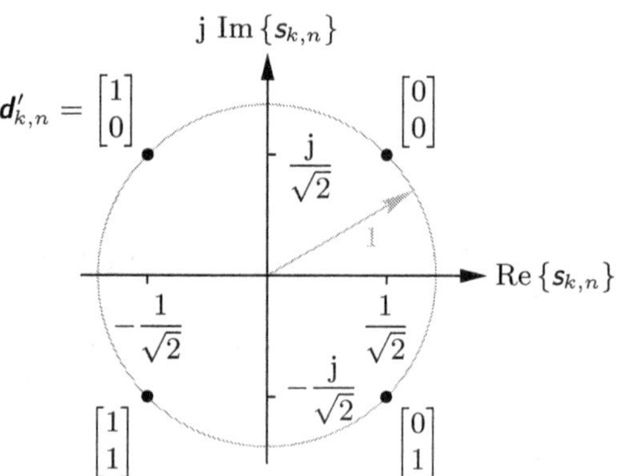

Fig. 5.4. QPSK symbols with $\varrho_s = 1$

The QPSK symbol mapper M_{G} based on the *Gray code* is defined via the mapping rules in Table 5.1 (e.g., [193]). As can be seen in Figure 5.4, one characteristic of the Gray code is that the bit tuples corresponding to nearby symbols differ only in one binary digit.

Table 5.1. QPSK symbol mapper based on the Gray code and with $\varrho_s = 1$

$d'_{k,n}$	$[0,0]^{\mathrm{T}}$	$[0,1]^{\mathrm{T}}$	$[1,0]^{\mathrm{T}}$	$[1,1]^{\mathrm{T}}$
$s_{k,n} = M_{\mathrm{G}}(d'_{k,n})$	$\dfrac{1}{\sqrt{2}} + \dfrac{j}{\sqrt{2}}$	$\dfrac{1}{\sqrt{2}} - \dfrac{j}{\sqrt{2}}$	$-\dfrac{1}{\sqrt{2}} + \dfrac{j}{\sqrt{2}}$	$-\dfrac{1}{\sqrt{2}} - \dfrac{j}{\sqrt{2}}$

Note that Gray mapping for a modulation alphabet with 2^Q symbols is one of $2^Q!$, i.e., $4! = 24$ in the case of QPSK modulation, possible symbol mappings. In fact, non-Gray mappings in combination with the *soft symbol demapping* of Subsection 5.3.2 which exploits *a priori* information available from the decoder in the Turbo process (cf. Section 5.3) can outperform the Gray code (see, e. g., [22, 21, 41]). Examples of these mappings are based on *set partitioning* [243], the *effective free distance criterion* [95], or the optimization as proposed in [217]. In single-user systems with frequency-flat channels, the gain of performing Turbo iterations is even zero when using the Gray code (e. g., [41]). The reason for the superiority of non-Gray mappings in Turbo systems is as follows: If we have full *a priori* information over all bits except from the one which has to be determined, we get the lowest BER if the symbols which differ only in this remaining bit have the largest distance in the complex plane. Clearly, this is contradictory to the Gray code where symbols which differ only in one bit are closest to each other (cf. Figure 5.4). Thus, if we choose non-Gray mappings instead of the Gray code, we get a performance improvement if we have full *a priori* knowledge. However, if there is no *a priori* information available, i. e., before the first iteration of the Turbo process, the system performance degrades due to the fact that Gray mapping is still optimal in this case. Hence, there is a fixed number of Turbo iterations where the system with non-Gray mapping beats the one with the Gray code. Compared to the single-user case with the frequency-flat channel where this number is one, systems with frequency-selective channels need generally more than one Turbo iteration such that non-Gray mappings outperform the Gray code (cf., e. g., [41]). Clearly, the more Turbo iterations the higher is the computational complexity of the receiver. Since we are finally interested in computationally cheap solutions, we restrict ourselves to the Gray mapping as introduced in Example 5.2 for QPSK modulation. Note that another drawback of using non-Gray mappings in Turbo systems is the increase of the *Signal to Noise Ratio* (SNR) *threshold*, i. e., the SNR value of the waterfall region.

5.1.4 Spreading

In order to separate the users in a DS-CDMA system (e. g., [216, 186, 111]), their symbol streams are multiplied by orthogonal code sequences. Each code sequence consists of $\chi \in \mathbb{N}$ so-called *chips* which can be either $+1$ or -1.

The rate $1/T_c$ of the resulting chip sequences is by a factor of χ larger than the rate $1/T_s$ of the symbol sequences, i.e., $T_s = \chi T_c$. Since this results in a spreading of the signal spectra, DS-CDMA is a *spread spectrum technique*. Here, $T_c \in \mathbb{R}_+$ denotes the duration of one chip, $T_s \in \mathbb{R}_+$ the duration of one symbol, and χ the *spreading factor*.

Mathematically speaking, with the code vector $\boldsymbol{c}_k \in \{-1,+1\}^\chi$ of user $k \in \{0,1,\ldots,K-1\}$, which is orthogonal to the code vectors of all other users, i.e., $\boldsymbol{c}_k^{\mathrm{T}}\boldsymbol{c}_i = \delta_{k,i}$, $i \in \{0,1,\ldots,K-1\}$, the symbol block of user k is *spread* according to

$$\boxed{\bar{\boldsymbol{s}}_k = (\boldsymbol{I}_S \otimes \boldsymbol{c}_k)\,\boldsymbol{s}_k \in \mathbb{M}^{\chi S},} \tag{5.17}$$

where '$\otimes$' denotes the Kronecker product and $\delta_{k,i}$ the Kronecker delta as defined in Chapter 2. Here, we use *Orthogonal Variable Spreading Factor* (OVSF) *codes* (e.g., [111]) based on a code tree generated according to Figure 5.5(a). Figure 5.5(b) depicts exemplarily the resulting code tree with the orthogonal codes $\boldsymbol{c}_k \in \{-1,+1\}^4$, $k \in \{0,1,2,3\}$, for $\chi = 4$. Note that two OVSF codes are orthogonal if and only if neither of them lies on the path from the other code to the root. Unfortunately, these codes do not have good auto-correlation or cross-correlation properties leading to severe *InterSymbol Interference* (ISI) and *Multiple-Access Interference* (MAI) in frequency-selective channels (see, e.g., [111]).

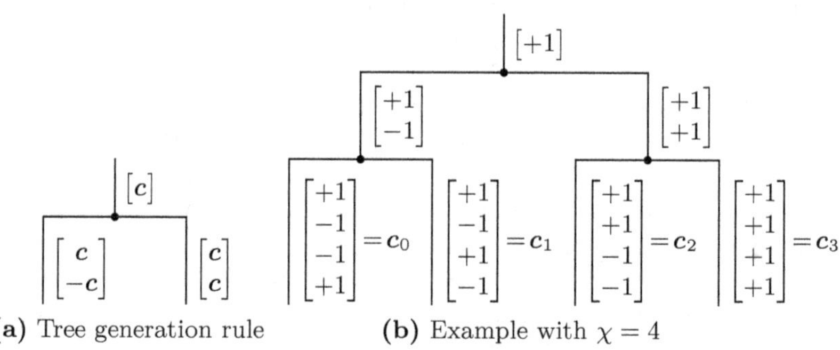

(a) Tree generation rule   (b) Example with $\chi = 4$

Fig. 5.5. Orthogonal Variable Spreading Factor (OVSF) code tree

For the derivation of the equalizer in Subsection 5.3.1, we need a convolutional matrix-vector model of the spreader. First, we define the sequences $\boldsymbol{s}[n] \in \mathbb{M}^K$, $\bar{\boldsymbol{s}}[n] \in \bar{\mathbb{M}}^K$, and $\boldsymbol{C}[n] \in \{-1,0,+1\}^{K \times K}$ such that they comprise the symbols, chips, or codes of all users, i.e.,

$$\boldsymbol{s}[n] = \begin{bmatrix} s_0[n] \\ s_1[n] \\ \vdots \\ s_{K-1}[n] \end{bmatrix} = \sum_{i=0}^{S-1} \begin{bmatrix} s_{0,i} \\ s_{1,i} \\ \vdots \\ s_{K-1,i} \end{bmatrix} \delta[n-i], \tag{5.18}$$

$$\bar{\boldsymbol{s}}[\bar{n}] = \begin{bmatrix} \bar{s}_0[\bar{n}] \\ \bar{s}_1[\bar{n}] \\ \vdots \\ \bar{s}_{K-1}[\bar{n}] \end{bmatrix} = \sum_{i=0}^{\chi S-1} \begin{bmatrix} \bar{s}_{0,i} \\ \bar{s}_{1,i} \\ \vdots \\ \bar{s}_{K-1,i} \end{bmatrix} \delta[\bar{n}-i], \tag{5.19}$$

$$\boldsymbol{C}[n] = \sum_{i=0}^{\chi-1} \boldsymbol{C}_i \delta[n-i], \quad \boldsymbol{C}_i = \operatorname{diag}\left\{c_{0,i}, c_{1,i}, \ldots, c_{K-1,i}\right\}, \tag{5.20}$$

where $s_{k,i} = \boldsymbol{e}_{i+1}^{\mathrm{T}} \boldsymbol{s}_k \in \mathbb{M}$, $\bar{s}_{k,i} = \boldsymbol{e}_{i+1}^{\mathrm{T}} \bar{\boldsymbol{s}}_k \in \bar{\mathbb{M}}$, and $c_{k,i} = \boldsymbol{e}_{i+1}^{\mathrm{T}} \boldsymbol{c}_k \in \{-1,+1\}$ are the $(i+1)$th element of the vectors $\boldsymbol{s}_k$, $\bar{\boldsymbol{s}}_k$, and $\boldsymbol{c}_k$, respectively. Then, the spreading of all users is represented by the block diagram in Figure 5.6.

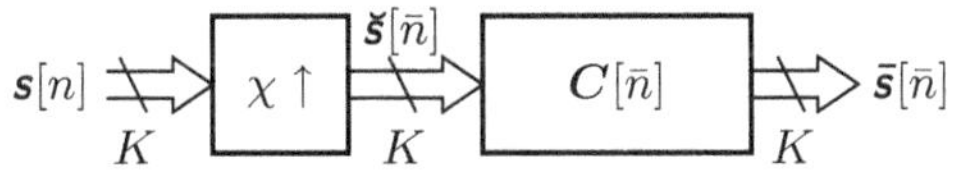

Fig. 5.6. Spreading for DS-CDMA

The first block in Figure 5.6 symbolizes *zero padding*, i.e., the insertion of zeros into the symbol sequence $\boldsymbol{s}[n]$ such that the intermediate sequence $\check{\boldsymbol{s}}[\bar{n}] \in \mathbb{M}^K$ reads as

$$\check{\boldsymbol{s}}[\bar{n}] = \sum_{n=-\infty}^{\infty} \boldsymbol{s}[n]\delta[\bar{n} - \chi n]. \tag{5.21}$$

Note that the different time indices, i.e., n before and $\bar{n}$ after zero padding, shall indicate the different rates of the sequences. In other words, the sequence $\check{\boldsymbol{s}}[\bar{n}]$ is sampled with chip rate $1/T_\mathrm{c}$ whereas $\boldsymbol{s}[n]$ is sampled with the symbol rate $1/T_\mathrm{s} = 1/(\chi T_\mathrm{c})$. Then, the convolution of $\check{\boldsymbol{s}}[\bar{n}]$ with the code sequence $\boldsymbol{C}[\bar{n}]$ yields the spread symbol sequence

$$\bar{\boldsymbol{s}}[\bar{n}] = \boldsymbol{C}[\bar{n}] * \check{\boldsymbol{s}}[\bar{n}] = \sum_{i=0}^{\chi-1} \boldsymbol{C}[i]\check{\boldsymbol{s}}[\bar{n} - i]. \tag{5.22}$$

Note that Equations (5.21) and (5.22) with the definitions of the sequences in Equations (5.18) to (5.20) is an alternative spreading representation compared to the block notation in Equation (5.17).

To get finally the convolutional matrix-vector model of the spreading procedure, we define the vectors[4]

---

[4] The index 'c' as an abbreviation for "composite" indicates that the corresponding vector is composed by adjacent samples of a vector sequence.

$$\bar{s}_{\mathrm{c}}[\bar{n}] = \begin{bmatrix} \bar{s}[\bar{n}] \\ \bar{s}[\bar{n} - 1] \\ \vdots \\ \bar{s}[\bar{n} - G + 1] \end{bmatrix} \in \bar{\mathbb{M}}^{KG}, \quad s_{\mathrm{c}}[n] = \begin{bmatrix} s[n] \\ s[n - 1] \\ \vdots \\ s[n - g + 1] \end{bmatrix} \in \mathbb{M}^{Kg}, \quad (5.23)$$

where $g = \lceil (G + \chi - 1)/\chi \rceil$ is the maximal number of samples which have an influence on the entries in $\bar{s}_{\mathrm{c}}[\bar{n}]$. Eventually, the spreading of all users for the samples needed in Subsection 5.3.1, i. e., for $\bar{n} = \chi n$, is given by

$$\boxed{\bar{s}_{\mathrm{c}}[\chi n] = C s_{\mathrm{c}}[n],} \qquad (5.24)$$

if we define additionally the matrix

$$\left| C = S_{(K(\chi-1),KG,K(\chi g-G))} \left(I_g \otimes \begin{bmatrix} C_{\chi-1} \\ C_{\chi-2} \\ \vdots \\ C_0 \end{bmatrix} \right) \in \{-1, 0, +1\}^{KG \times Kg}, \right| \quad (5.25)$$

which is the convolutional matrix according to Equation (5.22) where the block columns of width K which belong to the M-dimensional zero vectors in the intermediate sequence $\check{s}[\bar{n}]$ (cf. Equation 5.21) have been removed. Here, the selection matrix $S_{(K(\chi-1),KG,K(\chi g-G))} = [\mathbf{0}_{KG \times K(\chi-1)}, I_{KG}, \mathbf{0}_{KG \times K(\chi(g-1)-G+1))}] \in \{0, 1\}^{KG \times K\chi g}$.

5.2 Channel Model

In this section, we introduce a time-dispersive *Multiple-Access* (MA) *Single-Input Multiple-Output* (SIMO) channel as depicted in Figure 5.7. This is a common channel model for the *uplink* of a multiuser system as the one simulated in Chapter 6.

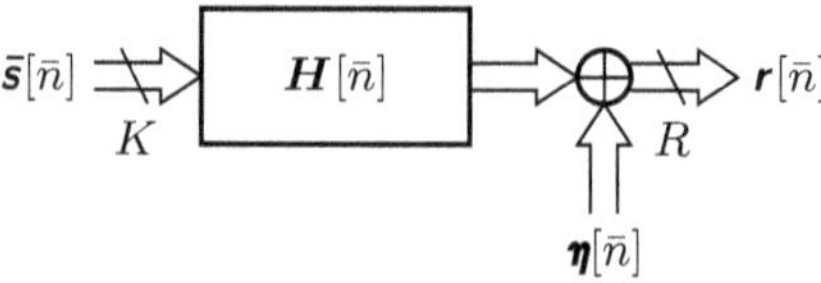

Fig. 5.7. Time-dispersive Multiple-Access (MA) Single-Input Multiple-Output (SIMO) channel

The spread symbol sequence $\bar{s}[\bar{n}]$ which comprises the signals of all users (see Equation 5.19) is transmitted over the frequency-selective multipath MA-SIMO channel with the impulse response

$$H[\bar{n}] = \sum_{\ell=0}^{L-1} H_\ell \delta[\bar{n} - \ell] \in \mathbb{C}^{R \times K}, \tag{5.26}$$

of length L and the coefficients $H_\ell \in \mathbb{C}^{R \times K}$, $\ell \in \{0, 1, \dots, L-1\}$. When receiving the distorted sequence by R antennas, it is perturbed by stationary AWGN $\boldsymbol{\eta}[\bar{n}] \in \mathbb{C}^R$ with the complex normal distribution $\mathcal{N}_{\mathbb{C}}(\mathbf{0}_R, C_{\boldsymbol{\eta}})$,[5] i.e., $\boldsymbol{\eta}[\bar{n}]$ is zero-mean and its auto-covariance matrix reads as $C_{\boldsymbol{\eta}} \in \mathbb{C}^{R \times R}$. The channel impulse response is assumed to be constant during one block, i.e., for the transmission of S symbols, but varies from block to block. The received signal sequence $\boldsymbol{r}[\bar{n}] \in \mathbb{C}^R$ can be finally written as

$$\boldsymbol{r}[\bar{n}] = H[\bar{n}] * \bar{\boldsymbol{s}}[\bar{n}] + \boldsymbol{\eta}[\bar{n}] = \sum_{\ell=0}^{L-1} H_\ell \bar{\boldsymbol{s}}[\bar{n} - \ell] + \boldsymbol{\eta}[\bar{n}]. \tag{5.27}$$

Again, for the derivation of the optimal linear equalizer filter with F taps in Subsection 5.3.1, we introduce the matrix-vector model

$$\boxed{\boldsymbol{r}_{\mathrm{c}}[\bar{n}] = H \bar{\boldsymbol{s}}_{\mathrm{c}}[\bar{n}] + \boldsymbol{\eta}_{\mathrm{c}}[\bar{n}],} \tag{5.28}$$

with the observation vector $\boldsymbol{r}_{\mathrm{c}}[\bar{n}] \in \mathbb{C}^{RF}$ and the noise vector $\boldsymbol{\eta}_{\mathrm{c}}[\bar{n}] \in \mathbb{C}^{RF}$ defined as

$$\boldsymbol{r}_{\mathrm{c}}[\bar{n}] = \begin{bmatrix} \boldsymbol{r}[\bar{n}] \\ \boldsymbol{r}[\bar{n} - 1] \\ \vdots \\ \boldsymbol{r}[\bar{n} - F + 1] \end{bmatrix}, \quad \boldsymbol{\eta}_{\mathrm{c}}[\bar{n}] = \begin{bmatrix} \boldsymbol{\eta}_{\mathrm{c}}[\bar{n}] \\ \boldsymbol{\eta}_{\mathrm{c}}[\bar{n} - 1] \\ \vdots \\ \boldsymbol{\eta}_{\mathrm{c}}[\bar{n} - F + 1] \end{bmatrix}, \tag{5.29}$$

the spread symbol vector $\bar{\boldsymbol{s}}_{\mathrm{c}}[\bar{n}] \in \mathbb{C}^{K(F+L-1)}$ of Equation (5.23) where $G = F + L - 1$, and the time-invariant channel convolutional matrix $H \in \mathbb{C}^{RF \times K(F+L-1)}$ defined as

$$\boxed{H = \sum_{\ell=0}^{L-1} S_{(\ell, F, L-1)} \otimes H_\ell.} \tag{5.30}$$

Since $G = F + L - 1$, the spreader as described by Equation (5.24) and (5.25) for $\bar{n} = \chi n$ and the channel in Equations (5.28) and (5.30) can be combined such that we get the transmission over the composite channel

$$\boxed{\boldsymbol{r}_{\mathrm{c}}[\chi n] = HC \boldsymbol{s}_{\mathrm{c}}[n] + \boldsymbol{\eta}_{\mathrm{c}}[\chi n].} \tag{5.31}$$

---

[5] The probability density function of the complex multivariate normal distribution $\mathcal{N}_{\mathbb{C}}(\boldsymbol{m}_{\boldsymbol{x}}, C_{\boldsymbol{x}})$ with mean $\boldsymbol{m}_{\boldsymbol{x}}$ and the auto-covariance matrix $C_{\boldsymbol{x}}$ reads as $\mathrm{p}_{\boldsymbol{x}}(\boldsymbol{x}) = \exp(-(\boldsymbol{x} - \boldsymbol{m}_{\boldsymbol{x}})^{\mathrm{H}} C_{\boldsymbol{x}}^{-1}(\boldsymbol{x} - \boldsymbol{m}_{\boldsymbol{x}}))/(\pi^n \det(C_{\boldsymbol{x}}))$ if the random variable $\boldsymbol{x}$ has n entries (e.g., [183]).

Note that with $G = F + L - 1$, it holds: $g = \lceil (F + L + \chi - 2)/\chi \rceil$. Throughout this book, we assume that the transmitted symbols are uncorrelated to the noise, i.e., $\mathrm{E}\{\boldsymbol{s}_{\mathrm{c}}^{\mathrm{H}}[n]\boldsymbol{\eta}_{\mathrm{c}}[\chi n]\} = \boldsymbol{0}_{Kg \times RF}$, and that the noise $\boldsymbol{\eta}_{\mathrm{c}}[\chi n]$ is stationary and zero-mean, i.e., $\boldsymbol{C}_{\boldsymbol{\eta}_{\mathrm{c}}}[\chi n] = \boldsymbol{C}_{\boldsymbol{\eta}_{\mathrm{c}}}$ and $\boldsymbol{m}_{\boldsymbol{\eta}_{\mathrm{c}}}[\chi n] = \boldsymbol{0}_{RF}$, respectively. Besides, the auto-covariance matrix of $\boldsymbol{s}_{\mathrm{c}}[n]$ is diagonal, i.e., $\boldsymbol{C}_{\boldsymbol{s}_{\mathrm{c}}}[n] = \mathrm{bdiag}\{\boldsymbol{C}_{\boldsymbol{s}}[n], \boldsymbol{C}_{\boldsymbol{s}}[n-1], \ldots, \boldsymbol{C}_{\boldsymbol{s}}[n-g+1]\} \in \mathbb{R}_{0,+}^{Kg \times Kg}$ with $\boldsymbol{C}_{\boldsymbol{s}}[n] = \mathrm{diag}\{c_{s_0}[n], c_{s_1}[n], \ldots, c_{s_{K-1}}[n]\} \in \mathbb{R}_{0,+}^{K \times K}$ (cf. Equations 5.23 and 5.18) because the symbols of each user are temporally uncorrelated due to the interleaver and the symbols of different users are uncorrelated anyway.

Example 5.3 (Uncorrelated channel coefficients with exponential or uniform power delay profile). If the channel coefficients $\boldsymbol{H}_\ell \in \mathbb{C}^{R \times K}$ are spatially and temporally uncorrelated, they are realization of the random variables $\boldsymbol{H}_\ell \in \mathbb{C}^{R \times K}$, $\ell \in \{0, 1, \ldots, L-1\}$, where all entries are statistically independent and distributed according to the complex normal distributions $\mathcal{N}_{\mathbb{C}}(0, c_{h,\ell})$. In the case of an *exponential power delay profile*, we choose the variance $c_{h,\ell}$ of the $(\ell + 1)$th channel tap entries as

$$c_{h,\ell} = \frac{\exp(-\ell)}{\sum_{i=0}^{L-1} \exp(-i)}, \tag{5.32}$$

and in the case of a *uniform power delay profile*, we set

$$c_{h,\ell} = 1/L, \quad \forall \ell \in \{0, 1, \ldots, L-1\}. \tag{5.33}$$

In any case, $\sum_{\ell=0}^{L-1} c_{h,\ell} = 1$.

5.3 Iterative Receiver Structure

The aim of the iterative multiuser receiver as depicted in Figure 5.8 is to process the received signal block $\boldsymbol{r} \in \mathbb{C}^{R(\chi S + L - 1)}$ such that each of its outputs $\tilde{\boldsymbol{b}}_k \in \{0, 1\}^B$, $k \in \{0, 1, \ldots, K-1\}$, is best possible equal to the transmitted data block $\boldsymbol{b}_k \in \{0, 1\}^B$ of the corresponding user k. Note that the received vector block $\boldsymbol{r}$ is seen as a realization of the vector random variable $\boldsymbol{r} \in \mathbb{C}^{R(\chi S + L - 1)}$ which comprises the received signal sequence $\boldsymbol{r}[\bar{n}] \in \mathbb{C}^R$ at the output of the channel model in Figure 5.7 (cf. also Equation 5.27) according to

$$\boldsymbol{r} = \begin{bmatrix} \boldsymbol{r}[0] \\ \boldsymbol{r}[1] \\ \vdots \\ \boldsymbol{r}[\chi S + L - 2] \end{bmatrix}, \tag{5.34}$$

as well as the data blocks $\boldsymbol{b}_k$ as realizations of $\boldsymbol{b}_k \in \{0, 1\}^B$ at the input of the transmitter models in Figure 5.1.

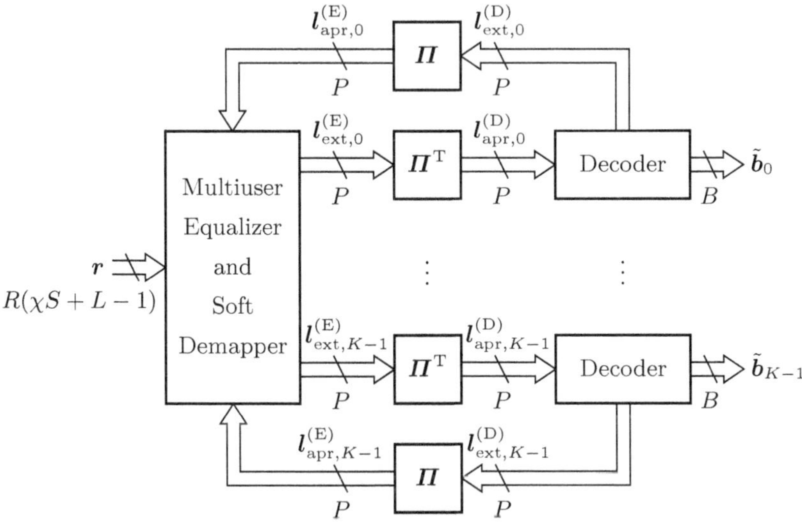

Fig. 5.8. Iterative multiuser receiver for a coded DS-CDMA system

Best possible detection is approximately achieved by exchanging soft information about the coded data bits between the receiver part performing *linear multiuser equalization* (cf. Subsection 5.3.1) and *soft symbol demapping*, and the *decoders* of each user in an iterative process. Here, soft information about a binary random variable $u \in \{0, 1\}$ is measured in terms of the *Log-Likelihood Ratio* (LLR) [107]

$$l = \ln \frac{\mathrm{P}\,(u = 0)}{\mathrm{P}\,(u = 1)} \in \mathbb{R}, \tag{5.35}$$

where $\mathrm{P}(u = u)$ denotes the probability that a realization of the random variable u is equal to $u \in \{0, 1\}$. If u depends statistically on another random variable, the probability function has to be replaced by the respective conditional probability function. Note that the sign of l is equal to the hard information about u obtained via hard decision and its magnitude tells you something about the reliability of this decision. As will be discussed more detailed in Subsection 5.3.2, the soft demapper of the $(k + 1)$th user calculates the extrinsic information $l_{\mathrm{ext},k}^{(\mathrm{E})} \in \mathbb{R}^P$ about the interleaved and coded data block $\boldsymbol{d}_k' \in \{0, 1\}^P$ using the observation signal block $\boldsymbol{r}$ and the *a priori* information $l_{\mathrm{apr},k}^{(\mathrm{E})} \in \mathbb{R}^P$ about $\boldsymbol{d}_k'$ obtained by interleaving of the extrinsic information $l_{\mathrm{ext},k}^{(\mathrm{D})} \in \mathbb{R}^P$ about the coded data block $\boldsymbol{d}_k \in \{0, 1\}^P$ computed by the decoder at the previous iteration step, i.e.,

$$l_{\mathrm{apr},k}^{(\mathrm{E})} = \boldsymbol{\Pi} l_{\mathrm{ext},k}^{(\mathrm{D})}, \quad k \in \{0, 1, \ldots, K - 1\}. \tag{5.36}$$

Further, the *Maximum A Posteriori* (MAP) *decoder* as introduced more detailed in Subsection 5.3.3 computes the extrinsic information $l_{\mathrm{ext},k}^{(\mathrm{D})}$ about the coded data block $\boldsymbol{d}_k$ of user k from its *a posteriori Log-Likelihood Ratio* (LLR) vector $\boldsymbol{l}_{\mathrm{apo},k}^{(\mathrm{D})} \in \mathbb{R}^P$ using its *a priori* information $\boldsymbol{l}_{\mathrm{apr},k}^{(\mathrm{D})} \in \mathbb{R}^P$ which is the deinterleaved extrinsic information $\boldsymbol{l}_{\mathrm{ext},k}^{(\mathrm{E})}$ about the interleaved and coded data block $\boldsymbol{d}_k'$ at the output of the soft demapper k, i. e.,

$$\boldsymbol{l}_{\mathrm{apr},k}^{(\mathrm{D})} = \boldsymbol{\Pi}^{\mathrm{T}} \boldsymbol{l}_{\mathrm{ext},k}^{(\mathrm{E})}, \quad k \in \{0, 1, \ldots, K-1\}. \tag{5.37}$$

For the remainder of this book, let $\boldsymbol{l}_k \in \mathbb{R}^P$ be the vector composed of the LLRs $l_{k,i} = \boldsymbol{e}_{i+1}^{\mathrm{T}} \boldsymbol{l}_k \in \mathbb{R}$, $i \in \{0, 1, \ldots, P-1\}$, which can be *a posteriori*, *a priori*, or extrinsic information at the equalizer or decoder, respectively. Clearly, these informations depend on the number of the current iteration. Nevertheless, an additional index is omitted due to simplicity. During the iterative process, the *a priori* LLRs $l_{\mathrm{apr},k,i}^{(\mathrm{D})}$ of the coded data bits $d_{k,i} = \boldsymbol{e}_{i+1}^{\mathrm{T}} \boldsymbol{d}_k \in \{0, 1\}$ improve from iteration to iteration. The asymptotic values are $+\infty$ or $-\infty$ depending on the detected coded data bits $\tilde{d}_{k,i}$, i. e., $l_{\mathrm{apr},k,i}^{(\mathrm{D})} \to +\infty$ if $\tilde{d}_{k,i} = 0$ and $l_{\mathrm{apr},k,i}^{(\mathrm{D})} \to -\infty$ if $\tilde{d}_{k,i} = 1$. In other words, the probability $\mathrm{P}(d_{k,i} = \tilde{d}_{k,i})$ converges to one. Note that $\mathrm{P}(d_{k,i} = \tilde{d}_{k,i}) \to 1$ does not mean that $\tilde{d}_{k,i}$ was actually transmitted but the receiver *assumes* that $\tilde{d}_{k,i}$ is the transmitted coded data bit $d_{k,i}$.

5.3.1 A Priori Based Linear Equalization

In order to remove MAI and ISI caused by the composite MA-SIMO channel as introduced in Section 5.2, we apply the *Matrix Wiener Filter* (MWF) of Chapter 2 or reduced-rank approximations thereof (cf. Section 2.3 and Chapter 4), respectively, to estimate the symbol vector sequence $\boldsymbol{s}[n] \in \mathbb{M}^K$ based on the observation vector sequence $\boldsymbol{r}_{\mathrm{c}}[\chi n] \in \mathbb{C}^{RF}$ and the *a priori* LLRs $\boldsymbol{l}_{\mathrm{apr},k}^{(\mathrm{E})} \in \mathbb{R}^P$, $k \in \{0, 1, \ldots, K-1\}$, of all K users. Since $\boldsymbol{s}[n]$ comprises the symbol streams of all users (see Equation 5.18) which are jointly estimated, the resulting filter is a multiuser equalizer whose block representation is depicted in Figure 5.9.

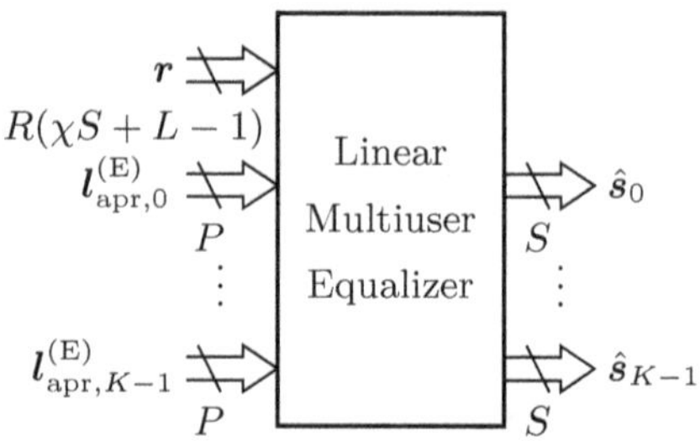

Fig. 5.9. A priori based linear multiuser equalizer

Before explaining how the *a priori* information affects the linear multiuser equalizer, we derive the MWF coefficients for the composite channel model of Equation (5.31). With the substitutions $\boldsymbol{x}[n] = \boldsymbol{s}[n - \nu] \in \mathbb{M}^M$, $M = K$, and $\boldsymbol{y}[n] = \boldsymbol{r}_{\mathrm{c}}[\chi n] \in \mathbb{C}^N$, $N = RF$, the MWF $\boldsymbol{W}[n] \in \mathbb{C}^{N \times M}$ and $\boldsymbol{a}[n] \in \mathbb{C}^M$ of Equation (2.7) can be used as the linear matrix filter which achieves the MMSE between the output

$$\hat{\boldsymbol{s}}[n] = \boldsymbol{W}^{\mathrm{H}}[n]\boldsymbol{r}_{\mathrm{c}}[\chi n] + \boldsymbol{a}[n], \tag{5.38}$$

and the desired vector random sequence $\boldsymbol{s}[n - \nu]$ where ν denotes the *latency time*. Recalling Equations (2.7) and (5.31), the MWF can be written as

$$\begin{aligned}
\boldsymbol{W}[n] &= \boldsymbol{C}_{\boldsymbol{y}}^{-1}[n]\boldsymbol{C}_{\boldsymbol{y},\boldsymbol{x}}[n] \\
&= \left(\boldsymbol{H}\boldsymbol{C}\boldsymbol{C}_{\boldsymbol{s}_{\mathrm{c}}}[n]\boldsymbol{C}^{\mathrm{T}}\boldsymbol{H}^{\mathrm{H}} + \boldsymbol{C}_{\boldsymbol{\eta}_{\mathrm{c}}}\right)^{-1}\boldsymbol{H}\boldsymbol{C}\boldsymbol{C}_{\boldsymbol{s}_{\mathrm{c}}}[n]\boldsymbol{S}_{\nu+1}, \\
\boldsymbol{a}[n] &= \boldsymbol{m}_{\boldsymbol{x}}[n] - \boldsymbol{W}^{\mathrm{H}}[n]\boldsymbol{m}_{\boldsymbol{y}}[n] \\
&= \boldsymbol{m}_{\boldsymbol{s}}[n - \nu] - \boldsymbol{W}^{\mathrm{H}}[n]\boldsymbol{H}\boldsymbol{C}\boldsymbol{m}_{\boldsymbol{s}_{\mathrm{c}}}[n],
\end{aligned} \tag{5.39}$$

where the matrix $\boldsymbol{S}_{\nu+1}^{\mathrm{T}} = \boldsymbol{S}_{(K\nu, K, K(g-1))} = [\boldsymbol{0}_{K \times K\nu}, \boldsymbol{I}_K, \boldsymbol{0}_{K \times K(g-\nu-1)}] \in \{0,1\}^{K \times Kg}$ selects the vector $\boldsymbol{s}[n - \nu] \in \mathbb{M}^K$ in $\boldsymbol{s}_{\mathrm{c}}[n] \in \mathbb{M}^{Kg}$ as defined in Equation (5.23), i.e., $\boldsymbol{s}[n - \nu] = \boldsymbol{S}_{\nu+1}^{\mathrm{T}}\boldsymbol{s}_{\mathrm{c}}[n]$. Remember that we assumed the symbols in $\boldsymbol{s}_{\mathrm{c}}[n]$ to be uncorrelated to the noise $\boldsymbol{\eta}_{\mathrm{c}}[n]$ and the noise to be stationary and zero-mean. Besides, the latency time ν is assumed to be fixed and not optimized with respect to the cost function in Equation (2.4) of Section 2.1 although this would improve the performance (cf., e. g., [154, 58]).

Further, the *a priori* LLRs $l_{\mathrm{apr},k}^{(\mathrm{E})}$, $k \in \{0, 1, \ldots, K-1\}$, are used to compute the time-variant statistics $\boldsymbol{C}_{\boldsymbol{s}_{\mathrm{c}}}[n] \in \mathbb{R}_{0,+}^{Kg \times Kg}$, $\boldsymbol{m}_{\boldsymbol{s}}[n - \nu] \in \mathbb{C}^K = \boldsymbol{S}_{\nu+1}^{\mathrm{T}}\boldsymbol{m}_{\boldsymbol{s}_{\mathrm{c}}}[n]$, and $\boldsymbol{m}_{\boldsymbol{s}_{\mathrm{c}}}[n] \in \mathbb{C}^{Kg}$ in Equation (5.39). In the following, $l_{k,nQ+q}^{(\mathrm{E})} = \boldsymbol{e}_{nQ+q}^{\mathrm{T}}\boldsymbol{l}_k^{(\mathrm{E})}$, $n \in \{0, 1, \ldots, S - 1\}$, $q \in \{0, 1, \ldots, Q - 1\}$, is the LLR of the interleaved and coded data bit $d'_{k,nQ+q} = \boldsymbol{e}_{q+1}^{\mathrm{T}}\boldsymbol{d}'_{k,n}$ which can be *a posteriori*, *a priori*, or extrinsic information (cf. Figure 5.8). In order to get the diagonal auto-covariance matrix (cf. end of Section 5.2)

$$\begin{aligned}
\boldsymbol{C}_{\boldsymbol{s}_{\mathrm{c}}}[n] &= \mathrm{bdiag}\left\{\boldsymbol{C}_{\boldsymbol{s}}[n], \boldsymbol{C}_{\boldsymbol{s}}[n - 1], \ldots, \boldsymbol{C}_{\boldsymbol{s}}[n - g + 1]\right\}, \\
\boldsymbol{C}_{\boldsymbol{s}}[n] &= \mathrm{diag}\left\{c_{s_0}[n], c_{s_1}[n], \ldots, c_{s_{K-1}}[n]\right\} \in \mathbb{R}_{0,+}^{K \times K},
\end{aligned} \tag{5.40}$$

and the mean (cf. Equations 5.18 and 5.23)

$$\boldsymbol{m}_{\boldsymbol{s}_{\mathrm{c}}}[n] = \begin{bmatrix} \boldsymbol{m}_{\boldsymbol{s}}[n] \\ \boldsymbol{m}_{\boldsymbol{s}}[n - 1] \\ \vdots \\ \boldsymbol{m}_{\boldsymbol{s}}[n - g + 1] \end{bmatrix}, \quad \boldsymbol{m}_{\boldsymbol{s}}[n] = \begin{bmatrix} m_{s_0}[n] \\ m_{s_1}[n] \\ \vdots \\ m_{s_{K-1}}[n] \end{bmatrix} \in \mathbb{C}^K, \tag{5.41}$$

we have to find expressions for $m_{s_k}[n]$ and $r_{s_k}[n]$, $k \in \{0, 1, \ldots, K - 1\}$, if we remember additionally that $c_{s_k}[n] = r_{s_k}[n] - |m_{s_k}[n]|^2$. It holds (e. g., [183])

$$m_{\boldsymbol{s}_k}[n] = \sum_{s\in\mathbb{M}} s\,\mathrm{P}\left(s_k[n] = s\right), \tag{5.42}$$

$$r_{\boldsymbol{s}_k}[n] = \sum_{s\in\mathbb{M}} |s|^2\,\mathrm{P}\left(s_k[n] = s\right), \tag{5.43}$$

with

$$\mathrm{P}\left(s_k[n] = s\right) = \mathrm{P}\left(\boldsymbol{d}'_{k,n} = \boldsymbol{d}\left(s\right)\right) = \prod_{q=0}^{Q-1} \mathrm{P}\left(d'_{k,nQ+q} = \boldsymbol{e}_{q+1}^{\mathrm{T}}\boldsymbol{d}\left(s\right)\right), \tag{5.44}$$

where

$$\boldsymbol{d}: \mathbb{M} \to \{0,1\}^Q, \quad s \mapsto \boldsymbol{d}'_{k,n} = \boldsymbol{d}\left(s\right), \tag{5.45}$$

is the inverse function of the mapper M, i.e., $\boldsymbol{d}(M(\boldsymbol{d}'_{k,n})) = \boldsymbol{d}'_{k,n}$, and the elements of $\boldsymbol{d}'_{k,n}$ are statistically independent due to the interleaver. Finally, the probability $\mathrm{P}(d'_{k,nQ+q} = \boldsymbol{e}_{q+1}^{\mathrm{T}}\boldsymbol{d}(s))$ can be expressed using the *a priori* information $l^{(\mathrm{E})}_{\mathrm{apr},k,nQ+q}$ in the following way:[6]

$$\mathrm{P}\left(d'_{k,nQ+q} = \boldsymbol{e}_{q+1}^{\mathrm{T}}\boldsymbol{d}\left(s\right)\right) = \frac{1}{2}\left(1 + \left(1 - 2\boldsymbol{e}_{q+1}^{\mathrm{T}}\boldsymbol{d}\left(s\right)\right)\tanh\frac{l^{(\mathrm{E})}_{\mathrm{apr},k,nQ+q}}{2}\right). \tag{5.46}$$

Summing up Equations (5.40) to (5.46), $\boldsymbol{C}_{\boldsymbol{s}_c}[n]$ and $\boldsymbol{m}_{\boldsymbol{s}_c}[n]$ and after all, the filter coefficients $\boldsymbol{W}[n]$ and $\boldsymbol{a}[n]$ in Equation (5.39) can be computed based on the *a priori* LLRs $\boldsymbol{l}^{(\mathrm{E})}_{\mathrm{apr},k}$, $k \in \{0,1,\ldots,K-1\}$.

Example 5.4 (Statistics based on a priori LLRs for QPSK). In case of QPSK modulation as presented in Example 5.2, we get the simplified expressions [240]

$$\left| m_{\boldsymbol{s}_k}[n] = \frac{1}{\sqrt{2}}\left(\tanh\frac{l^{(\mathrm{E})}_{\mathrm{apr},k,2n}}{2} + \mathrm{j}\tanh\frac{l^{(\mathrm{E})}_{\mathrm{apr},k,2n+1}}{2}\right), \quad r_{\boldsymbol{s}_k}[n] = 1. \right| \tag{5.47}$$

Revealing the influence of the *a priori* LLRs on the statistics explains why the MWF coefficients $\boldsymbol{W}[n]$ and $\boldsymbol{a}[n]$ must be time-variant. Besides, it gives a reason for the need of the additive term $\boldsymbol{a}[n]$ when estimating $\boldsymbol{s}[n-\nu]$ according to Equation (5.38). More precisely speaking, the fact that the *a priori* information differs from symbol to symbol and that it is generally unequal to zero involves that the underlying random sequences are generally non-stationary and non-zero mean, i.e., the statistics $\boldsymbol{C}_{\boldsymbol{s}_c}[n]$ and $\boldsymbol{m}_{\boldsymbol{s}_c}[n]$ depend on the times index n and almost surely $\boldsymbol{m}_{\boldsymbol{s}_c}[n] \neq \boldsymbol{0}_{Kg}$, respectively. Note that using the *a priori* LLRs to compute the probability of each element of the symbol alphabet $\mathbb{M}$ leads to a non-circular distribution. Although *widely*

---

[6] It holds for the LLR l of the binary random variable $x \in \{0,1\}$: $\tanh(l/2) = \mathrm{P}(x=0) - \mathrm{P}(x=1) = 2\,\mathrm{P}(x=0) - 1 = 1 - 2\,\mathrm{P}(x=1)$.

linear processing [185] improves generally the equalizer performance for such distributions, it was shown in [50, 54] that the gain is only marginal in Turbo systems. This is why we do not consider widely linear processing in this book.

In order to ensure the adherence of the *turbo principle*, we choose the filter coefficients such that they do not depend on the *a priori* information corresponding to the symbol vector $\boldsymbol{s}[n - \nu]$ which is currently estimated (e. g., [239]). In other words, we assume each element in $\boldsymbol{s}[n-\nu]$ to be uniformly distributed with zero-mean, i. e., $\boldsymbol{m_s}[n-\nu] = \boldsymbol{0}_K$ and $\boldsymbol{C_s}[n-\nu] = \boldsymbol{R_s}[n-\nu] = 2^{-Q} \sum_{s\in\mathrm{M}} |s|^2 \boldsymbol{I}_K = \varrho_s \boldsymbol{I}_K$ (cf. Equations 5.13 and 5.14 in Subsection 5.1.3). Let the matrix

$$
\begin{aligned}
\boldsymbol{\Gamma_{s_c}}[n] &= \boldsymbol{C_{s_c}}[n] + \boldsymbol{S}_{\nu+1} \left(\varrho_s \boldsymbol{I}_K - \boldsymbol{C_s}[n - \nu]\right) \boldsymbol{S}_{\nu+1}^{\mathrm{T}} \\
&= \mathrm{bdiag}\left\{\boldsymbol{C_s}[n], \ldots, \boldsymbol{C_s}[n - \nu + 1], \varrho_s \boldsymbol{I}_K,\right. \\
&\qquad \left. \boldsymbol{C_s}[n - \nu - 1], \ldots, \boldsymbol{C_s}[n - g + 1]\right\} \in \mathbb{R}_{0,+}^{Kg \times Kg},
\end{aligned}
\tag{5.48}
$$

be the *adjusted auto-covariance matrix*[7] of the symbol vector $\boldsymbol{s}_c[n]$ which is the $Kg \times Kg$ diagonal auto-covariance matrix $\boldsymbol{C_{s_c}}[n]$ where the K diagonal elements beginning at the $(K\nu + 1)$th position are replaced by ϱ_s, and the vector

$$
\boldsymbol{\mu_{s_c}}[n] = \boldsymbol{m_{s_c}}[n] - \boldsymbol{S}_{\nu+1} \boldsymbol{m_s}[n - \nu] =
\begin{bmatrix}
\boldsymbol{m_s}[n] \\
\vdots \\
\boldsymbol{m_s}[n - \nu + 1] \\
\boldsymbol{0}_K \\
\boldsymbol{m_s}[n - \nu - 1] \\
\vdots \\
\boldsymbol{m_s}[n - g + 1]
\end{bmatrix}
\in \mathbb{C}^{Kg},
\tag{5.49}
$$

the *adjusted mean* of $\boldsymbol{s}_c[n]$ which is the mean $\boldsymbol{m_{s_c}}[n]$ where the K elements beginning from the $(K\nu+1)$th element are set to zero. Then, the *adjusted MWF coefficients* $\boldsymbol{\Omega}[n] \in \mathbb{C}^{RF \times K}$ and $\boldsymbol{\alpha}[n] \in \mathbb{C}^K$ finally used as the linear multiuser equalizer in the Turbo system can be computed by verifying Equation (5.39) according to

$$
\begin{aligned}
\boldsymbol{\Omega}[n] &= \boldsymbol{\Gamma_y}^{-1}[n] \boldsymbol{\Gamma_{y,x}}[n] \\
&= \left(\boldsymbol{H}\boldsymbol{C}\boldsymbol{\Gamma_{s_c}}[n]\boldsymbol{C}^{\mathrm{T}}\boldsymbol{H}^{\mathrm{H}} + \boldsymbol{C_{\eta_c}}\right)^{-1} \boldsymbol{H}\boldsymbol{C}\boldsymbol{S}_{\nu+1}\varrho_s, \\
\boldsymbol{\alpha}[n] &= -\boldsymbol{\Omega}^{\mathrm{H}}[n]\boldsymbol{\mu_y}[n] = -\boldsymbol{\Omega}^{\mathrm{H}}[n]\boldsymbol{H}\boldsymbol{C}\boldsymbol{\mu_{s_c}}[n].
\end{aligned}
\tag{5.50}
$$

Clearly, the linear multiuser equalizer of the Turbo receiver as depicted in Figure 5.9, i. e., the MWF in Equation (5.50), can be efficiently implemented

---

[7] Throughout this book, the adjusted statistics and the resulting filter coefficients are denoted by Greek letters.

based on the Cholesky factorization as presented in Subsection 2.2.1. Moreover, we can also use the reduced-complexity approach of Subsection 2.2.2 because the auto-covariance matrix $C_y[n] = C_{r_c}[\chi n] \in \mathbb{C}^{RF \times RF}$ has the desired time-dependent block structure as defined in Equation (2.29). This can be easily verified by recalling the composite channel model of Equation (5.31) and the fact that the last $K(g-1)$ diagonal elements of $C_{s_c}[n]$ are equal to the first $K(g-1)$ diagonal elements of $C_{s_c}[n-1]$ (cf. Equation 5.40). It remains to show how the inverse of the adjusted auto-covariance matrix $\boldsymbol{\Gamma}_y[n] = HC\boldsymbol{\Gamma}_{s_c}[n]C^T H^H + C_{\boldsymbol{\eta}_c} \in \mathbb{C}^{RF \times RF}$ depends on the inverse $C_y^{-1}[n]$ computed based on the computationally efficient Algorithm 2.4.. With Equations (5.48) and (5.39), we get

$$\boldsymbol{\Gamma}_y[n] = C_y[n] + HCS_{\nu+1}\left(\varrho_s I_K - C_s[n-\nu]\right)S_{\nu+1}^T C^T H^H. \qquad (5.51)$$

Then, using the matrix inversion lemma of Appendix A.1.1 and the cross-covariance matrix $\boldsymbol{\Gamma}_{y,x}[n] = \varrho_s HCS_{\nu+1} \in \mathbb{C}^{RF \times K}$ (cf. Equation 5.50) for abbreviation yields

$$\boldsymbol{\Gamma}_y^{-1}[n] = C_y^{-1}[n] + \frac{1}{\varrho_s^2}C_y^{-1}[n]\boldsymbol{\Gamma}_{y,x}[n]K^{-1}[n]\boldsymbol{\Gamma}_{y,x}^H[n]C_y^{-1}[n],$$

$$K[n] = \left(\varrho_s I_K - C_s[n-\nu]\right)^{-1} - \frac{1}{\varrho_s^2}\boldsymbol{\Gamma}_{y,x}^H[n]C_y^{-1}[n]\boldsymbol{\Gamma}_{y,x}[n] \in \mathbb{C}^{K \times K}. \qquad (5.52)$$

We see that the inverse $\boldsymbol{\Gamma}_y^{-1}[n]$ can be easily computed based on the inverse of the $RF \times RF$ matrix $C_y[n]$ which we obtain by applying Algorithm 2.4., and the inverses of the $K \times K$ matrices $\varrho_s I_K - C_s[n-\nu]$ and $K[n]$, respectively. Note that the computational complexity of the latter two inversions as well as the one of the remaining matrix-matrix products is negligible if $RF \gg K$.

Besides the application of the Cholesky based implementation of the MWF and the RCMWF, the application of the suboptimal reduced-rank approaches of Section 2.3 and Chapter 4 decreases computational complexity as well. The filter coefficients of the *adjusted reduced-rank MWFs* read as (cf. Equation 2.48)

$$\left| \begin{array}{l} \boldsymbol{\Omega}^{(D)}[n] = T^{(D)}[n]\left(T^{(D),H}[n]\boldsymbol{\Gamma}_y[n]T^{(D)}[n]\right)^{-1}T^{(D),H}[n]\boldsymbol{\Gamma}_{y,x}[n] \\[2mm] \boldsymbol{\alpha}^{(D)}[n] = -\boldsymbol{\Omega}^{(D),H}[n]\boldsymbol{\mu}_y[n], \end{array} \right. \qquad (5.53)$$

where the prefilter matrix $T^{(D)} \in \mathbb{C}^{RF \times D}$ can be the one of the eigensubspace based *Principal Component* (PC) or *Cross-Spectral* (CS) method as derived in Subsections 2.3.1 and 2.3.2 but based on the eigenvectors of the adjusted auto-covariance matrix $\boldsymbol{\Gamma}_y[n]$, or the one of the Krylov subspace based *MultiStage MWF* (MSMWF) applied to $\boldsymbol{\Gamma}_y[n]$ and the cross-covariance matrix $\boldsymbol{\Gamma}_{y,x}[n]$, whose different implementations are derived and investigated in Chapter 4. Remember the most important implementations of the MSMWF, viz., the

Block Lanczos–Ruhe (BLR) *procedure* in Algorithm 4.2. and the *Block Conjugate Gradient* (BCG) *method* in Algorithm 3.4..

Further reduction in computational complexity can be achieved if we avoid the calculation of the time-variant filter coefficients at each time index n via the approximation of the adjusted auto-covariance matrix $\boldsymbol{\Gamma}_{\boldsymbol{s}_c}[n]$ by its time-average

$$\bar{\boldsymbol{\Gamma}}_{\boldsymbol{s}_c} = \frac{1}{S} \sum_{n=0}^{S-1} \boldsymbol{\Gamma}_{\boldsymbol{s}_c}[n], \tag{5.54}$$

for each symbol block of length S [240]. This method can be applied to either the optimal MWF implementation (cf. Subsection 2.2.1) or the suboptimal reduced-rank solutions of Section 2.3 or Chapter 4, respectively. However, it is not applicable to the RCMWF implementation of Subsection 2.2 due to its strong dependency on the time variance of $\boldsymbol{\Gamma}_{\boldsymbol{s}_c}[n]$. Note that an even worse approximation is obtained if the adjusted auto-covariance $\boldsymbol{\Gamma}_{\boldsymbol{s}_c}[n]$ is replaced by the scaled identity matrix $\varrho_s \boldsymbol{I}_{RF}$, i.e., the available *a priori* information is solely used to compute the filter coefficient $\boldsymbol{\alpha}[n]$ by means of $\boldsymbol{\mu}_{\boldsymbol{s}_c}[n]$ and no longer for the calculation of $\boldsymbol{\Omega}[n]$ (cf. also Subsection 5.3.4).

After all, it remains to mention that in the case of spatially and temporally white noise where $\boldsymbol{C}_{\boldsymbol{\eta}_c} = c_\eta \boldsymbol{I}_{RF}$, the matrix inversion lemma of Appendix A.1.1 can be used to simplify Equation (5.50) to[8]

$$\boldsymbol{\Omega}[n] = \boldsymbol{HC} \left(c_\eta \boldsymbol{\Gamma}_{\boldsymbol{s}_c}^{-1}[n] + \boldsymbol{C}^{\mathrm{T}} \boldsymbol{H}^{\mathrm{H}} \boldsymbol{HC} \right)^{-1} \boldsymbol{\Gamma}_{\boldsymbol{s}_c}^{-1}[n] \boldsymbol{S}_{\nu+1} \varrho_s. \tag{5.55}$$

With this simplification, the linear equation system with the $RF \times RF$ system matrix $\boldsymbol{HC\Gamma}_{\boldsymbol{s}_c}[n]\boldsymbol{C}^{\mathrm{T}}\boldsymbol{H}^{\mathrm{H}} + \boldsymbol{C}_{\boldsymbol{\eta}_c}$ (cf. Equation 5.50) can be efficiently solved by inverting the in general smaller $Kg \times Kg$ matrix $c_\eta \boldsymbol{\Gamma}_{\boldsymbol{s}_c}^{-1}[n] + \boldsymbol{C}^{\mathrm{T}}\boldsymbol{H}^{\mathrm{H}}\boldsymbol{HC}$. Note that the inversion of $\boldsymbol{\Gamma}_{\boldsymbol{s}_c}[n]$ is no challenge due to its diagonal structure. Although the simulations in Chapter 6 cover only spatially and temporally white noise, we do no longer consider the simplification of the MWF coefficient in Equation (5.55) because we are eventually interested in reduced-complexity methods which could be applied to any noise scenario.

It remains to mention that in CDMA systems with *scrambling* which is not considered here, neither the time-average approach nor the matrix inversion lemma can be applied to reduce the computational complexity required to compute the filter coefficients, because in this case, the auto-covariance matrix $\boldsymbol{\Gamma}_{\boldsymbol{s}_c}[n]$ is not only time-varying due to the *a priori* information but also due to the time-varying scrambling code. However, the complexity can be still reduced via the application of reduced-rank filters.

---

[8] With the matrix inversion lemma of Appendix A.1.1, it can be easily shown that $\left(\boldsymbol{ADA}^{\mathrm{H}} + \alpha \boldsymbol{I}_n\right)^{-1}\boldsymbol{A} = \boldsymbol{A}\left(\alpha\boldsymbol{D}^{-1} + \boldsymbol{A}^{\mathrm{H}}\boldsymbol{A}\right)^{-1}\boldsymbol{D}^{-1}$ where $\boldsymbol{D} \in \mathbb{C}^{m \times m}$ is invertible, $\boldsymbol{A} \in \mathbb{C}^{n \times m}$, and $\alpha \in \mathbb{C}$.

5.3.2 Soft Symbol Demapping

The $(k+1)$th *soft symbol demapper* as depicted in Figure 5.10 calculates the extrinsic LLR $\boldsymbol{l}_{\mathrm{ext},k}^{(\mathrm{E})} \in \mathbb{R}^P$, $k \in \{0,1,\ldots,K-1\}$, about the interleaved and coded data bits based on their *a priori* information $\boldsymbol{l}_{\mathrm{apr},k}^{(\mathrm{E})} \in \mathbb{R}^P$ delivered from the decoder at the previous iteration step, and the linear multiuser equalizer's estimate $\hat{\boldsymbol{s}}_k \in \mathbb{C}^S$ of the transmitted symbol block $\boldsymbol{s}_k \in \mathrm{M}^S$. Again, it is essential for the functionality of the Turbo receiver that we forward only the extrinsic LLRs and not the complete *a posteriori* information. Here, soft symbol demapping is performed for each user k separately.

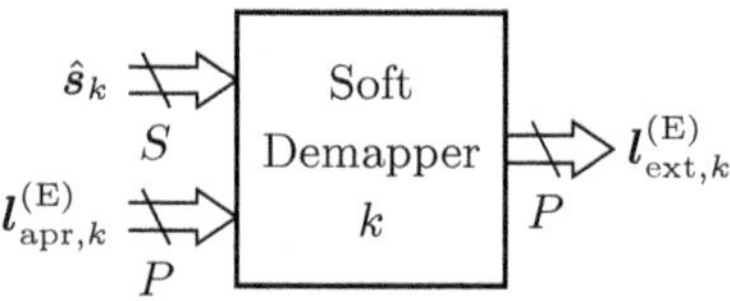

Fig. 5.10. Soft symbol demapper for user k

The $(nQ+q)$th element $l_{\mathrm{ext},k,nQ+q}^{(\mathrm{E})} \in \mathbb{R}$, $n \in \{0,1,\ldots,S-1\}$, $q \in \{0,1,\ldots,Q-1\}$, of the extrinsic LLR vector $\boldsymbol{l}_{\mathrm{ext},k}^{(\mathrm{E})}$, i.e., the soft extrinsic information about the $(q+1)$th interleaved and coded data bit $d_{k,nQ+q}' = \boldsymbol{e}_{q+1}^{\mathrm{T}} \boldsymbol{d}_{k,n}' \in \{0,1\}$ of the vector $\boldsymbol{d}_{k,n}' = \boldsymbol{S}_{(nQ,Q,P-Q)} \boldsymbol{d}_k' \in \{0,1\}^Q$ which comprises the bits belonging to the $(n+1)$th element $s_{k,n} = s_k[n] \in \mathrm{M}$ of the $(k+1)$th user's symbol block $\boldsymbol{s}_k \in \mathrm{M}^S$ (cf. Subsection 5.1.3), can be derived based on the *a posteriori LLR*

$$l_{\mathrm{apo},k,nQ+q}^{(\mathrm{E})} = \ln \frac{\mathrm{P}\left(d_{k,nQ+q}' = 0 \,\middle|\, \hat{s}_{k,n} = \hat{s}_{k,n}\right)}{\mathrm{P}\left(d_{k,nQ+q}' = 1 \,\middle|\, \hat{s}_{k,n} = \hat{s}_{k,n}\right)} \in \mathbb{R}, \qquad (5.56)$$

where $\hat{s}_{k,n} \in \mathbb{C}$ is the $(n+1)$th element of the block estimate $\hat{\boldsymbol{s}}_k$, i.e., $\hat{s}_{k,n} = \boldsymbol{e}_{n+1}^{\mathrm{T}} \hat{\boldsymbol{s}}_k$. Using Bayes' theorem (e.g., [183]) and the fact that elements of the vector $\boldsymbol{d}_{k,n}'$ are statistically independent due to the interleaver (cf. Equation 5.44), we get further

$$l_{\mathrm{apo},k,nQ+q}^{(\mathrm{E})} = \ln \frac{\displaystyle\sum_{\substack{\boldsymbol{d}\in\{0,1\}^Q \\ \boldsymbol{e}_{q+1}^{\mathrm{T}}\boldsymbol{d}=0}} \mathrm{p}_{\hat{s}_{k,n}}\!\left(\hat{s}_{k,n}\middle|\boldsymbol{d}_{k,n}'=\boldsymbol{d}\right) \mathrm{P}\left(\boldsymbol{d}_{k,n}'=\boldsymbol{d}\right)}{\displaystyle\sum_{\substack{\boldsymbol{d}\in\{0,1\}^Q \\ \boldsymbol{e}_{q+1}^{\mathrm{T}}\boldsymbol{d}=1}} \mathrm{p}_{\hat{s}_{k,n}}\!\left(\hat{s}_{k,n}\middle|\boldsymbol{d}_{k,n}'=\boldsymbol{d}\right) \mathrm{P}\left(\boldsymbol{d}_{k,n}'=\boldsymbol{d}\right)}, \qquad (5.57)$$

or

$$l^{(\mathrm{E})}_{\mathrm{apo},k,nQ+q} = \ln \frac{\displaystyle\sum_{\substack{d\in\{0,1\}^Q \\ e_{q+1}^{\mathrm{T}}d=0}} \mathrm{p}_{\hat{s}_{k,n}}\!\left(\hat{s}_{k,n}\big|d'_{k,n}=d\right) \prod_{\ell=0}^{Q-1} \mathrm{P}\left(d'_{k,nQ+\ell}=e_{\ell+1}^{\mathrm{T}}d\right)}{\displaystyle\sum_{\substack{d\in\{0,1\}^Q \\ e_{q+1}^{\mathrm{T}}d=1}} \mathrm{p}_{\hat{s}_{k,n}}\!\left(\hat{s}_{k,n}\big|d'_{k,n}=d\right) \prod_{\ell=0}^{Q-1} \mathrm{P}\left(d'_{k,nQ+\ell}=e_{\ell+1}^{\mathrm{T}}d\right)} . \tag{5.58}$$

Finally, the extrinsic LLR which is defined as the *a posteriori* LLR $l^{(\mathrm{E})}_{\mathrm{apo},k,nQ+q}$ where the *a priori* LLR

$$l^{(\mathrm{E})}_{\mathrm{apr},k,nQ+q} = \ln \frac{\mathrm{P}\left(d'_{k,nQ+q}=0\right)}{\mathrm{P}\left(d'_{k,nQ+q}=1\right)} \in \mathbb{R}, \tag{5.59}$$

about the interleaved and coded data bit $d'_{k,nQ+q}$ has been subtracted, i.e.,

$$l^{(\mathrm{E})}_{\mathrm{ext},k,nQ+q} = l^{(\mathrm{E})}_{\mathrm{apo},k,nQ+q} - l^{(\mathrm{E})}_{\mathrm{apr},k,nQ+q} \tag{5.60}$$

$$= \ln \frac{\displaystyle\sum_{\substack{d\in\{0,1\}^Q \\ e_{q+1}^{\mathrm{T}}d=0}} \mathrm{p}_{\hat{s}_{k,n}}\!\left(\hat{s}_{k,n}\big|d'_{k,n}=d\right) \prod_{\substack{\ell=0 \\ \ell\neq q}}^{Q-1} \mathrm{P}\left(d'_{k,nQ+\ell}=e_{\ell+1}^{\mathrm{T}}d\right)}{\displaystyle\sum_{\substack{d\in\{0,1\}^Q \\ e_{q+1}^{\mathrm{T}}d=1}} \mathrm{p}_{\hat{s}_{k,n}}\!\left(\hat{s}_{k,n}\big|d'_{k,n}=d\right) \prod_{\substack{\ell=0 \\ \ell\neq q}}^{Q-1} \mathrm{P}\left(d'_{k,nQ+\ell}=e_{\ell+1}^{\mathrm{T}}d\right)} .$$

From Equation (5.46), we know that

$$\mathrm{P}\left(d'_{k,nQ+\ell}=e_{\ell+1}^{\mathrm{T}}d\right) = \frac{1}{2}\left(1 + \left(1 - 2e_{\ell+1}^{\mathrm{T}}d\right)\tanh\frac{l^{(\mathrm{E})}_{\mathrm{apr},k,nQ+\ell}}{2}\right). \tag{5.61}$$

The remaining part in calculating $l^{(\mathrm{E})}_{\mathrm{ext},k,nQ+q}$ according to Equation (5.60) is the unknown probability density function $\mathrm{p}_{\hat{s}_{k,n}}(\hat{s}_{k,n}\,|\,d'_{k,n}=d)$ which we approximate assuming a normal distribution,[9] i.e.,

$$\mathrm{p}_{\hat{s}_{k,n}}\!\left(\hat{s}_{k,n}\big|d'_{k,n}=d\right) = \frac{1}{\pi\gamma_{(d,k,n)}}\,\exp\left(-\frac{\left|\hat{s}_{k,n}-\mu_{(d,k,n)}\right|^2}{\gamma_{(d,k,n)}}\right), \tag{5.62}$$

with the adjusted mean[10]

---

[9] In [189], it was shown that the sum of interference and noise at the output of a linear MMSE multiuser detector is well approximated by a normal distribution.

[10] Note that the adjusted mean $\mu_{(d,k,n)}$ and the adjusted variance $\gamma_{(d,k,n)}$ do not denote the adjusted mean and variance of d_k, respectively. To avoid misunderstandings, the index is enclosed in brackets.

$$\mu_{(\boldsymbol{d},k,n)} = \mathrm{E}\left\{\hat{s}_{k,n}\big|s_k[n-\nu]=M(\boldsymbol{d})\right\}\Big|_{\substack{\boldsymbol{m_s}[n-\nu]=\boldsymbol{0}_K \\ \boldsymbol{C_s}[n-\nu]=\varrho_{\boldsymbol{s}}\boldsymbol{I}_K}}$$
$$= M(\boldsymbol{d})\boldsymbol{e}_{k+1}^{\mathrm{T}}\boldsymbol{\Omega}^{\mathrm{H}}[n]\boldsymbol{H}\boldsymbol{C}\boldsymbol{e}_{K\nu+k+1}, \tag{5.63}$$

and the adjusted variance

$$\gamma_{(\boldsymbol{d},k,n)} = \mathrm{E}\left\{|\hat{s}_{k,n}|^2\big|s_k[n-\nu]=M(\boldsymbol{d})\right\}\Big|_{\substack{\boldsymbol{m_s}[n-\nu]=\boldsymbol{0}_K \\ \boldsymbol{C_s}[n-\nu]=\varrho_{\boldsymbol{s}}\boldsymbol{I}_K}} - \big|\mu_{(\boldsymbol{d},k,n)}\big|^2$$
$$= \boldsymbol{e}_{k+1}^{\mathrm{T}}\boldsymbol{\Omega}^{\mathrm{H}}[n]\left(\boldsymbol{H}\boldsymbol{C}\left(\boldsymbol{\Gamma}_{\boldsymbol{s}_{\mathrm{c}}}[n]-\varrho_s\boldsymbol{e}_{K\nu+k+1}\boldsymbol{e}_{K\nu+k+1}^{\mathrm{T}}\right)\boldsymbol{C}^{\mathrm{T}}\boldsymbol{H}^{\mathrm{H}}\right.$$
$$\left.+\boldsymbol{C}_{\boldsymbol{\eta}_{\mathrm{c}}}\right)\boldsymbol{\Omega}[n]\boldsymbol{e}_{k+1}, \tag{5.64}$$
$$= \varrho_s\left(\frac{\mu_{(\boldsymbol{d},k,n)}}{M(\boldsymbol{d})}-\frac{\big|\mu_{(\boldsymbol{d},k,n)}\big|^2}{|M(\boldsymbol{d})|^2}\right).$$

For the derivation of Equations (5.63) and (5.64), we used the expression

$$\hat{s}_{k,n} = \hat{s}_k[n] = \boldsymbol{e}_{k+1}^{\mathrm{T}}\left(\boldsymbol{\Omega}^{\mathrm{H}}[n]\boldsymbol{r}_{\mathrm{c}}[n]+\boldsymbol{\alpha}[n]\right)$$
$$= \boldsymbol{e}_{k+1}^{\mathrm{T}}\left(\boldsymbol{\Omega}^{\mathrm{H}}[n]\boldsymbol{H}\boldsymbol{C}\underbrace{\left(\boldsymbol{s}_{\mathrm{c}}[n]-\boldsymbol{\mu}_{\boldsymbol{s}_{\mathrm{c}}}[n]\right)}_{\boldsymbol{s}'_{\mathrm{c}}[n]}+\boldsymbol{\Omega}^{\mathrm{H}}[n]\boldsymbol{\eta}_{\mathrm{c}}[\chi n]\right), \tag{5.65}$$

where we combined the adjusted counterpart of Equation (5.38) with the filter coefficient $\boldsymbol{\alpha}[n]$ from Equation (5.50) and the channel model in Equation (5.31). Note that with the assumptions $\boldsymbol{m_s}[n-\nu]=\boldsymbol{0}_K$ and $\boldsymbol{C_s}[n-\nu]=\varrho_s\boldsymbol{I}_K$ which have been necessary for the adherence of the turbo principle, $\mathrm{E}\{\boldsymbol{s}'_{\mathrm{c}}[n]\,|\,s_k[n-\nu]=M(\boldsymbol{d})\}=M(\boldsymbol{d})\boldsymbol{e}_{K\nu+k+1}\in\mathbb{C}^{Kg}$ where $\boldsymbol{s}'_{\mathrm{c}}[n]=\boldsymbol{s}_{\mathrm{c}}[n]-\boldsymbol{\mu}_{\boldsymbol{s}_{\mathrm{c}}}[n]\in\mathbb{C}^{Kg}$ and $\boldsymbol{e}_{K\nu+k+1}=\boldsymbol{S}_{\nu+1}\boldsymbol{e}_{k+1}\in\{0,1\}^{Kg}$, and $\mathrm{E}\{\boldsymbol{s}'_{\mathrm{c}}[n]\boldsymbol{s}'^{,\mathrm{H}}_{\mathrm{c}}[n]\,|\,s_k[n-\nu]=M(\boldsymbol{d})\}=\boldsymbol{\Gamma}_{\boldsymbol{s}_{\mathrm{c}}}[n]-\varrho_s\boldsymbol{e}_{K\nu+k+1}\boldsymbol{e}_{K\nu+k+1}^{\mathrm{T}}\in\mathbb{C}^{Kg\times Kg}$. The last equality in Equation (5.64) can be verified by exploiting the expression of the filter coefficient $\boldsymbol{\Omega}[n]$ in Equation (5.50) and the one of the adjusted mean $\mu_{(\boldsymbol{d},k,n)}$ in Equation (5.63).

Due to the fact that we use the *a priori* information in order to compute the extrinsic LLRs (cf. Equations 5.60 and 5.61), the presented soft symbol demapper can be categorized as an *iterative demapper* which has been introduced in [22, 21].

Example 5.5 (Soft demapping for QPSK). For QPSK modulation as defined in Example 5.2, the elements $l^{(\mathrm{E})}_{\mathrm{ext},k,2n+q}$, $n\in\{0,1,\dots,S-1\}$, $q\in\{0,1\}$, of the soft extrinsic information $\boldsymbol{l}^{(\mathrm{E})}_{\mathrm{ext},k}$, $k\in\{0,1,\dots,K-1\}$, are given by [240]

$$\left|l^{(\mathrm{E})}_{\mathrm{ext},k,2n}+\mathrm{j}\,l^{(\mathrm{E})}_{\mathrm{ext},k,2n+1}=\frac{\sqrt{8}}{1-\boldsymbol{e}_{k+1}^{\mathrm{T}}\boldsymbol{\Omega}^{\mathrm{H}}[n]\boldsymbol{H}\boldsymbol{C}\boldsymbol{e}_{K\nu+k+1}}\hat{s}_{k,n}\right.\Bigg| \tag{5.66}$$

5.3.3 Decoding

The Turbo receiver in Figure 5.8 decodes each user $k \in \{0, 1, \ldots, K-1\}$ separately using a *soft-input soft-output Maximum A Posteriori* (MAP) *decoder* (see, e.g., [163]) whose basics are briefly reviewed in this subsection. Recall from the introduction to Section 5.3 that the $(i+1)$th element of the LLR vector $\boldsymbol{l}_k^{(D)} \in \mathbb{R}^P$ is denoted as $l_{k,i}^{(D)} \in \mathbb{R}$, $i \in \{0, 1, \ldots, P-1\}$, which can be *a posteriori*, *a priori*, or extrinsic information. Then, based on the *a priori* information $\boldsymbol{l}_{\mathrm{apr},k}^{(D)} \in \mathbb{R}^P$ which is the deinterleaved extrinsic information obtained from the soft demapper of user k (cf. Figure 5.8 and Equation 5.37), the optimal soft-input soft-output MAP decoder calculates the *a posteriori* probabilities $\mathrm{P}(d_{k,i} = d \,|\, \boldsymbol{l}_{\mathrm{apr},k}^{(D)} = \boldsymbol{l}_{\mathrm{apr},k}^{(D)})$, $d \in \{0,1\}$, or the corresponding *a posteriori* LLRs

$$l_{\mathrm{apo},k,i}^{(D)} = \ln \frac{\mathrm{P}\left(d_{k,i} = 0 \,\Big|\, \boldsymbol{l}_{\mathrm{apr},k}^{(D)} = \boldsymbol{l}_{\mathrm{apr},k}^{(D)}\right)}{\mathrm{P}\left(d_{k,i} = 1 \,\Big|\, \boldsymbol{l}_{\mathrm{apr},k}^{(D)} = \boldsymbol{l}_{\mathrm{apr},k}^{(D)}\right)}, \tag{5.67}$$

of the code bits $d_{k,i} = \boldsymbol{e}_{i+1}^{\mathrm{T}} \boldsymbol{d}_k \in \{0,1\}$ (see Subsection 5.1.1). Here, we assume that the *a priori* LLR vector $\boldsymbol{l}_{\mathrm{apr},k}^{(D)}$ is a realization of the vector random variable $\boldsymbol{l}_{\mathrm{apr},k}^{(D)} \in \mathbb{R}^P$.[11] Further, with the *a priori* information

$$l_{\mathrm{apr},k,i}^{(D)} = \ln \frac{\mathrm{P}\left(d_{k,i} = 0\right)}{\mathrm{P}\left(d_{k,i} = 1\right)}, \tag{5.68}$$

about the $(i+1)$th code bit $d_{k,i}$ of user k, the extrinsic LLR $l_{\mathrm{ext},k,i}^{(D)}$ which is fed back by the decoder to the multiuser equalizer via the interleaver (cf. Figure 5.8), computes as

$$\left|\begin{aligned} l_{\mathrm{ext},k,i}^{(D)} &= l_{\mathrm{apo},k,i}^{(D)} - l_{\mathrm{apr},k,i}^{(D)} \\ &= \ln \frac{\mathrm{P}\left(d_{k,i} = 0 \,\Big|\, \boldsymbol{l}_{\mathrm{apr},k}^{(D)} = \boldsymbol{l}_{\mathrm{apr},k}^{(D)}\right)}{\mathrm{P}\left(d_{k,i} = 1 \,\Big|\, \boldsymbol{l}_{\mathrm{apr},k}^{(D)} = \boldsymbol{l}_{\mathrm{apr},k}^{(D)}\right)} - \ln \frac{\mathrm{P}\left(d_{k,i} = 0\right)}{\mathrm{P}\left(d_{k,i} = 1\right)}, \end{aligned}\right| \tag{5.69}$$

where we subtracted again the *a priori* information $l_{\mathrm{apr},k,i}^{(D)}$ computed by the soft demapper k, from the *a posteriori* LLR $l_{\mathrm{apo},k,i}^{(D)}$ offered by the decoder at the current iteration step, similar to the computation of the extrinsic information in Equation (5.60). Again, it is essential that only the extrinsic

---

[11] The modeling of the *a priori* LLR by a random variable is necessary to design the decoder based on the MAP criterion. Certainly, this model is reasonable since the actual *a priori* knowledge is estimated via the deinterleaved extrinsic information offered by the soft demapper k.

information $l_{\text{ext},k}^{(D)}$ is fed back to the equalizer in order to ensure the adherence of the Turbo principle (e. g., [106]).

After the last iteration, the $(k+1)$th MAP decoder provides also the decoded data bits $\tilde{b}_{k,i} = \boldsymbol{e}_{i+1}^{\text{T}} \tilde{\boldsymbol{b}}_k \in \{0,1\}$, $i \in \{0, 1, \ldots, B-1\}$, of user k, i. e.,

$$\tilde{b}_{k,i} = \underset{b \in \{0,1\}}{\operatorname{argmax}} \operatorname{P} \left(b_{k,i} = b \,\middle|\, \boldsymbol{l}_{\text{apr},k}^{(D)} = \boldsymbol{l}_{\text{apr},k}^{(D)} \right), \tag{5.70}$$

by using all available *a priori* information $\boldsymbol{l}_{\text{apr},k}^{(D)}$.

A well-known implementation of the soft-input soft-output MAP decoder is the *Bahl–Cocke–Jelinek–Raviv* (BCJR) *algorithm* [7] introduced by Lalit R. Bahl, John Cocke, Frederick Jelinek, and Josef Raviv in 1974. Similar to the *Viterbi algorithm* [248] which has been presented by Andrew J. Viterbi in 1967, the BCJR decoder is based on a Trellis structure (see, e. g., [163]). However, whereas the BCJR decoding algorithm minimizes the bit error probability, G. David Forney showed in [70, 71] that the Viterbi algorithm is a *Maximum Likelihood* (ML) *decoder* for convolutional codes choosing the output as the code word which maximizes the conditional probability of the received sequence or, in other words, minimizes the code word error probability. Although the Viterbi algorithm is easier to implement, thus, often used in practical applications, the BCJR algorithm is preferred for receivers with Turbo processing. For a detailed derivation of the BCJR algorithm, we refer to either the original publication or the book [163].

5.3.4 Relationship to Iterative Soft Interference Cancellation

In order to reveal the relationship between the linear multiuser equalizer of Subsection 5.3.1 and *soft interference cancellation* [254, 198] or *parallel decision feedback detection* [62, 127], we rewrite the estimate $\hat{\boldsymbol{s}}[n] = \boldsymbol{\Omega}^{\text{H}}[n]\boldsymbol{r}_{\text{c}}[n] + \boldsymbol{\alpha}[n] \in \mathbb{C}^K$ such that it no longer depends on $\boldsymbol{\alpha}[n] \in \mathbb{C}^K$. Similar to Equation (2.9) of Section 2.1, use Equation (5.50) to get the expression

$$\hat{\boldsymbol{s}}[n] = \boldsymbol{\Omega}^{\text{H}}[n] \left(\boldsymbol{r}_{\text{c}}[n] - \boldsymbol{H}\boldsymbol{C}\boldsymbol{\mu}_{\boldsymbol{s}_{\text{c}}}[n] \right), \tag{5.71}$$

which is represented by the block diagram of Figure 5.11.

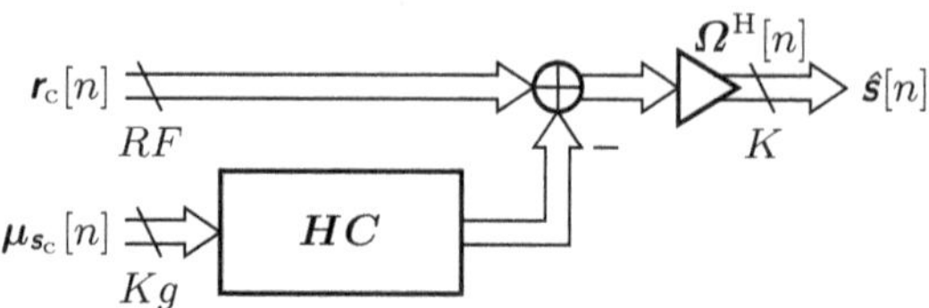

Fig. 5.11. Alternative representation of linear multiuser equalizer

It can be seen that the adjusted mean vector $\boldsymbol{\mu}_{\boldsymbol{s}_c}[n] \in \mathbb{C}^{Kg}$ based on the *a priori* information in the way described by Equations (5.42), (5.44), and (5.46), is used to subtract its influence after the transmission over the composite channel $\boldsymbol{HC} \in \mathbb{C}^{RF \times Kg}$ from the observation vector $\boldsymbol{r}_c[n] \in \mathbb{C}^{RF}$ to increase the performance of the following linear MWF $\boldsymbol{\Omega}[n] \in \mathbb{C}^{RF \times K}$. This procedure is strongly related to the *decision feedback equalizer* [5, 51] or *interference canceler* [78, 170] where not the adjusted mean vector is used to reconstruct the interference which is subtracted from the observation, but the already decided symbols. Based on this analogy, the elements of the adjusted mean vector $\boldsymbol{\mu}_{\boldsymbol{s}_c}[n]$ defined via Equation (5.42) can be interpreted as *soft decisions* or *soft symbols*. Note that the soft symbol is no soft information in terms of LLRs but a non-linear transformation thereof (cf. Equations 5.42, 5.44, and 5.46). By the way, besides the exploitation of the *a priori* LLRs to compute the soft symbols, they are still used to compute the statistics of the MWF coefficient $\boldsymbol{\Omega}[n]$ according to Equation (5.50). Only in the case where the adjusted auto-covariance matrix $\boldsymbol{\Gamma}_{\boldsymbol{s}_c}[n]$ of the symbol vector is approximated by the scaled identity matrix $\varrho_s \boldsymbol{I}_{RF}$ as mentioned in Subsection 5.3.1, the dependency on the *a priori* information does no longer influence the filter coefficient $\boldsymbol{\Omega}[n]$.

Furthermore, we see that the interference cancellation system of Figure 5.11 does not cancel the MAI between the users at time index $n - \nu$ because we assume the symbols of *all* K users to be zero-mean at this time instance which is necessary for the application of the MWF framework. However, the remaining MAI is suppressed via the orthogonal codes of the MA scheme. In DS-CDMA systems with non-orthogonal codes, it is recommended to use single-user equalization where the MWF estimating all K users jointly is replaced by K *Vector Wiener Filters* (VWFs), i. e., MWFs with $M = K = 1$, estimating each user separately, in order to exploit the *a priori* information also to remove the MAI at time index $n - \nu$. The simulation results in Section 6.3 (see Figure 6.16) confirm that the performances of the MWF and the K VWFs are very close to each other if we use OVSF codes.

The multiuser equalizer can be also derived by assuming the interference cancellation structure of Figure 5.11 from the first (cf., e. g., [254, 198, 127]). Nevertheless, doing so, it is hard to explain why the soft symbols are computed exactly based on the non-linear transformation of the *a priori* LLRs as given in Equations (5.42), (5.44), and (5.46), and on no other heuristical chosen transformation. This is the reason why we prefer the linear or *affine* MWF based approach of Subsection 5.3.1 where the *a priori* information is used by its definition to compute the statistics which is needed for the design of the filter coefficients $\boldsymbol{\Omega}[n]$ and $\boldsymbol{\alpha}[n]$.

6

System Performance

This chapter is to investigate the performance of the proposed linear multiuser equalizers when used for iterative soft interference cancellation of a coded *Direct Sequence Code Division Multiple-Access* (DS-CDMA) system (cf. Subsection 5.3.1). Besides the conventional performance evaluation based on the *Bit Error Rate* (BER) obtained via *Monte Carlo* simulations, we apply *EXtrinsic Information Transfer* (EXIT) *charts* to estimate either the *Mutual Information* (MI) between the *Log-Likelihood Ratio* (LLR) at the output of the soft symbol demapper and the corresponding interleaved and coded data bit, or the *Mean Square Error* (MSE) at the output of the equalizer, both for a given number of Turbo iterations. Note that Stephan ten Brink [19, 20] introduced the EXIT charts originally to analyze the convergence properties of Turbo schemes by simulating each block, e. g., equalizer inclusive soft demapper or decoder, separately. Although the EXIT based performance measure is computationally cheaper than simulating the BER via *Monte Carlo* methods, the computational burden is still very high because the MI has to be determined for a large number of LLRs. Therefore, we present an alternative method where the EXIT characteristic of the equalizer-demapper combination is obtained by numerical integration instead of histogram measurements. The resulting approach extends the contributions of Valéry Ramon and Luc Vandendorpe [195] from a single-user system with *Binary Phase Shift Keying* (BPSK) *modulation* to a multiuser system with *QuadriPhase Shift Keying* (QPSK) *modulation*. Note that we calculate only the EXIT characteristic of the linear equalizer and soft demapper in the proposed semianalytical way, the decoder's EXIT curve is still computed based on the standard histogram measurement method. Besides the BER analysis and the EXIT based MSE analysis of the DS-CDMA system, we introduce so-called *complexity–performance charts* to investigate whether reduced-rank filters or full-rank filters with a reduced order are preferable in Turbo systems. Unless otherwise stated, the *Channel State Information* (CSI) is assumed to be perfectly known at the receiver.

6.1 Extrinsic Information Transfer Charts

6.1.1 Principles

The basic idea of *EXtrinsic Information Transfer* (EXIT) *charts* is to treat the equalizer-demapper combination and the decoder as independent blocks which is justified by the interleaver and deinterleaver both providing statistically independent *Log-Likelihood Ratios* (LLRs). Thus, the performance of the iterative system can be evaluated by firstly simulating each of the blocks separately, and subsequently investigating the resulting EXIT characteristics. Note that this procedure is computationally much more efficient than simulating the whole iterative system via explicitly transmitting bits.

Both the equalizer-demapper combination and the decoder receive *a priori* information in form of the LLR $l_{\mathrm{apr}} \in \mathbb{R}$ at their input and map it to the extrinsic LLR $l_{\mathrm{ext}} \in \mathbb{R}$ at their output (cf. Figure 5.8). As in [20], we assume that each element of the blocks $\boldsymbol{l}_{\mathrm{apr},k} \in \mathbb{R}^P$, $k \in \{0, 1, \ldots K - 1\}$, of *a piori* LLRs either at the input of the linear multiuser equalizer or the inputs of the decoders are realizations of *independent and identically distributed* (i. i. d.) random variables which are real normally distributed according to $\mathcal{N}_{\mathbb{R}}(m_{\mathrm{apr}}, c_{\mathrm{apr}})$ with the probability density function

$$\mathrm{p}_{l_{\mathrm{apr}}}\left(l_{\mathrm{apr}} \middle| u = u\right) = \frac{1}{\sqrt{2\pi c_{\mathrm{apr}}}} \exp\left(-\frac{(l_{\mathrm{apr}} - m_{\mathrm{apr}})^2}{2 c_{\mathrm{apr}}}\right), \qquad (6.1)$$

the variance $c_{\mathrm{apr}} \in \mathbb{R}_+$, and the mean $m_{\mathrm{apr}} = (1 - 2u)c_{\mathrm{apr}}/2 = \pm c_{\mathrm{apr}}/2 \in \mathbb{R}$.[1] The magnitude of m_{apr} is $c_{\mathrm{apr}}/2$ and its sign depends on the realization $u \in \{0, 1\}$ of the binary random variable $u \in \{0, 1\}$ which can be interpreted as the coded data bit or interleaved coded data bit, i. e., an element in $\boldsymbol{d}_k \in \{0, 1\}^P$ or $\boldsymbol{d}'_k \in \{0, 1\}^P$, respectively (cf. Figure 5.1).[2] Since the magnitude of an LLR l determines the reliability of the information provided by the LLR about the respective bit, the magnitude of the mean m_{apr} determines the average quality of the LLRs. Note that the assumption of i. i. d. *a priori* LLRs is only valid for blocks with an infinite length and perfect interleaving which is only approximately fulfilled in real iterative systems.

The EXIT characteristic is measured in terms of the *Mutual Information* (MI, e. g., [20])

---

[1] Note that the mean m_{apr} depends on the current realization u of the random variable u even though this is not explicitly expressed by the notation. We avoided an additional index due to simplicity.

[2] If a binary random variable $x \in \{-1, +1\}$ is perturbed by zero-mean real-valued *Additive White Gaussian Noise* (AWGN) n with the distribution $\mathcal{N}_{\mathbb{R}}(0, c_n)$, the LLR of the resulting random variable $z = x + n$ reads as $l_z = \ln(\mathrm{p}_z(z \,|\, x = +1)/\mathrm{p}_z(z \,|\, x = -1)) = 2(x + n)/c_n$. Thus, the distribution of $l_z = 2(x + n)/c_n$ for a fixed realization x can be modeled via $\mathcal{N}_{\mathbb{R}}(m, c)$ with the variance $c = 4/c_n$ and the mean $m = 2x/c_n = xc/2$.

$$\left| I(l; u) = \frac{1}{2} \sum_{u \in \{0,1\}} \int_{\mathbb{R}} \mathrm{p}_l \left(l \big| u = u\right) \log_2 \frac{2\,\mathrm{p}_l \left(l \big| u = u\right)}{\mathrm{p}_l \left(l \big| u = 0\right) + \mathrm{p}_l \left(l \big| u = 1\right)} \, \mathrm{d}l, \right| \quad (6.2)$$

between the binary random variable $u \in \{0,1\}$ representing the (interleaved) coded data bit and the corresponding LLR $l \in \mathbb{R}$ which can be either *a priori* information at the input or extrinsic information at the output of the block under consideration. Roughly speaking, the MI $I(l; u) \in [0, 1]$ shows how much information of the sent random variable u is included in the corresponding soft information l. The closer $I(l; u)$ is to one, the more reliable is the information delivered by l about u. A gain in reliability which the equalizer-demapper combination or decoder achieves by computing the extrinsic LLRs based on the available *a priori* LLRs is represented by an increase in MI, i.e., the MI $I_{\mathrm{ext}} - I(l_{\mathrm{ext}}; u)$ at the output is larger than the MI $I_{\mathrm{apr}} = I(l_{\mathrm{apr}}; u)$ at the input. This input-output characteristic with respect to MI is exactly represented by the EXIT chart which depicts the output MI I_{ext} for values of the input MI I_{apr} between zero and one.

To get the EXIT characteristic of the $(k + 1)$th user's decoder, $k \in \{0, 1, \ldots, K - 1\}$, we calculate several tuples $(I_{\mathrm{apr},k}^{(D)}, I_{\mathrm{ext},k}^{(D)})$ for different values of the variance c_{apr}. Since all users have the same decoder, these EXIT characteristics are independent of k. However, for the sake of generality, we keep the user index k in the notation. Whereas the input MI $I_{\mathrm{apr},k}^{(D)}$ can be easily obtained via numerical integration according to Equation (6.2) based on the probability density function of Equation (6.1), the output MI $I_{\mathrm{ext},k}^{(D)}$ must be determined by the following procedure:

1. The elements of the data bit block $\boldsymbol{b}_k \in \{0, 1\}^B$ are chosen as realizations of the random variable $b \in \{0, 1\}$ with $\mathrm{P}(b = 0) = \mathrm{P}(b = 1) = 1/2$.
2. After *Forward Error Correction* (FEC) channel coding as described in Subsection 5.1.1, we get the coded data bit block $\boldsymbol{d}_k \in \{0, 1\}^P$.
3. Each element in the *a priori* LLR block $\boldsymbol{l}_{\mathrm{apr},k}^{(D)} \in \mathbb{R}^P$ is chosen as the sum of $+c_{\mathrm{apr}}/2$ if the corresponding element in $\boldsymbol{d}_k$ is 0, or $-c_{\mathrm{apr}}/2$ if the corresponding element is 1, and a random term which is drawn from the distribution $\mathcal{N}_{\mathbb{R}}(0, c_{\mathrm{apr}})$. Doing so, we get *a priori* LLRs which are distributed according to Equation (6.1).
4. The *a priori* LLR block $\boldsymbol{l}_{\mathrm{apr},k}^{(D)}$ is then used as the input for the decoder whose EXIT characteristic is unknown. After decoding as briefly reviewed in Subsection 5.3.3, we obtain the extrinsic LLR block $\boldsymbol{l}_{\mathrm{ext},k}^{(D)} \in \mathbb{R}^P$.
5. Finally, the output MI $I_{\mathrm{ext},k}^{(D)}$ is obtained via numerical evaluation of Equation (6.2) where the probability density function $\mathrm{p}_{l_{\mathrm{ext},k}^{(D)}} (l_{\mathrm{ext},k}^{(D)} \,|\, u = u)$ is approximated based on histogram measurements on the extrinsic LLRs at the output of the decoder.

With this procedure, the EXIT characteristic of the *Bahl–Cocke–Jelinek–Raviv* (BCJR) *algorithm* (cf. Subsection 5.3.3) decoding the $(7, 5)$-

convolutionally encoded data (see Example 5.1) calculates as the solid line with circles in Figure 6.1. Note that in Figure 6.1, the input MI $I^{(D)}_{\mathrm{apr},k}$ is the ordinate and the output MI $I^{(D)}_{\mathrm{ext},k}$ is the abscissa. The reason for this choice is given later on. The EXIT chart of the decoder visualizes that for low *a priori* information, i.e., $I^{(D)}_{\mathrm{apr},k} < 1/2$, the decoding process even lowers the output MI $I^{(D)}_{\mathrm{ext},k}$ whereas for $I^{(D)}_{\mathrm{apr},k} > 1/2$, $I^{(D)}_{\mathrm{ext},k}$ is larger than $I^{(D)}_{\mathrm{apr},k}$.

Contrary to the decoder whose inputs are solely *a priori* LLRs, the equalizer-demapper combination has the observation vector $\boldsymbol{r} \in \mathbb{C}^{R(\chi S+L-1)}$ as a second source of information (cf. Figure 5.8). Therefore, its EXIT characteristic is influenced by the convolutional spreading matrix $\boldsymbol{C} \in \{-1,0,+1\}^{KG \times Kg}$, the convolutional channel matrix $\boldsymbol{H} \in \mathbb{C}^{RF \times KG}$, and the *Signal-to-Noise Ratio* (SNR) defined as

$$\frac{E_{\mathrm{b}}}{N_0} = \frac{E_{\mathrm{s}}}{rQN_0} = \frac{\varrho_s \chi}{rQc_\eta}, \tag{6.3}$$

where E_{b} is the bit energy, $E_{\mathrm{s}} = \varrho_s T_{\mathrm{s}} = \varrho_s \chi T_{\mathrm{c}}$ the symbol energy, and $N_0 = c_\eta T_{\mathrm{c}}$ the noise power density. For the SNR definition in Equation (6.3), we assumed spatially and temporally white noise with $\boldsymbol{C}_{\boldsymbol{\eta}_{\mathrm{c}}} = c_\eta \boldsymbol{I}_{RF}$ as used in all simulations of this chapter. Remember that the noise $\boldsymbol{\eta}[\bar{n}]$ has been sampled with the chip rate $1/T_{\mathrm{c}}$ whereas the symbol sequence $\boldsymbol{s}[n]$ with the symbol rate $1/T_{\mathrm{s}} = 1/(\chi T_{\mathrm{c}})$. The simulation procedure to obtain the EXIT characteristic of the equalizer-demapper combination is very similar to the one of the decoder which we described above. First, we choose each element of the interleaved coded data blocks $\boldsymbol{d}'_k \in \{0,1\}^P$, $k \in \{0,1,\ldots,K-1\}$, as realizations of the uniformly distributed random variable $d \in \{0,1\}$ and compute the corresponding *a priori* LLR blocks $\boldsymbol{l}^{(E)}_{\mathrm{apr},k} \in \mathbb{R}^P$ where we assume the same variance c_{apr} for all users (see Step 3 of the above procedure). After symbol mapping of $\boldsymbol{d}'_k$ according to Subsection 5.1.3, the resulting symbol streams are transmitted over the composite channel to get the received signal vector $\boldsymbol{r}$ as described in Section 5.2. Then, the extrinsic LLR block $\boldsymbol{l}^{(E)}_{\mathrm{ext},k} \in \mathbb{R}^P$ of user k is computed based on $\boldsymbol{r}$ and the available *a priori* information, i.e., $\boldsymbol{l}^{(E)}_{\mathrm{apr},k}$, $k \in \{0,1,\ldots,K-1\}$ (cf. Subsections 5.3.1 and 5.3.2). Finally, the tuples $(I^{(E)}_{\mathrm{apr},k}, I^{(E)}_{\mathrm{ext},k})$ of the EXIT characteristic of the equalizer-demapper combination can be calculated by evaluating Equations (6.1) and (6.2) in the same way as described above for the decoder.

Figure 6.1 depicts exemplarily the EXIT characteristic of a *Wiener Filter* (WF), i.e., a *Matrix Wiener Filter* (MWF) with $M = 1$,[3] which has been applied to equalize the Proakis b channel (cf. Table 6.1) at a log-SNR of $10\log_{10}(E_{\mathrm{b}}/N_0) = 5\,\mathrm{dB}$ in a system with one user, one receive antenna, and no spreading, i.e., $K = R = \chi = 1$. The filter length has been chosen to

---

[3] Remember from Chapter 2 that the term VWF denotes a MWF with $M = 1$ if it is part of a filter bank estimating a vector signal with $M > 1$ whereas it is denoted WF when used alone to estimate a single scalar signal.

be $F = 7$, i.e., $N = RF = 7$, the latency time is set to $\nu = 5$, and the size of the data block which is mapped to *QuadriPhase Shift Keying* (QPSK) symbols ($Q = 2$, cf. Example 5.2) is chosen to be $B = 2048$. We see that even for zero *a priori* information, i.e., for $I_{\mathrm{apr},0}^{(E)} = 0$, the equalizer-demapper combination delivers non-zero MI $I_{\mathrm{ext},0}^{(E)}$ at its output. This is due to the fact that the observation vector can still be exploited although no further *a priori* knowledge is available.

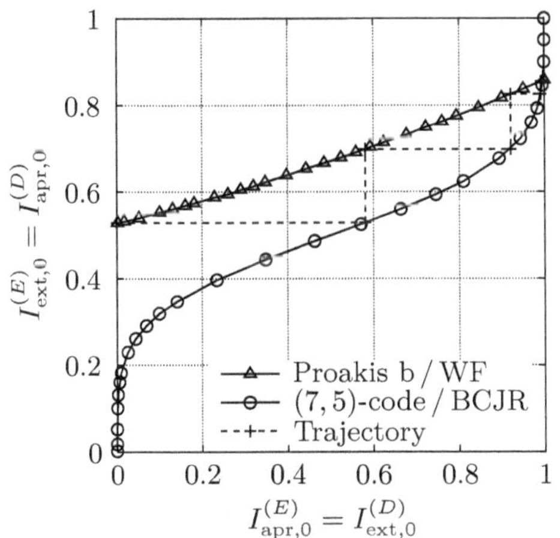

Fig. 6.1. EXIT chart example for single-user system with one receive antenna ($K = R = \chi = 1$, $F = 7$) at $10\log_{10}(E_\mathrm{b}/N_0) = 5\,\mathrm{dB}$

In a Turbo receiver, the equalizer-demapper combination and the decoder exchange soft information which can be also seen as an exchange of MI between both components. More precisely speaking, the output MI $I_{\mathrm{ext},k}^{(E)}$ of the equalizer-demapper acts as the input MI $I_{\mathrm{apr},k}^{(D)}$ of the decoder, and the output MI $I_{\mathrm{ext},k}^{(D)}$ of the decoder as the input MI $I_{\mathrm{apr},k}^{(E)}$ of the equalizer-demapper. Thus, $I_{\mathrm{apr},k}^{(E)} = I_{\mathrm{ext},k}^{(D)}$ and $I_{\mathrm{ext},k}^{(E)} = I_{\mathrm{apr},k}^{(D)}$, and the EXIT characteristics of the equalizer-demapper and decoder can be drawn in one diagram as given in Figure 6.1. Finally, the iterative Turbo process resulting in an increase of MI can be visualized by the dashed trajectory of Figure 6.1. This trajectory is used in the analysis of Section 6.2 to compute the output MI $I_{\mathrm{ext},k}^{(E)}$ or an estimate of the *Mean Square Error* (MSE) for an arbitrary number of Turbo iterations. Moreover, the MI at the intersection of both EXIT characteristics provides the performance bound of the iterative receiver if an infinite number of Turbo iterations is performed.

It remains to mention that the output MI at the decoder is related to the *Bit Error Rate* (BER) of the iterative system, i.e., the higher the MI at the output of the decoder, the lower the BER of the system. This relationship has been already presented in [20] for the iterative decoding of parallel concatenated codes but holds also for systems with Turbo equalization as shown in, e. g., [177, 160].

6.1.2 Semianalytical Calculation of Extrinsic Information Transfer Characteristics for Linear Equalizers

The computation of the equalizer-demapper EXIT characteristic as briefly discussed in the previous subsection requires the simulated transmission of the data bits based on *Monte Carlo* techniques which can be a time-consuming issue, especially if the block size is very large. Therefore, we derive in the following a method which calculates semianalytically the EXIT characteristic of the equalizer-demapper combination in a computationally efficient way. To do so, we assume not only the *a priori* LLRs to be real normal distributed but also the extrinsic LLRs at the output of the soft symbol demapper (cf. Subsection 5.3.2). This approach has been firstly proposed by Valéry Ramon and Luc Vandendorpe in [195] for a single-user system with BPSK modulation. Here, we will extend it to multiuser systems with QPSK modulation as introduced in Example 5.2. To obtain finally the parameters of the probability density function of the extrinsic LLRs, we must additionally assume that the symbol estimates of all users at the output of the linear multiuser equalizer (cf. Subsection 5.3.1) are complex normal distributed. However, this assumption holds only if we replace the MWF with K outputs from Equation (5.50), i. e.,

$$
\begin{aligned}
\boldsymbol{\Omega}[n] &= \left(\boldsymbol{HC\Gamma}_{\boldsymbol{s}_{\mathrm{c}}}[n]\boldsymbol{C}^{\mathrm{T}}\boldsymbol{H}^{\mathrm{H}} + \boldsymbol{C}_{\boldsymbol{\eta}_{\mathrm{c}}}\right)^{-1}\boldsymbol{HCS}_{\nu+1}\varrho_{\boldsymbol{s}} \in \mathbb{C}^{RF \times K}, \\
\boldsymbol{\alpha}[n] &= -\boldsymbol{\Omega}^{\mathrm{H}}[n]\boldsymbol{HC\mu}_{\boldsymbol{s}_{\mathrm{c}}}[n] \in \mathbb{C}^{K},
\end{aligned}
\tag{6.4}
$$

by the bank of K *Vector Wiener Filters* (VWFs)

$$
\left|
\begin{aligned}
\boldsymbol{\omega}_{k}[n] &= \left(\boldsymbol{HC\Gamma}_{\boldsymbol{s}_{\mathrm{c}}}^{(k)}[n]\boldsymbol{C}^{\mathrm{T}}\boldsymbol{H}^{\mathrm{H}} + \boldsymbol{C}_{\boldsymbol{\eta}_{\mathrm{c}}}\right)^{-1}\boldsymbol{HCe}_{K\nu+k+1}\varrho_{\boldsymbol{s}} \in \mathbb{C}^{RF}, \\
\alpha_{k}[n] &= -\boldsymbol{\omega}_{k}^{\mathrm{H}}[n]\boldsymbol{HC\mu}_{\boldsymbol{s}_{\mathrm{c}}}^{(k)}[n] \in \mathbb{C},
\end{aligned}
\right.
\tag{6.5}
$$

$k \in \{0, 1, \ldots, K - 1\}$, each providing the estimate

$$
\hat{s}_{k}[n] = \boldsymbol{\omega}_{k}^{\mathrm{H}}[n]\boldsymbol{r}_{\mathrm{c}}[\chi n] + \alpha_{k}[n] \in \mathbb{C},
\tag{6.6}
$$

of the $(k+1)$th user's symbol $s_{k}[n-\nu] \in \mathbb{M}$ at time index $n-\nu$, such that the *a priori* information about the interleaved and coded data bits of all users can be also exploited to remove the *Multiple-Access Interference* (MAI) at time index $n - \nu$ (cf. Subsection 5.3.4). This is realized by choosing the adjusted auto-covariance matrix as $\boldsymbol{\Gamma}_{\boldsymbol{s}_{\mathrm{c}}}^{(k)}[n] = \boldsymbol{C}_{\boldsymbol{s}_{\mathrm{c}}}[n] + (\varrho_{\boldsymbol{s}} - c_{s_{k}}[n - \nu])\boldsymbol{e}_{K\nu+k+1}\boldsymbol{e}_{K\nu+k+1}^{\mathrm{T}} \in \mathbb{R}_{0,+}^{Kg \times Kg}$ where only the $(K\nu + k + 1)$th diagonal element is set to $\varrho_{\boldsymbol{s}}$, and the adjusted mean vector as $\boldsymbol{\mu}_{\boldsymbol{s}_c}^{(k)}[n] = \boldsymbol{m}_{\boldsymbol{s}_c}[n] - m_{s_k}[n-\nu]\boldsymbol{e}_{K\nu+k+1} \in \mathbb{C}^{Kg}$ with a zero at the $(K\nu + k + 1)$th position, instead of the adjusted statistics as defined in Equations (5.48) and (5.49), respectively. Remember from Equation (5.18) that $s_k[n] = \boldsymbol{e}_{k+1}^T \boldsymbol{s}[n]$. Due to the fact that all inputs and outputs of the $(k+1)$th user's equalizer and soft symbol demapper are modeled by normal distributions, their processing can be described by the respective transformations of the statistical parameters, i.e., the mean and the variance, as depicted in Figure 6.2.

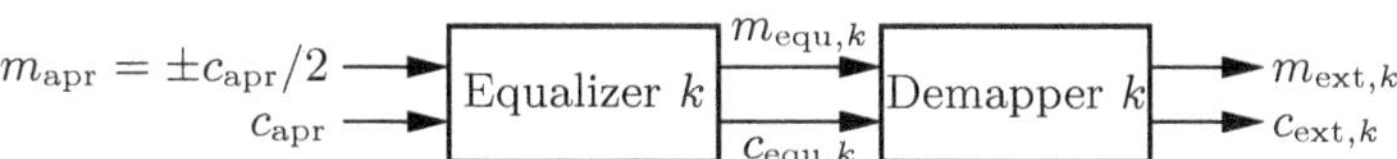

Fig. 6.2. Statistical model of linear equalizer and soft symbol demapper of user k

In the sequel, we restrict ourselves to QPSK modulation, i.e., $Q = 2$ and $\varrho_s = 1$ (cf. Example 5.2). In order to compute the EXIT chart without simulating the transmission of symbols over the channel, we have to find a way how the *a priori* LLRs distributed according to $\mathcal{N}_{\mathbb{R}}(\pm c_{\mathrm{apr}}/2, c_{\mathrm{apr}})$ influence the statistical parameters, i.e., the mean $m_{\mathrm{equ},k}$ and variance $c_{\mathrm{equ},k}$, of the symbol estimate $\hat{s}_k[n]$ of user k which is assumed to be distributed according to $\mathcal{N}_{\mathbb{C}}(m_{\mathrm{equ},k}, c_{\mathrm{equ},k})$.[4] Such an influence can only be found if we approximate the adjusted auto-covariance matrix $\boldsymbol{\Gamma}_{\boldsymbol{s}_c}^{(k)}[n]$ by the diagonal matrix

$$\tilde{\boldsymbol{\Gamma}}_{\boldsymbol{s}_c}^{(k)} = \tilde{c}_s^{(k)} \boldsymbol{I}_{Kg} + \left(1 - \tilde{c}_s^{(k)}\right) \boldsymbol{e}_{K\nu+k+1}\boldsymbol{e}_{K\nu+k+1}^T \in \mathbb{R}_{0,+}^{Kg \times Kg}, \qquad (6.7)$$

where the $(K\nu + k + 1)$th element is one and $\tilde{c}_s^{(k)}$ is the average variance of $s_k[n]$ over the real normal distributed *a priori* LLRs at the input of the equalizer, i.e.,

$$\tilde{c}_s^{(k)} = \int_{\mathbb{R}} \int_{\mathbb{R}} \left(1 - \frac{1}{2}\left(\tanh^2 \frac{l_0}{2} + \tanh^2 \frac{l_1}{2}\right)\right) p_{l_{\mathrm{apr},0},l_{\mathrm{apr},1}}(l_0,l_1)\,\mathrm{d}l_0\mathrm{d}l_1, \quad (6.8)$$

where

$$p_{l_{\mathrm{apr},0},l_{\mathrm{apr},1}}(l_0,l_1) = \frac{1}{8\pi c_{\mathrm{apr}}} \sum_{u \in \{0,1\}} \sum_{v \in \{0,1\}} \exp\left(-\frac{\left(l_0 - (1-2u)\frac{c_{\mathrm{apr}}}{2}\right)^2}{2c_{\mathrm{apr}}}\right)$$

$$\cdot \exp\left(-\frac{\left(l_1 - (1-2v)\frac{c_{\mathrm{apr}}}{2}\right)^2}{2c_{\mathrm{apr}}}\right), \qquad (6.9)$$

---

[4] The probability density function of the complex normal distribution $\mathcal{N}_{\mathbb{C}}(m_x, c_x)$ with mean m_x and variance c_x reads as $p_x(x) = \exp(-|x - m_x|^2/c_x)/(\pi c_x)$ (e.g., [183]).

is the joint probability density function of the random variables $l_{\mathrm{apr},0} \in \mathbb{R}$ and $l_{\mathrm{apr},1} \in \mathbb{R}$ from which the *a priori* LLRs $l^{(\mathrm{E})}_{\mathrm{apr},k,2n}$ and $l^{(\mathrm{E})}_{\mathrm{apr},k,2n+1}$ are drawn. Recall that $c_{s_k}[n] = r_{s_k}[n] - |m_{s_k}[n]|^2$ where for QPSK modulation, $m_{s_k}[n]$ and $r_{s_k}[n]$ are given in Equation (5.47), and note that we assume equiprobable and independent interleaved code bits which is justified due to the interleaver. Besides, $l_{\mathrm{apr},0}$ and $l_{\mathrm{apr},1}$ are independent of the time index n and the user index k because we assume the same statistics for all *a priori* LLRs. Finally, the approximated filter coefficients which estimate the transmitted QPSK symbols ($\varrho_s = 1$) read as

$$
\left|
\begin{aligned}
\tilde{\boldsymbol{\omega}}_k &= \left(\boldsymbol{HC}\tilde{\boldsymbol{\Gamma}}^{(k)}_{\boldsymbol{s}_{\mathrm{c}}}\boldsymbol{C}^{\mathrm{T}}\boldsymbol{H}^{\mathrm{H}} + \boldsymbol{C}_{\boldsymbol{\eta}_{\mathrm{c}}} \right)^{-1} \boldsymbol{HC}\boldsymbol{e}_{K\nu+k+1} \in \mathbb{C}^{RF}, \\
\tilde{\alpha}_k[n] &= -\tilde{\boldsymbol{\omega}}_k^{\mathrm{H}}\boldsymbol{HC}\boldsymbol{\mu}^{(k)}_{\boldsymbol{s}_{\mathrm{c}}}[n] \in \mathbb{C},
\end{aligned}
\right.
\tag{6.10}
$$

where the additive scalar $\tilde{\alpha}_k[n]$ still depends on the time index n due to the time-variant *a priori* information in $\boldsymbol{\mu}^{(k)}_{\boldsymbol{s}_{\mathrm{c}}}[n]$. As we will see in the following, this is no problem for deriving the mean $m_{\mathrm{equ},k}$ and variance $c_{\mathrm{equ},k}$ of $\hat{s}_k[n]$ solely based on the variance c_{apr}.

With the composite channel model of Equation (5.31) and the definition of $\boldsymbol{s}'_{\mathrm{c},k}[n] = \boldsymbol{s}_{\mathrm{c}}[n] - \boldsymbol{\mu}^{(k)}_{\boldsymbol{s}_{\mathrm{c}}}[n] \in \mathbb{C}^{Kg}$ (cf. Equation 5.65), the estimate $\hat{s}_k[n] \in \mathbb{C}$ of the $(k+1)$th user's symbol $s_k[n-\nu]$ can be written as

$$
\begin{aligned}
\hat{s}_k[n] &= \tilde{\boldsymbol{\omega}}_k^{\mathrm{H}}\boldsymbol{r}_{\mathrm{c}}[\chi n] + \tilde{\alpha}_k[n] = \tilde{\boldsymbol{\omega}}_k^{\mathrm{H}}\left(\boldsymbol{HC}\boldsymbol{s}'_{\mathrm{c},k}[n] + \boldsymbol{\eta}_{\mathrm{c}}[\chi n] \right) \\
&= \tilde{\boldsymbol{\omega}}_k^{\mathrm{H}}\boldsymbol{HC}\boldsymbol{e}_{K\nu+k+1}s_k[n-\nu] + v_k[n],
\end{aligned}
\tag{6.11}
$$

where the composite noise $v_k[n] \in \mathbb{C}$ is given by

$$
v_k[n] = \tilde{\boldsymbol{\omega}}_k^{\mathrm{H}}\left(\boldsymbol{HC}\left(\boldsymbol{s}'_{\mathrm{c},k}[n] - \boldsymbol{e}_{K\nu+k+1}s_k[n-\nu] \right) + \boldsymbol{\eta}_{\mathrm{c}}[\chi n] \right).
\tag{6.12}
$$

In order to ensure that the estimate $\hat{s}_k[n]$ in Equation (6.11) is complex normal distributed for the specific symbol realization $s_k[n-\nu] = s_k[n-\nu]$, we must assume that $v_k[n]$ is complex normal distributed according to $\mathcal{N}_{\mathbb{C}}(0, c_{v_k}[n])$. This assumption makes sense due to the fact that $\boldsymbol{\eta}_{\mathrm{c}}[\chi n]$ is distributed according to $\mathcal{N}_{\mathbb{C}}(\boldsymbol{0}_{RF}, \boldsymbol{C}_{\boldsymbol{\eta}_{\mathrm{c}}})$ (cf. Section 5.2) and $\mathrm{E}\{\boldsymbol{s}'_{\mathrm{c},k}[n] - \boldsymbol{e}_{K\nu+k+1}s_k[n-\nu]\} = \boldsymbol{0}_{Kg}$ because $\boldsymbol{s}'_{\mathrm{c},k}[n]$ is zero-mean apart from its $(K\nu + k + 1)$th entry which is set to zero anyway. With Equation (6.12) and the fact that $\mathrm{E}\{\boldsymbol{s}_{\mathrm{c}}^{\mathrm{H}}[n]\boldsymbol{\eta}_{\mathrm{c}}[\chi n]\} = \boldsymbol{0}_{Kg\times RF}$ (cf. Section 5.2), the variance of $v_k[n] \sim \mathcal{N}_{\mathbb{C}}(0, c_{v_k}[n])$ computes as

$$
\begin{aligned}
c_{v_k}[n] &= \mathrm{E}\left\{ |v_k[n]|^2 \right\} \\
&= \tilde{\boldsymbol{\omega}}_k^{\mathrm{H}}\left(\boldsymbol{HC}\left(\boldsymbol{\Gamma}^{(k)}_{\boldsymbol{s}_{\mathrm{c}}}[n] - \boldsymbol{e}_{K\nu+k+1}\boldsymbol{e}^{\mathrm{T}}_{K\nu+k+1} \right) \boldsymbol{C}^{\mathrm{T}}\boldsymbol{H}^{\mathrm{H}} + \boldsymbol{C}_{\boldsymbol{\eta}_{\mathrm{c}}} \right) \tilde{\boldsymbol{\omega}}_k.
\end{aligned}
\tag{6.13}
$$

Due to Equation (6.11), the variance of $\hat{s}_k[n]$ in the case where $s_k[n-\nu] = s_k[n-\nu]$ is also $c_{v_k}[n]$. However, we approximate the distribution of $\hat{s}_k[n]$ by the time-invariant complex normal distribution $\mathcal{N}_{\mathbb{C}}(m_{\mathrm{equ},k}, c_{\mathrm{equ},k})$ where the variance $c_{\mathrm{equ},k}$ is obtained from the variance $c_{v_k}[n]$ by replacing again the time-variant auto-covariance matrix $\boldsymbol{\Gamma}_{s_\mathrm{c}}^{(k)}[n]$ with $\tilde{\boldsymbol{\Gamma}}_{s_\mathrm{c}}^{(k)}$ (cf. Equation 6.7), i. e.,

$$
\begin{aligned}
c_{\mathrm{equ},k} &= \mathrm{E}\left\{\left|\hat{s}_k[n]\right|^2 \Big| s_k[n-\nu] = s_k[n-\nu]\right\}\Bigg|_{\boldsymbol{\Gamma}_{s_\mathrm{c}}^{(k)}[n]=\tilde{\boldsymbol{\Gamma}}_{s_\mathrm{c}}^{(k)}} - \left|m_{\mathrm{equ},k}\right|^2 \\
&= c_{v_k}[n]\Big|_{\boldsymbol{\Gamma}_{s_\mathrm{c}}^{(k)}[n]=\tilde{\boldsymbol{\Gamma}}_{s_\mathrm{c}}^{(k)}} \\
&= \tilde{\boldsymbol{\omega}}_k^{\mathrm{H}}\left(\boldsymbol{HC}\left(\tilde{\boldsymbol{\Gamma}}_{s_\mathrm{c}}^{(k)} - \boldsymbol{e}_{K\nu+k+1}\boldsymbol{e}_{K\nu+k+1}^{\mathrm{T}}\right)\boldsymbol{C}^{\mathrm{T}}\boldsymbol{H}^{\mathrm{H}} + \boldsymbol{C}_{\boldsymbol{\eta}_\mathrm{c}}\right)\tilde{\boldsymbol{\omega}}_k,
\end{aligned}
\tag{6.14}
$$

where the mean $m_{\mathrm{equ},k}$ of $\hat{s}_k[n]$ in Equation (6.11) for $s_k[n-\nu] = s_k[n-\nu]$ computes as

$$
\begin{aligned}
m_{\mathrm{equ},k} &= \mathrm{E}\left\{\hat{s}_k[n]\big| s_k[n-\nu] = s_k[n-\nu]\right\} \\
&= \tilde{\boldsymbol{\omega}}_k^{\mathrm{H}}\boldsymbol{HC}\boldsymbol{e}_{K\nu+k+1}s_k[n-\nu].
\end{aligned}
\tag{6.15}
$$

Before deriving the mean $m_{\mathrm{ext},k}$ and the variance $c_{\mathrm{ext},k}$ of the extrinsic LLRs in terms of the statistical parameters $m_{\mathrm{equ},k}$ and $c_{\mathrm{equ},k}$ at the output of the $(k+1)$th user's equalizer (cf. Figure 6.2), we present an interesting application of the AWGN channel model in Equation (6.11). In fact, it can be used to approximate the MSE between the QPSK symbol $s_k[n-\nu]$ of user k and its estimate $\hat{s}_k[n]$ according to (cf. also Equation 2.4)

$$
\begin{aligned}
\tilde{\xi}^{(k)} &= \mathrm{E}\left\{\left|s_k[n-\nu] - \hat{s}_k[n]\right|^2\right\}\Bigg|_{\boldsymbol{\Gamma}_{s_\mathrm{c}}^{(k)}[n]=\tilde{\boldsymbol{\Gamma}}_{s_\mathrm{c}}^{(k)}} \\
&= \mathrm{E}\left\{\left|\left(1-\tilde{\boldsymbol{\omega}}_k^{\mathrm{H}}\boldsymbol{HC}\boldsymbol{e}_{K\nu+k+1}\right)s_k[n-\nu] - v_k[n]\right|^2\right\}\Bigg|_{\boldsymbol{\Gamma}_{s_\mathrm{c}}^{(k)}[n]=\tilde{\boldsymbol{\Gamma}}_{s_\mathrm{c}}^{(k)}} \\
&= \left(1-\tilde{\boldsymbol{\omega}}_k^{\mathrm{H}}\boldsymbol{HC}\boldsymbol{e}_{K\nu+k+1}\right)^2 + c_{\mathrm{equ},k},
\end{aligned}
\tag{6.16}
$$

where we used Equation (6.14) and the fact that $|s_k[n-\nu]| = 1$ in case of QPSK modulation (cf. Example 5.2). Note that the estimate $\tilde{\xi}^{(k)}$ in Equation (6.16) reveals the access to the MSE approximation of the optimal linear equalizer $\boldsymbol{\omega}_k[n]$ and $\alpha_k[n]$ for an arbitrary input MI which has been parameterized by c_{apr}. In other words, for each tuple $(I_{\mathrm{apr},k}^{(E)}, I_{\mathrm{ext},k}^{(F)})$ of the EXIT characteristic, there exists a corresponding MSE estimate. Thus, we can estimate the MSE for an arbitrary number of Turbo iterations by using the trajectory as exemplarily depicted in Figure 6.1, as well as the MSE at an infinite number of Turbo iterations if we consider the intersection point in the EXIT chart, both without performing any *Monte Carlo* simulations. Note that an information-theoretic discussion about the connection between MI and MSE in Gaussian channels can be found in [103].

Finally, we have to express the mean $m_{\mathrm{ext},k}$ and the variance $c_{\mathrm{ext},k}$ of the extrinsic LLRs at the output of the soft symbol demapper in terms of $m_{\mathrm{equ},k}$ and $c_{\mathrm{equ},k}$ (cf. Figure 6.2). To do so, we assume that the extrinsic LLRs $l_{\mathrm{ext},k,2n}^{(E)}$ and $l_{\mathrm{ext},k,2n+1}^{(E)}$ are drawn from the random variables $l_{\mathrm{ext},k,0}$ and $l_{\mathrm{ext},k,1}$ both distributed according to the same real normal distribution $\mathcal{N}_{\mathbb{R}}(m_{\text{ext}}, c_{\text{ext}})$. Recall the soft symbol demapper for QPSK modulation given in Equation (5.66) of Example 5.5 which results in the following transformation of the random variable $\hat{s}_k[n]$:

$$l_{\text{ext},k,0} + \mathrm{j}\, l_{\text{ext},k,1} = \frac{\sqrt{8}}{1 - \tilde{\omega}_k^{\mathrm{H}} \boldsymbol{H}\boldsymbol{C}\boldsymbol{e}_{K\nu+k+1}} \hat{s}_k[n]. \tag{6.17}$$

Since this transformation is only a scaling by the real-valued factor $\sqrt{8}/(1 - \tilde{\omega}_k^{\mathrm{H}} \boldsymbol{H}\boldsymbol{C}\boldsymbol{e}_{K\nu+k+1})$, we get the mean

$$\left| \begin{aligned} m_{\text{ext},k} &= \mathrm{E}\left\{ l_{\text{ext},k,0} \big| s_k[n-\nu] = s_k[n-\nu] \right\} \\ &= \frac{\sqrt{8}}{1 - \tilde{\omega}_k^{\mathrm{H}} \boldsymbol{H}\boldsymbol{C}\boldsymbol{e}_{K\nu+k+1}} \, \mathrm{Re}\left\{ m_{\text{equ},k} \right\} \\ &= \frac{\sqrt{8}\, \tilde{\omega}_k^{\mathrm{H}} \boldsymbol{H}\boldsymbol{C}\boldsymbol{e}_{K\nu+k+1}}{1 - \tilde{\omega}_k^{\mathrm{H}} \boldsymbol{H}\boldsymbol{C}\boldsymbol{e}_{K\nu+k+1}} \, \mathrm{Re}\left\{ s_k[n-\nu] \right\}, \end{aligned} \right| \tag{6.18}$$

of $l_{\text{ext},k,0}$ for $s_k[n-\nu] = s_k[n-\nu]$ if we apply Equation (6.15), and its variance

$$\left| \begin{aligned} c_{\text{ext},k} &= \mathrm{E}\left\{ |l_{\text{ext},k,0}|^2 \big| s_k[n-\nu] = s_k[n-\nu] \right\} - m_{\text{ext},k}^2 \\ &= \frac{8}{\left(1 - \tilde{\omega}_k^{\mathrm{H}} \boldsymbol{H}\boldsymbol{C}\boldsymbol{e}_{K\nu+k+1}\right)^2} \frac{c_{\text{equ},k}}{2}. \end{aligned} \right| \tag{6.19}$$

Note that for QPSK modulation as introduced in Example 5.2, $\mathrm{Re}\{s_k[n-\nu]\} = \pm 1/\sqrt{2}$. Again, the mean and variance of $l_{\text{ext},k,1}$ is also $m_{\text{ext},k}$ and $c_{\text{ext},k}$, respectively.

The procedure to semianalytically calculate the tuples $(I_{\text{apr},k}^{(E)}, I_{\text{ext},k}^{(E)})$ of the equalizer-demapper EXIT characteristic for different values of the variance c_{apr} can be summarized as follows:

1. The input MI $I_{\text{apr},k}^{(E)}$ is computed via numerical integration of Equation (6.2) based on the probability density function of Equation (6.1) with mean $m_{\text{apr}} = (1 - 2u)c_{\text{apr}}/2$ and variance c_{apr}.
2. The approximated filter coefficient $\tilde{\omega}_k$ is calculated in terms of c_{apr} with Equations (6.9), (6.8), (6.7), and (6.10).
3. The mean $m_{\text{equ},k}$ and the variance $c_{\text{equ},k}$ are given by Equations (6.15) and (6.14), respectively.
4. These statistical parameters are then used to compute the mean $m_{\text{ext},k}$ and the variance $c_{\text{ext},k}$ of the extrinsic LLRs according to Equations (6.18) and (6.19), respectively, where $\mathrm{Re}\{s_k[n-\nu]\} = (1 - 2u)/\sqrt{2}$, $u \in \{0,1\}$.
5. Finally, the output MI $I_{\text{ext},k}^{(E)}$ is obtained via numerical evaluation of Equation (6.2) based on the probability density function

$$\mathrm{p}_{l_{\mathrm{ext},k}}\left(l_{\mathrm{ext},k}\middle|u=u\right)=\frac{1}{\sqrt{2\pi c_{\mathrm{ext},k}}}\exp\left(-\frac{(l_{\mathrm{ext},k}-m_{\mathrm{ext},k})^2}{2c_{\mathrm{ext},k}}\right).\quad(6.20)$$

We see that the given procedure requires neither a simulated transmission of symbols over a channel nor the evaluation of histogram measurements. However, the equalizer-demapper EXIT characteristic cannot be calculated strictly analytically because the integration in Equation (6.2) still involves numerical methods. This is the reason why we denote the presented calculation as *semianalytical*.

Figure 6.3 is to validate the quality of the semianalytical calculation of the equalizer-demapper EXIT characteristic by comparing it to the one obtained via *Monte Carlo* simulations as described in Subsection 6.1.1. First, we investigate the single-user system with the Proakis b channel and the parameters as already used for the EXIT chart example in Figure 6.1. Besides using the WF for equalization, we consider also the approximation of the WF based on the *Conjugate Gradient* (CG) *implementation* of the *MultiStage WF*, i.e., the *Block Conjugate Gradient* (BCG) *implementation* of the *MultiStage Matrix WF* (MSMWF) given by Algorithm 4.3. if we set $M=1$. Except from the WF where we also present the results for *Time-Variant* (TV) filter coefficients with the auto-covariance matrix $\boldsymbol{\Gamma}_{\boldsymbol{s}_{\mathrm{c}}}[n]$ (cf. Equation 6.5),[5] all equalizers are based on time-invariant filter coefficients where $\boldsymbol{\Gamma}_{\boldsymbol{s}_{\mathrm{c}}}[n]$ has been replaced by $\bar{\boldsymbol{\Gamma}}_{\boldsymbol{s}}$ of Equation (5.54).[6] It can be observed that the semianalytically calculated WF EXIT characteristic is closer to the simulated EXIT characteristic of the time-invariant WF than to the one of the TV WF. This can be explained if we recall that the semianalytical calculation of the equalizer-demapper EXIT characteristic is based on the approximation of $\boldsymbol{\Gamma}_{\boldsymbol{s}_{\mathrm{c}}}[n]$ by the scaled identity matrix $\tilde{c}_{\boldsymbol{s}}\boldsymbol{I}_{Kg}$ where the $(K\nu+k+1)$th diagonal entry has been replaced by one, which is clearly a better approximation of $\bar{\boldsymbol{\Gamma}}_{\boldsymbol{s}}$ than of the TV auto-covariance matrix $\boldsymbol{\Gamma}_{\boldsymbol{s}_{\mathrm{c}}}[n]$. Anyway, Figure 6.3 shows that the semianalytically calculated EXIT characteristics are more or less identical to the simulated ones if we restrict ourselves to the computationally efficient time-invariant equalizers.

Finally, Figure 6.3 depicts also the first user's ($k=0$) equalizer-demapper EXIT characteristic at $10\log_{10}(E_{\mathrm{b}}/N_0)=5\,\mathrm{dB}$ in a multiuser system with $K=4$ users, a spreading factor of $\chi=4$, $R=2$ receive antennas, QPSK modulation ($Q=2$), and one channel realization with $L=5$ taps and an exponential power delay profile as defined in Example 5.3. Here, all users are equalized separately based on the time-invariant version of the VWF in Equation (6.5) with a length of $F=25$, i.e., $N=RF=50$, and a latency time of $\nu=2$. Again, the semianalytically calculated EXIT characteristic is almost

---

[5] For $K=M=1$, it holds: $\boldsymbol{\Gamma}_{\boldsymbol{s}_{\mathrm{c}}}^{(k)}[n]=\boldsymbol{\Gamma}_{\boldsymbol{s}_{\mathrm{c}}}[n]$.

[6] Note that throughout this chapter, the filters with the abbreviations as introduced in the previous chapters for the TV versions are equally used for their time-invariant approximations. It will be explicitly denoted if the filter is TV.

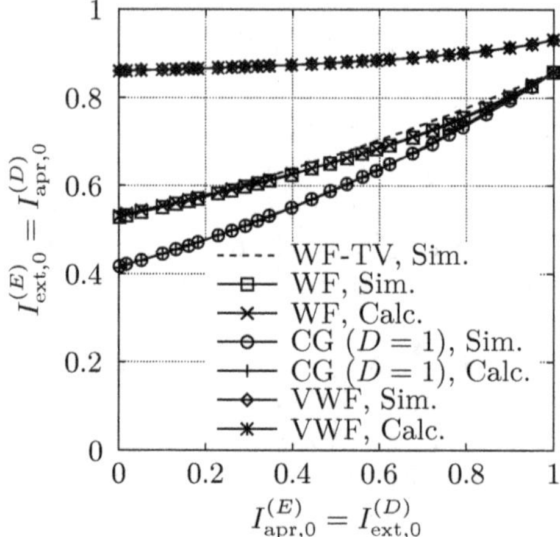

Fig. 6.3. Quality of semianalytical calculation of equalizer-demapper EXIT characteristics for single-user system with Proakis b channel ($K = R = \chi = 1$, $F = 7$) and multiuser system with random channel ($K = \chi = 4$, $R = 2$, $F = 25$) at $10 \log_{10}(E_\mathrm{b}/N_0) = 5\,\mathrm{dB}$

identical to the respective simulated EXIT characteristic which justifies all the assumptions having been necessary to end up in the semianalytical approach.

Although the semianalytical calculation of the equalizer-demapper EXIT characteristic is restricted to the case where the MWF is replaced by K VWFs, the result can also be used as an approximation for the EXIT characteristic of the MWF. Actually, this approximation must be very good if we consider the close BER performance of both methods in a DS-CDMA systems with orthogonal codes (cf. Figure 6.16) which can be explained if we recall the interpretation of the Turbo equalizer as an iterative soft interference canceler (see end of Subsection 5.3.4).

6.2 Analysis Based on Extrinsic Information Transfer Charts

6.2.1 Extrinsic Information Transfer Characteristics for Single-User System with Fixed Channels

Here and in the next subsection, we consider the application of the proposed equalizers to an iterative receiver (cf. Section 5.3) in a single-user system ($K = M = \chi = 1$) with a fixed channel of norm one. As usually done for the evaluation of Turbo systems, we choose the Proakis b, Proakis c [193], or the Porat [190] channel with the coefficients as given in Table 6.1. Thus, the system under consideration has $R = 1$ receive antenna and if we apply an equalizer filter of length $F = 7$, the dimension of the observation vector computes as $N = RF = 7$. Besides, we set the latency time to $\nu = 5$, the size of the data block to $B = 2048$, and transmit QPSK modulated symbols ($Q = 2$, see Example 5.2). With the rate $r = 1/2$ (7, 5)-convolutional encoder of Example 5.1, $rQ = 1$, and the length of the symbol block is the same as the length of the data block, i.e., $S = B/(rQ) = 2048$. Remember that we assume spatially and temporally white noise for all the simulations in this chapter, i.e., $\boldsymbol{C}_{\boldsymbol{\eta}_{\mathrm{c}}} = c_\eta \boldsymbol{I}_{RF}$. Whereas we apply the semianalytical calculation of the EXIT characteristics in the next three subsections, we present here simulated EXIT characteristics of time-variant equalizers, i.e., the statistics $\boldsymbol{\Gamma}_{\boldsymbol{s}_{\mathrm{c}}}[n]$ and $\boldsymbol{\mu}_{\boldsymbol{s}_{\mathrm{c}}}[n]$ are assumed to be time-varying.

Table 6.1. Fixed channels used for simulations of the single-user system

	Proakis b	Proakis c	Porat
L	3	5	5
h_0	0.408	0.227	$0.49 + \mathrm{j}\,0.10$
h_1	0.815	0.46	$0.36 + \mathrm{j}\,0.44$
h_2	0.408	0.688	0.24
h_3	0	0.46	$0.29 - \mathrm{j}\,0.32$
h_4	0	0.227	$0.19 + \mathrm{j}\,0.39$

Figure 6.4 shows the EXIT characteristics of the proposed equalizers for the real-valued and symmetric Proakis b channel at a log-SNR of $10 \log_{10}(E_{\mathrm{b}}/N_0) = 5\,\mathrm{dB}$ as well as the EXIT characteristic of the BCJR decoder from Figure 6.1. It can be seen that all investigated algorithms achieve the same output MI $I_{\mathrm{ext},0}^{(\mathrm{E})}$ in case of full *a priori* knowledge, i.e., for $I_{\mathrm{apr},0}^{(\mathrm{E})} = 1$, because with the *a priori* MI being one, all symbols except from the one that shall be detected are known with high reliability and the interference caused by these symbols can be removed completely. Therefore, the impulse response of the channel is irrelevant and only the norm of the channels and the SNR determine the extrinsic MI. Note that the common intersection point of the different EXIT characteristics at $I_{\mathrm{apr},0}^{(\mathrm{E})} = 1$ means that all considered equalizers perform equally for a sufficient high number of iterations if the SNR is high enough such that their EXIT characteristics do not intersect the one of the decoder. Nevertheless, the number of iterations which are necessary to achieve the optimum depends on the curvature of the corresponding EXIT graph and on the mutual information $I_{\mathrm{ext},0}^{(\mathrm{E})}$ at $I_{\mathrm{apr},0}^{(\mathrm{E})} = 0$ (no *a priori* information available) which corresponds to a coded system with a receiver performing no Turbo iterations. Consequently, the WF with the highest $I_{\mathrm{ext},0}^{(\mathrm{E})}$ for all $I_{\mathrm{apr},0}^{(\mathrm{E})}$ needs the lowest number of Turbo iterations. The proposed CG based MSWF implementation (cf. Algorithm 4.3.) with rank $D = 2$ has only a small degradation with respect to the extrinsic MI compared to the WF. Contrary to the Krylov subspace based MSWF, the eigensubspace based *Principal*

Component (PC) or *Cross-Spectral* (CS) *method* (cf. Subsections 2.3.1 and 2.3.2) with rank $D = 2$ behave even worse than the rank-one MSWF.

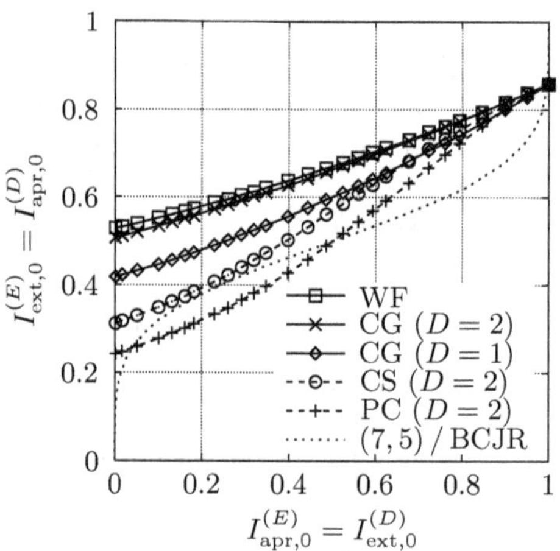

Fig. 6.4. EXIT characteristics of different time-variant equalizers in a single-user system with Proakis b channel ($K = R = \chi = 1$, $F = 7$) at $10 \log_{10}(E_b/N_0) = 5\,\mathrm{dB}$

Next, we consider the real-valued and symmetric Proakis c channel as given in Table 6.1. Compared to the real-valued and symmetric Proakis b channel which already causes severe *InterSymbol Interference* (ISI) due to its low-pass characteristic, the ISI introduced by the Proakis c channel is even higher because of the notch in the middle of the relevant spectrum [193]. Therefore, in Figure 6.5, the resulting EXIT characteristics at $10 \log_{10}(E_b/N_0) = 5\,\mathrm{dB}$ start at a lower MI $I_{\mathrm{ext},0}^{(\mathrm{E})}$ compared to the graphs in Figure 6.4 and their curvatures are positive for all $I_{\mathrm{apr},0}^{(\mathrm{E})}$, i. e., the number of iterations needed for convergence is higher compared to a non-positive curved graph. Again, the optimum can only be reached for high SNR values, e. g., at the investigated log-SNR of 5 dB, only the WF and the CG based MSWF with rank $D = 2$ do not intersect the decoder curve, thus, achieving optimal performance after several iterations. In case of the Proakis c channel, the rank $D = 2$ CS method as the MSE optimal eigensubspace algorithm is better than the rank-one MSWF but performs still worse than the MSWF with rank $D = 2$. Remember that the CS filter is only considered as a performance bound for eigensubspace methods since its computational complexity has the same order as the one of the WF (cf. Subsection 2.3.2). Again, the EXIT characteristic of the MSWF with rank $D = 2$ is almost equal to the one of the full-rank WF.

Finally, we investigate the complex-valued Porat channel (cf. Table 6.1) which is known to be of good nature. The obtained EXIT characteristics are very similar to the one of the Proakis b channel in Figure 6.4 except from

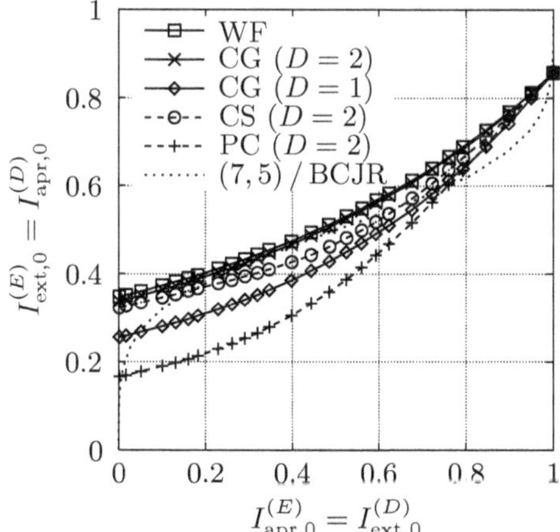

Fig. 6.5. EXIT characteristics of different time-variant equalizers in a single-user system with Proakis c channel ($K = R = \chi = 1$, $F = 7$) at $10\log_{10}(E_{\mathrm{b}}/N_0) = 5\,\mathrm{dB}$

the fact that the equalization of the Porat channel via a CS based reduced-rank WF achieves almost the same MI as the PC based reduced-rank WF. However, again, the CG based MSWF implementation with rank $D = 2$ is a very good approximation of the WF and the rank-one MSWF is even better than the rank $D = 2$ eigensubspace based equalizers.

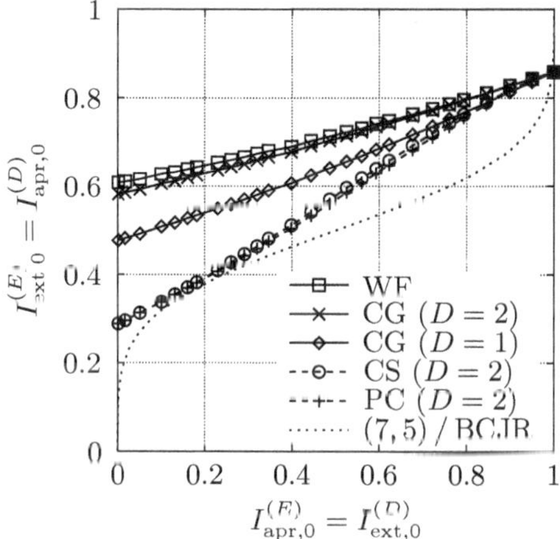

Fig. 6.6. EXIT characteristics of different time-variant equalizers in a single-user system with Porat channel ($K = R = \chi = 1$, $F = 7$) at $10\log_{10}(E_{\mathrm{b}}/N_0) = 5\,\mathrm{dB}$

6.2.2 Mean Square Error Analysis for Single-User System with Fixed Channels

The previous subsection investigated the EXIT characteristics of time-variant equalizers based on *Monte Carlo* simulations at one SNR value. In order to evaluate the EXIT charts for several SNRs, we approximate the equalizer's EXIT characteristics based on the semianalytical approach of Subsection 6.1.2. Recall that these approximations fit almost perfectly the EXIT characteristics of the time-invariant filter solutions (cf. Figure 6.3). Moreover, we use the resulting EXIT charts not only to determine the output MI $I_{\mathrm{ext},k}^{(E)}$ but also to estimate the MSE $\tilde{\xi}^{(k)}$ according to Equation (6.16) for a given number of Turbo iterations as described in Subsection 6.1.2. Here, we discuss especially the MSE performance if the Turbo receiver performs an infinite number of iterations. This theoretical MSE bound is obtained via the output MI at the intersection between the EXIT characteristic of the equalizer and the one of the decoder. The simulation parameters throughout this subsection are the same as those of the single-user system in Subsection 6.2.1.

Figure 6.7 depicts the SNR dependency of the MSE $\tilde{\xi}^{(0)}$ and the MI $I_{\mathrm{ext},0}^{(E)}$ at the output of the corresponding equalizer. We see the close connection between MI and MSE, i. e., the higher the extrinsic MI the smaller the MSE (cf. also [103]). Since a high MI results also in a low BER [20] which is the common performance criterion in coded systems, this observation justifies the choice of the MSE as the optimization criterion for the design of the equalizer filters. In Figure 6.7, we observe the *SNR threshold* (cf. Subsection 5.1.3) which is typical for Turbo systems. Again, the SNR threshold denotes the SNR value where the tunnel between the equalizer's and decoder's EXIT characteristic opens such that the output MI after an infinite number of Turbo iterations (cf. Figure 6.1) suddenly achieves the optimum, i. e., the output MI where the corresponding *a priori* information $I_{\mathrm{apr},0}^{(E)}$ is close to one. Note that the PC based rank $D = 2$ WF achieves this optimum at $10\log_{10}(E_{\mathrm{b}}/N_0) = 15\,\mathrm{dB}$ which is the highest SNR threshold of all considered equalizers. This coincides with the EXIT chart of Figure 6.4 where the EXIT characteristic of the PC algorithm has the smallest output MI for all possible input MIs. In contrast to the PC method, the CS based rank $D = 2$ WF already converges to the full-rank WF at $10\log_{10}(E_{\mathrm{b}}/N_0) = 5\,\mathrm{dB}$. However, the CG based implementation of the rank $D = 2$ MSWF performs identically to the full-rank WF for all given SNR values. Moreover, even the rank-one MSWF is better than the eigensubspace based rank $D = 2$ WFs. Recall from Subsection 4.1.2 that the MSWF with $D = M = 1$ is equal to an arbitrarily scaled cross-covariance vector followed by a scalar WF. Note that contrary to the proposed rank-one MSWF which is used in any turbo iteration, alternative matched filter approaches as introduced in [176] or [239] have a higher computational complexity since they are applied in a *hybrid manner*, i. e., the first iteration still applies a full-rank WF whereas only the following iterations are based on matched filter approximations.

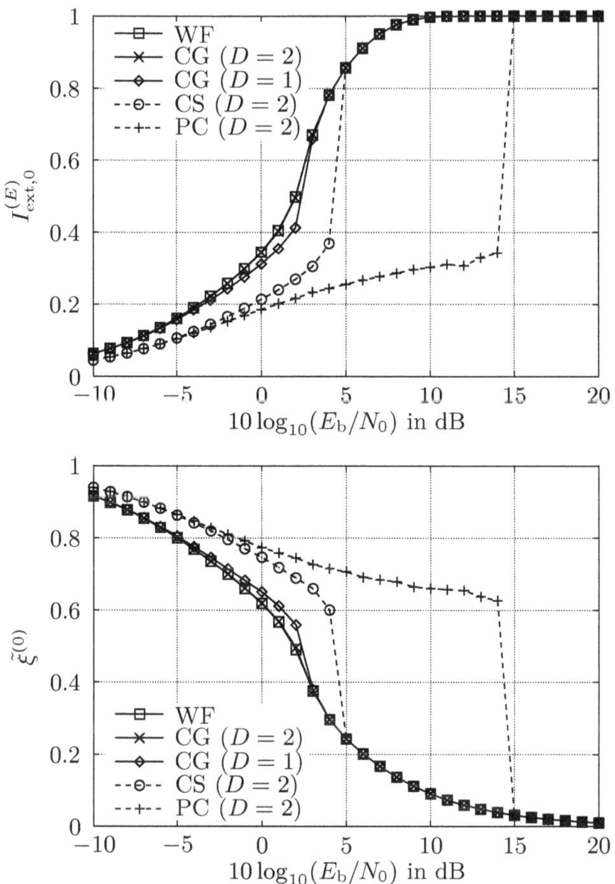

Fig. 6.7. Mutual Information (MI) and Mean Square Error (MSE) performance of different equalizers in a single-user system with Proakis b channel ($K = R = \chi = 1$, $\Gamma = 7$, Turbo iteration ∞)

Next, we consider the MSE performance of the proposed equalizers in case of the Proakis c channel. Since the investigation of the output MI and MSE performance lead again to the same insights, the SNR dependency of the MI is no longer presented. In Figure 6.8, we see again that similar EXIT characteristics in the EXIT chart of Figure 6.5 result in close MSE performances. In fact, the full-rank WF, the CG based MSWF with rank $D = 2$, and the CS based rank $D = 2$ WF have almost the same SNR threshold of $10 \log_{10}(E_\mathrm{b}/N_0) = 6\,\mathrm{dB}$. Only the rank-one MSWF and the PC based rank $D = 2$ WF do not achieve the optimal performance, i.e., the corresponding tunnels in the EXIT charts do not open at all. The close performance of the eigensubspace based CS method with $D = 2$ to the Krylov subspace based CG algorithm with $D = 1$ implies that in case of the Proakis c channel, reduced-rank approximations of the WF in eigensubspaces can achieve, at least in principal, almost the same performance as reduced-rank approximations in Krylov subspaces. However, this is only true for specific channel realizations as we will see when simulating random channels in the sequel of this chapter.

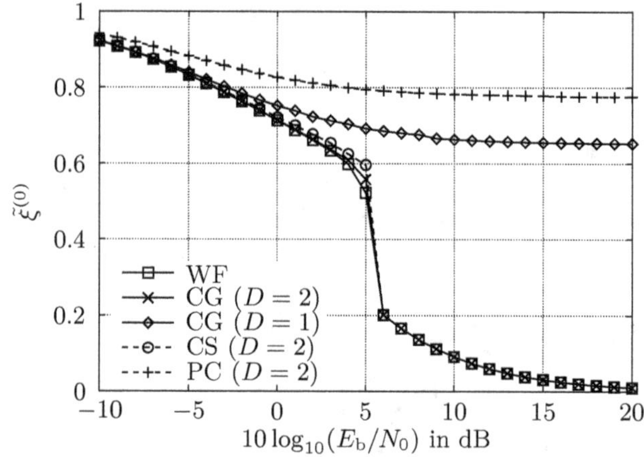

Fig. 6.8. MSE performance of different equalizers in a single-user system with Proakis c channel ($K = R = \chi = 1$, $F = 7$, Turbo iteration ∞)

Finally, Figure 6.9 depicts the MSE performance of the different equalizers if the Porat channel is present. In this case, the eigensubspace based PC and CS rank $D = 2$ WF performs almost equally, and again, worse than the Krylov subspace based WF approximations. However, if $10 \log_{10}(E_b/N_0) \geq 5\,\mathrm{dB}$, all equalizers achieve the same MSE, i. e., the one of the optimal full-rank WF.

To conclude, in the single-user system with the considered fixed channels, the available *a priori* information from the decoder leads to a good MSE performance even if the rank of the WF approximation is reduced drastically to two or even one. If this is still true in multiuser systems with random channels will be investigated in the following subsection.

6.2.3 Mean Square Error Analysis for Multiuser System with Random Channels

Recall the multiuser system as introduced at the end of Subsection 6.1.2. The data bits of each of the $K = 4$ users are encoded based on the $(7,5)$-convolutional channel coder with rate $r = 1/2$, QPSK modulated ($Q = 2$), spread with a factor of $\chi = 4$, transmitted over a frequency-selective channel with an exponential power delay profile of length $L = 5$ (cf. Example 5.3), and received at $R = 2$ antennas. At the receiver side, each of the $K = 4$ users is

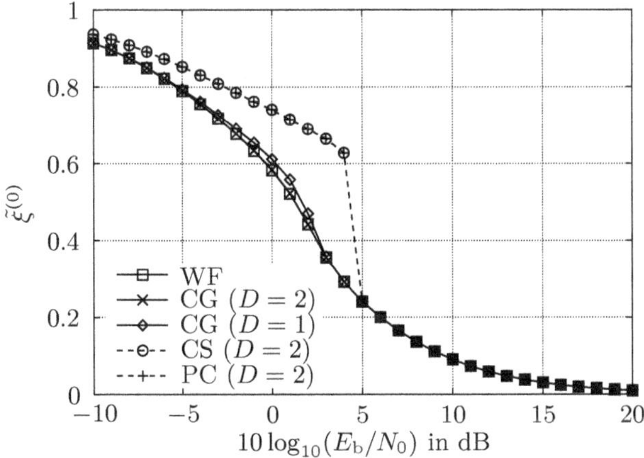

Fig. 6.9. MSE performance of different equalizers in a single-user system with Porat channel ($K = R = \chi = 1$, $F = 7$, Turbo iteration ∞)

equalized based on the VWF in Equation (6.5) of Subsection 6.1.2 with $F = 25$ taps and a latency time of $\nu = 2$, or based on the proposed reduced-rank approximations thereof, viz., the *Vector Conjugate Gradient* (VCG), the *Vector Principal Component* (VPC), or the *Vector Cross-Spectral* (VCS) *method.*[7] Hence, the dimension of the observation vector is again $N = RF = 50$. As in the previous subsection, we use the semianalytical approach to calculate the equalizer's EXIT characteristic for each channel realization and use the resulting EXIT chart with the BCJR algorithm as the decoder to get a MSE estimate. Thus, the following results can be used to discuss the time-invariant VWF where the auto-covariance matrix $\boldsymbol{\Gamma}_{\boldsymbol{s}_c}^{(k)}[n]$ has been replaced by its time-average $\bar{\boldsymbol{\Gamma}}_{\boldsymbol{s}_c}^{(k)}$ (cf. the discussion to Figure 6.3 at the end of Subsection 6.1.2). Without loss of generality, we restrict the following performance analysis to the first user ($k = 0$). The performance of the remaining users is the same due to the fact that the channels of all users have the same statistics (cf. Example 5.3) and the performance measure is averaged over 2000 channel realizations.

Figure 6.10 depicts the MSE $\tilde{\xi}^{(0)}$ of the first user for the proposed equalizers in a receiver without any Turbo feedback as well as a Turbo receiver performing one or two iterations. We see that the eigensubspace based approximations of the VWF with rank $D = 2$ perform worst of all reduced-rank equalizers whereas the Krylov subspace based reduced-rank *MultiStage Vector Wiener Filter* (MSVWF) with $D \in \{1, 2\}$ represent a very good approximation of the full-rank VWF up to certain SNR values. For example, the rank

---

[7] The vector versions of the reduced-rank approximations are obtained by applying the matrix methods of Subsections 2.3.1, 2.3.2, and 4.3.2 for $M = 1$ to the VWF in Equation (6.5).

$D = 2$ VCG in Figure 6.10(a) without any Turbo feedback has almost the same MSE as the VWF up to $10 \log_{10}(E_b/N_0) = 4$ dB, and in the case of one Turbo iteration as shown in Figure 6.10(b) even up to $10 \log_{10}(E_b/N_0) = 14$ dB. After two Turbo iterations, also the rank-one VCG is close to optimum for log-SNRs smaller than 8 dB. In other words, we observe a trade-off between the number of Turbo iterations and the rank of the VWF approximations, i. e., a higher number of Turbo iterations allows a smaller choice of the rank and vice versa. Therefore, the gain due to the Turbo feedback can be exploited to reduce computational complexity in the design of the equalizer as discussed next.

Table 6.2 summarizes the number of FLOPs of the different VWF implementations. Here, we consider the total computational complexity needed per Turbo iteration to equalize all $K = 4$ blocks each with $S = 512$ symbols based on a filter bank with $K = 4$ VWFs.[8] Besides, for the VCG based implementation of the reduced-rank VWF, we assume the same computational complexity as the one of the CG algorithm derived in the next subsection (cf. Equation 6.21). The computational complexity of the VWF is taken from Equation (6.22). Note that the given computational complexity of the full-rank VWF holds only if at least one Turbo iteration is performed because without any Turbo feedback, we can exploit the fact that the system matrix is the same for all users since the auto-covariance matrix $\boldsymbol{\Gamma}_{\boldsymbol{s}_c}^{(k)}[n]$ does not depend on k in this case. Table 6.2 confirms the tremendous reduction in computational complexity when using the VCG algorithm in Turbo receivers. In fact, the computational complexity of the rank $D = 2$ VCG is by factor of more than 4 smaller than the one of the full-rank VWF, and the rank-one VCG even by a factor of more than 9.

Table 6.2. Computational complexity of different VWF implementations per block and Turbo iteration ($N = RF = 50$, $M = K = 4$, $S = 512$)

	D	FLOPs
4 VWFs	50	191 700
4 VCGs	2	42 796
4 VCGs	1	20 996

In Figure 6.11, the theoretical MSE bound is plotted for an optimal Turbo receiver, i. e., a receiver which performs an infinite number of Turbo iterations. Again, we determine this value via the intersection point between the equalizer's and decoder's EXIT characteristics. We see that even for the optimal

---

[8] Note that the block length has not been specified so far because it has no influence on the semianalytical estimation of the MSE. However, for the computational complexity considerations, we assume the same block length as in the *Monte Carlo* simulations of Section 6.3.

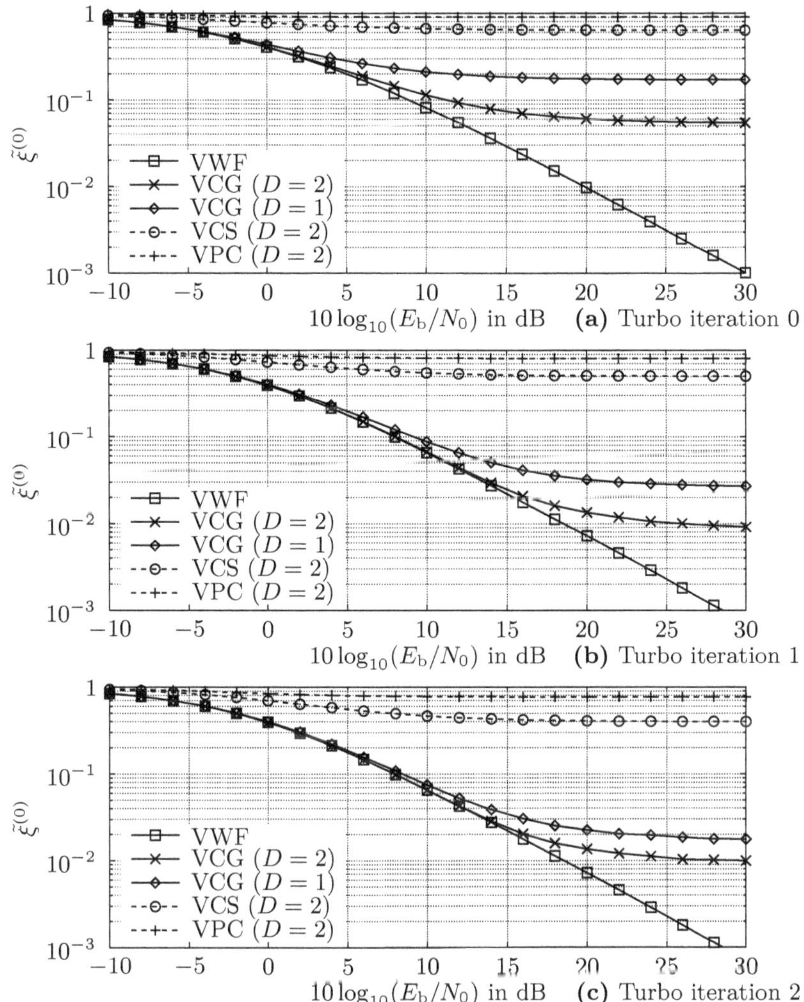

Fig. 6.10. MSE performance of different equalizers in a multiuser system with random channel of an exponential power delay profile ($K = \chi = 4$, $R = 2$, $F = 25$)

Turbo receiver, the approximation of the reduced-rank VWF in 2-dimensional eigensubspaces cannot achieve the performance of the full-rank VWF. Due to this poor average performance in case of random channels, they are no longer investigated in the sequel of this chapter. Considering the Krylov subspace based VCG implementations of the MSVWF, it is astonishing that their rank can be reduced from $N = 50$ to $D = 2$ without changing the performance in the whole given SNR range. Even, the rank-one VCG method is close to optimum up to approximately $10 \log_{10}(E_b/N_0) = 12\,\text{dB}$.

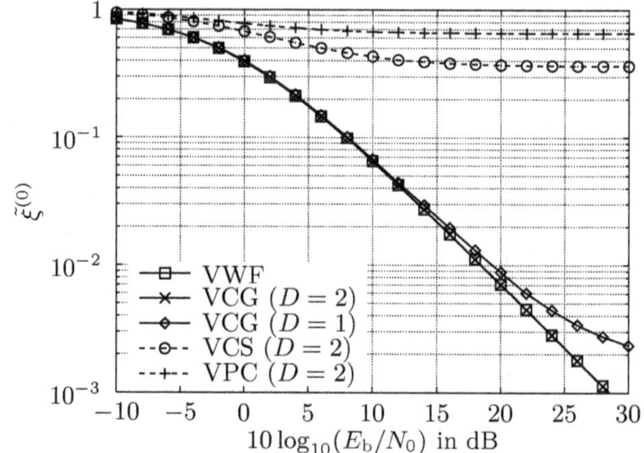

Fig. 6.11. MSE performance of different equalizers in a multiuser system with random channel of an exponential power delay profile ($K = \chi = 4$, $R = 2$, $F = 25$, Turbo iteration ∞)

6.2.4 Complexity Analysis – Rank Versus Order Reduction

In the multiuser system of the previous subsection, we chose a very large dimension N of the observation vector which could be also decreased to reduce computational complexity. The question remains whether it is better to reduce N or the rank D to obtain low-complexity equalizer solutions with an acceptable performance. We restrict this discussion to the special case where $M = 1$. Nevertheless, the generalization to arbitrary M is straightforward. To find out which combination of N and D associated with a given computational complexity yields the best performance according to a chosen criterion, we introduce in the following the *complexity–performance chart*.

Remember from Section 4.4 that the CG algorithm for $D = N$ is computationally more expensive than the Cholesky based WF implementation (cf. Subsection 2.2.1). With $M = 1$, the computational complexity of the BCG algorithm as given in Equation (4.108) reduces to the one of the CG algorithm, i. e.,

$$\zeta_{\mathrm{CG}}(N, D) = \zeta_{\mathrm{BCG}}(N, 1, D) - 2(2D - 1) = 2DN^2 + (9D - 4)N - 1, \quad (6.21)$$

where we considered the fact that solving the system of linear equations with a $M \times M$ system matrix and M right-hand sides which is done $2D - 1$ times in Algorithm 4.3. if $M = 1$,[9] is counted with a computational complexity of $\zeta_{\mathrm{MWF}}(M, M) = 3$ FLOPs for $M = 1$ (cf. Equation 2.28) although the division

---

[9] If $M = 1$, it follows that $d = D$ and $\mu = M$. Thus, $\zeta_{\mathrm{MWF}}(M, M)$ occurs $D - 1$ times in Line 7 of Algorithm 4.3., $D - 2$ times in Line 12 , once in Line 17, and once in Line 21, resulting in $2D - 1$ occurrences.

by a scalar requires actually 1 FLOP. Besides, the computational complexity
of the MWF of Equation (2.28) reduces to the one of the WF

$$\zeta_{\mathrm{WF}}(N) = \zeta_{\mathrm{MWF}}(N, 1) = \frac{1}{3}N^3 + \frac{5}{2}N^2 + \frac{1}{6}N. \tag{6.22}$$

If we use the CG based WF approximation for $D < N$ but the Cholesky based
implementation of the WF for $D = N$, calculating the rank D MSWF for an
arbitrary $D \in \{1, 2, \ldots, N\}$ involves the computational complexity

$$\zeta(N, D) = \begin{cases} \zeta_{\mathrm{CG}}(N, D), & N > D, \\ \zeta_{\mathrm{WF}}(N), & N = D. \end{cases} \tag{6.23}$$

Next, we define the β-ζ_0-*complexity isoset*[10] $\mathbb{I}_{\zeta_0}^{(\beta)}$ as the set of (N, D)-tuples
leading to the computational complexity ζ_0 with a tolerance of $\pm\beta\zeta_0$, i.e.,

$$\mathbb{I}_{\zeta_0}^{(\beta)} = \left\{ (N, D) \in \mathbb{N} \times \{1, 2, \ldots, N\} : \left| \frac{\zeta(N, D) - \zeta_0}{\zeta_0} \right| \leq \beta \right\}.$$

The complexity–performance chart depicts the β-ζ_0-*complexity isoset* in the
N-D-plane for a given computational complexity ζ_0. Finally, the performance
for each element of the β-ζ_0-*complexity isoset* is investigated in order to find
the best combination of the dimension N and the rank D with respect to the
chosen performance criterion and for a given computational complexity ζ_0.

We consider for the following simulations a single-user system ($K = M =
1$) with $R = 1$ antenna at the receiver. Thus, the dimension N of the obser-
vation vector is equal to the filter length F. The results are averaged over
several channel realizations each of length $L = 30$ and a uniform power delay profile as introduced in Example 5.3. Note that the uniform power delay
profile can be seen as a worst case scenario because the performance would
be clearly better for an exponential power delay profile. Figure 6.12 depicts
the complexity–performance chart for $10\log_{10}(E_\mathrm{b}/N_0) = 5$ dB. The tuples of
N and D plotted with the same marker correspond to the 10%-ζ_0-complexity
isoset $\mathbb{I}_{\zeta_0}^{(10\%)}$ with the computational complexity ζ_0 as given in the legend. The
size of a marker depends linearly on the MSE which we have chosen as the
performance criterion. As in the previous subsections, this MSE is estimated
based on the semianalytical approach of Subsection 6.1.2. The filled marker
denotes the combination of N and D with the optimal performance for a given
ζ_0. It can be seen that the reduced-rank method yields the best performance
if we are interested in low-complexity receivers. Only for $\zeta_0 = 50000$ FLOPs,
the full-rank WF achieves the lowest MSE. Moreover, if we perform Turbo
iterations, we observe that the MSE is drastically decreased. In other words,
the computational complexity of the linear equalizer can be further reduced
without changing its performance.

---

[10] Here, we extend the notation *isoset* by introducing a certain tolerance level. This
is necessary due to the fact that N and D are discrete.

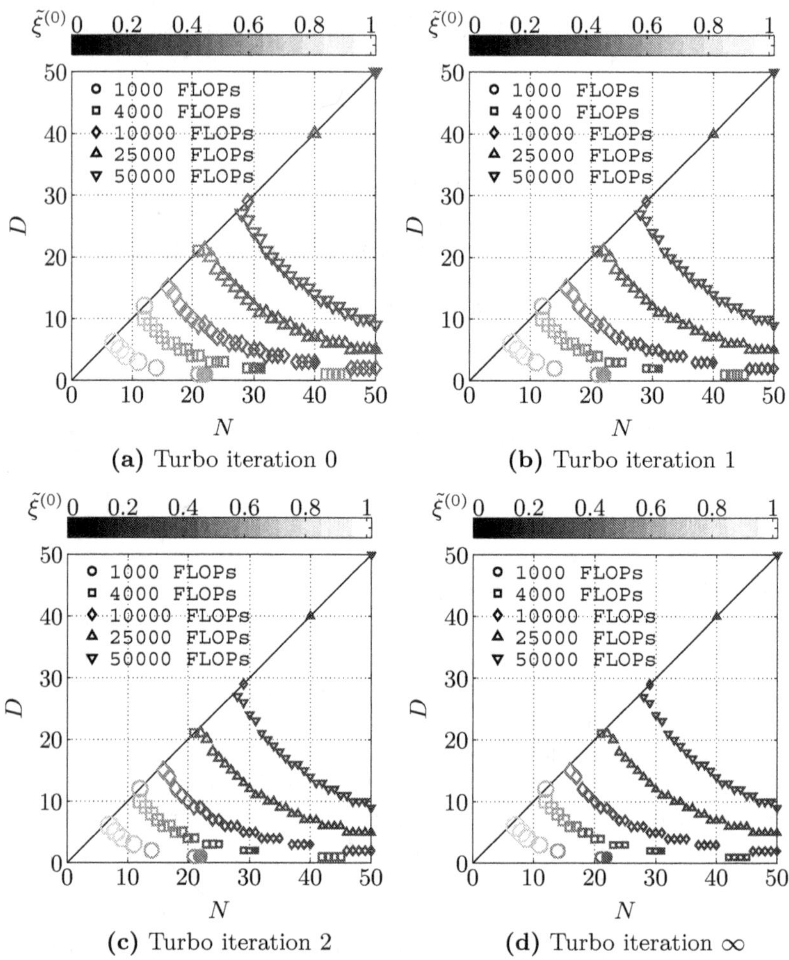

Fig. 6.12. Complexity–performance chart for single-user system with a random channel of a uniform power delay profile ($K = R = \chi = 1$, $F = N$) at $10 \log_{10}(E_b/N_0) = 5\,\mathrm{dB}$ assuming perfect Channel State Information (CSI)

Figure 6.13 presents the complexity–performance chart for a system where the channel is estimated based on 100 pilot symbols using the *Least Squares* (LS) *method* (e. g., [188, 53]). Although the MSE is greater than in the case of perfect channel state information, reducing the rank D should be again preferred to decreasing the filter order of the full-rank WF.

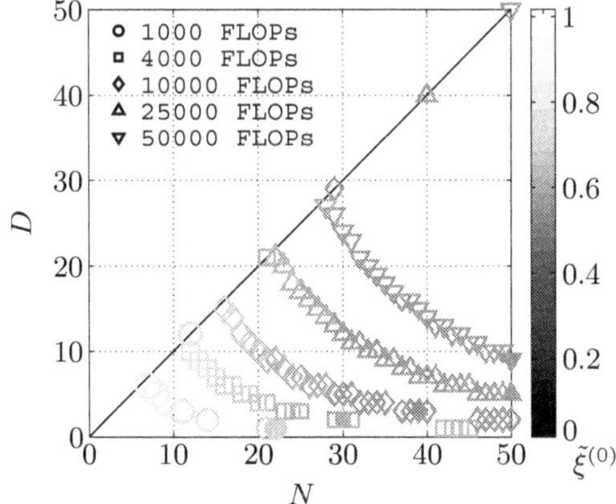

Fig. 6.13. Complexity–performance chart for single-user system with a random channel of a uniform power delay profile ($K = R = \chi = 1$, $F = N$) at $10\log_{10}(E_b/N_0) = 5\,\text{dB}$ assuming estimated CSI (Turbo iteration 0)

6.3 Bit Error Rate Performance Analysis

Again, we consider the multiuser system with $K = 4$ users whose data blocks consisting of $B = 512$ bits are $(7,5)$-convolutionally encoded with rate $r = 1/2$ (cf. Example 5.1), QPSK modulated ($Q = 2$, see Example 5.2), spread with *Orthogonal Variable Spreading Factor* (OVSF) *codes* of length $\chi = 4$ (cf. Subsection 5.1.4), and transmitted over a channel with an exponential power delay profile of length $L = 5$ (cf. Example 5.3). With this setting, the length of the symbol block is the same as the length of the data bit block, i.e., $S = B/(rQ) = 512$, because $rQ = 1$. Again, $C_{\eta_c} = c_\eta I_{RF}$. After receiving the signal at $R = 2$ antennas, it is processed as described in Section 5.3 where the length of the linear equalizer filter is chosen as $F = 25$ and the latency time as $\nu = 2$. Thus, the dimension of the observation vector computes as $N = RF = 50$. Compared to the MSE analysis in the previous section, we consider here the BER at the output of the BCJR decoder (cf. Subsection 5.3.3) obtained via simulating the whole chain of transmitters, channel, and receiver based on *Monte Carlo* methods. All BER results are averaged over several thousand channel realizations.

Figure 6.14 compares the time-invariant MWF and its reduced-rank BCG approximations with their Time-Variant (TV) counterparts. Besides the time-invariant equalizers where the auto-covariance matrix $\boldsymbol{\Gamma}_{s_c}[n]$ in the filter coefficients of Equation (5.50) has been replaced by its time-average $\bar{\boldsymbol{\Gamma}}_{s_c}$ of Equation (5.54), we investigate also the performance of time-invariant equalizers where the auto-covariance matrix has been approximated by the identity matrix (cf. Subsections 5.3.1 and 5.3.4). For receivers without any Turbo feedback, all versions of the auto-covariance matrices are equal to the identity matrix, i.e., $\boldsymbol{\Gamma}_{s_c}[n] = \bar{\boldsymbol{\Gamma}}_{s_c} = I_{RF}$, because all QPSK symbols have the variance $\varrho_s = 1$ if there is no *a priori* information available (cf. Example 5.2). Hence, all different versions of the MWF and the BCG algorithm, respectively, have the same BER performance in Figure 6.14(a). However, if the receiver performs Turbo iterations (cf. Figures 6.14(b) and (c)), we get different behaviors because the auto-covariance matrix $\boldsymbol{\Gamma}_{\boldsymbol{s}_{\mathrm{c}}}[n]$ and its time-average $\bar{\boldsymbol{\Gamma}}_{\boldsymbol{s}_{\mathrm{c}}}$ is no longer the identity matrix. Whereas the performances of the equalizers with the time-average approximation are still very close to those of the TV equalizers, the approximation with the identity matrix raises the BER more severely. For example, after two Turbo iterations in Figure 6.14(c), it achieves the BER of 10^{-4} at an approximately $1\,\mathrm{dB}$ higher log-SNR if compared to its opponents. Moreover, the approximation with $\boldsymbol{I}_{RF}$ degrades the performance of the BCG based implementations of the reduced-rank MSMWF. After performing two Turbo iterations, the rank $D = 4$ BCG has still not converged to the performance of the full-rank MWF which is the case for the BCG algorithm using $\boldsymbol{\Gamma}_{\boldsymbol{s}_{\mathrm{c}}}[n]$ as well as the one based on $\bar{\boldsymbol{\Gamma}}_{\boldsymbol{s}_{\mathrm{c}}}$. Thus, it should be preferred to use the time-average $\bar{\boldsymbol{\Gamma}}_{\boldsymbol{s}_{\mathrm{c}}}$ of the time-varying auto-covariance matrix $\boldsymbol{\Gamma}_{\boldsymbol{s}_{\mathrm{c}}}[n]$ instead of totally neglecting the *a priori* information as done when using the identity matrix. In the following, time-invariant equalizers denote solely the ones based on the time-average approximation $\bar{\boldsymbol{\Gamma}}_{\boldsymbol{s}_{\mathrm{c}}}$.

Note that the approximation of $\boldsymbol{\Gamma}_{\boldsymbol{s}_{\mathrm{c}}}[n]$ by the identity matrix $\boldsymbol{I}_{RF}$ is especially very interesting in the design of low-complexity equalizers. With $\boldsymbol{\Gamma}_{\boldsymbol{s}_{\mathrm{c}}}[n] = \boldsymbol{I}_{RF}$, the auto-covariance matrix of the observation vector reads as $\boldsymbol{\Gamma}_{\boldsymbol{y}}[n] = \boldsymbol{H}\boldsymbol{C}\boldsymbol{C}^{\mathrm{T}}\boldsymbol{H}^{\mathrm{H}} + \boldsymbol{C}_{\boldsymbol{\eta}_{\mathrm{c}}}$, i.e., it has a *block Toeplitz structure* [94] if we recall the block Toeplitz matrices $\boldsymbol{C}$ and $\boldsymbol{H}$ from Equations (5.25) and (5.30), respectively. The auto-covariance matrix $\boldsymbol{C}_{\boldsymbol{\eta}_{\mathrm{c}}}$ is also block Toeplitz due to the definition of the noise vector $\boldsymbol{\eta}_{\mathrm{c}}[\bar{n}]$ according to Equation (5.29) and the fact that the noise is stationary. Thus, computationally efficient *block Toeplitz methods* [2, 255] as mentioned in the introduction to Section 2.2 can be used for cheap implementations of the full-rank MWF in Equation (5.50). Nevertheless, since the performance degradation of the identity approximations is to large in the considered communication system, the block Toeplitz methods are not discussed in this book.

Table 6.3 summarizes the computational complexities of the different MWF implementations for the given simulation parameters based on Equations (2.28), (2.41), and (4.108). Again, we depict the number of FLOPs per Turbo iteration to equalize all $K = 4$ blocks each with $S = 512$ symbols. Considering Figure 6.14 and Table 6.3 together, we must conclude that the first step in the direction of low-complexity equalizers should be switching from TV to time-invariant filters. Whereas the performance degradation can be neglected, the reduction in computational complexity is immense. For example, using the time-invariant MWF instead of the TV MWF decreases the required number of FLOPs by a factor of more than 500. Even the reduced-complexity approach of Subsection 2.2.2 which exploits the time-dependency of the auto-covariance matrices is too computationally intense. Thus, we

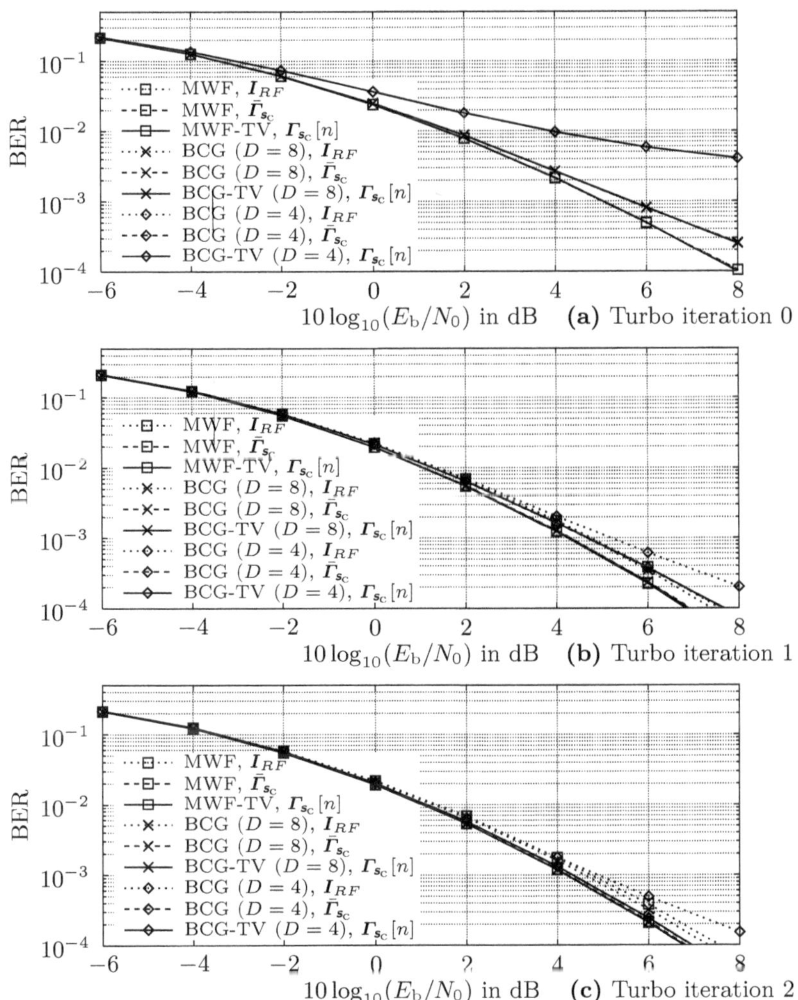

Fig. 6.14. BER performance of time-invariant and Time-Variant (TV) equalizers in a multiuser system with random channel of an exponential power delay profile ($K = \chi = 4$, $R = 2$, $F = 25$)

restrict ourselves to the discussion of time-invariant equalizers in the sequel of this chapter.

Next, we investigate the reduced-rank approximations of the MWF based on the BCG implementation of the MSMWF given by Algorithm 4.3.. In Table 6.3, it can be seen that the BCG with rank $D = 8$ has already a smaller computational complexity than the Cholesky based implementation of the full-rank MWF which is even further reduced if $D < 8$. Figure 6.15 presents

Table 6.3. Computational complexity of different MWF implementations per block and Turbo iteration ($N = RF = 50$, $M = K = 4$, $S = 512$)

	D	FLOPs
MWF-TV	50	32 217 600
RCMWF-TV	50	29 271 614
MWF	50	62 925
BCG	8	50 434
BCG	6	37 759
BCG	4	23 538

the BERs of the dimension-flexible BCG algorithm for $D \in \{4, 6, 8\}$. Note that the rank $D = 6$ requires the dimension-flexible extension of the BCG method as introduced in Subsection 3.4.4. We observe that the BCG with $D = 8$ is close to the full-rank MWF only up to 2 dB if we use a receiver without any Turbo feedback (cf. Figure 6.15(a)). In Turbo receivers though, the rank can be even decreased more drastically without loosing performance compared to the full-rank solution. With one Turbo iteration as depicted in Figure 6.15(b), the rank $D = 6$ BCG performs almost equally to the full-rank MWF down to a BER of 10^{-5}, and after two Turbo iterations in Figure 6.15(c), the rank can be even reduced to $D = 4$. This is again the trade-off between the equalizer's rank and the iterations in Turbo systems as already mentioned in the MSE analysis of Subsection 6.2.3.

In Figure 6.16, we compare the multiuser equalizer based on the MWF of Equation (5.50) with the one based on a filter bank with $K = 4$ VWFs (cf. Equation 6.5). For the Krylov subspace based reduced-rank approximations thereof, we choose the rank in the matrix case to be $D \in \{4, 8\}$ and in the vector case as $D \in \{1, 2\}$. This misadjustment is necessary to end up in comparable computational complexities (cf. Tables 6.3 and 6.2). As already explained at the end of Section 2.1, the MWF and the bank of VWFs perform equally in receivers without any Turbo feedback (see Figure 6.16(a)). However, if the extrinsic information from the decoder is used as *a priori* information for the equalizer, the VWF can also suppress the MAI from the other users at time instance $n - \nu$ (cf. Subsection 5.3.4) which results in a small log-SNR improvement up to approximately 1 dB at 10^{-5} compared to the MWF (cf. Figures 6.16(b) and (c)). Contrary to the full-rank case, the reduced-rank approximations of the VWF have a higher BER than the one of the MWF in receivers without any Turbo feedback as shown in Figure 6.16(a). After one Turbo iteration though (cf. Figure 6.16(b)), the rank $D = 2$ VCG is already better than the MWF but the rank $D = 1$ VCG performs still worse than the computationally comparable rank $D = 4$ BCG for $10 \log_{10}(E_b/N_0) > 6$ dB. In Figure 6.16(c), i.e., after two Turbo iterations, we see that the convergence of the MWF approximations is better than the one of the VWF approximations. Whereas the BER of the rank $D = 4$ BCG is almost the same as the one of the

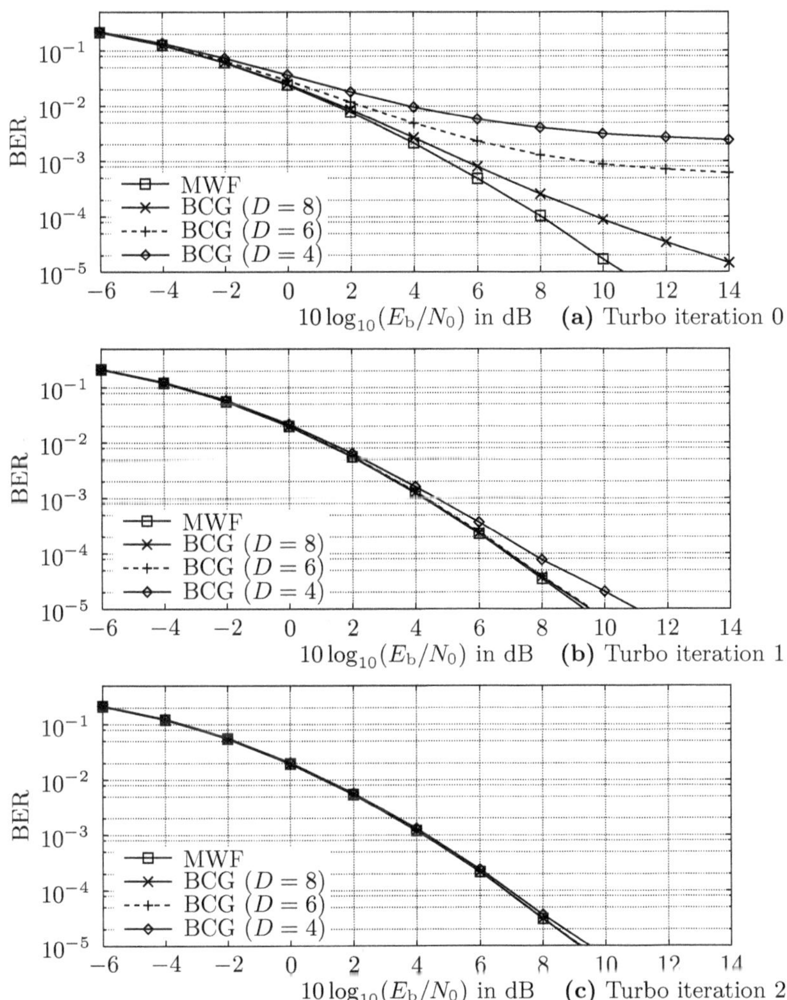

Fig. 6.15. BER performance of the dimension-flexible BCG in a multiuser system with random channel of an exponential power delay profile ($K = \chi = 4$, $R = 2$, $F = 25$)

full-rank MWF, the BER of the rank $D = 1$ VCG is still higher than the one of the full-rank VWF. Nevertheless, the rank $D = 1$ VCG performs already better than the full-rank MWF in the given SNR range. If one is interested in a Turbo system where the number of iterations must be adaptable, the BCG implementation of the MSMWF is preferable compared to the VCG implementation of the MSVWF because its gain in the case without any feedback is tremendous and its loss when performing Turbo iterations is still acceptable.

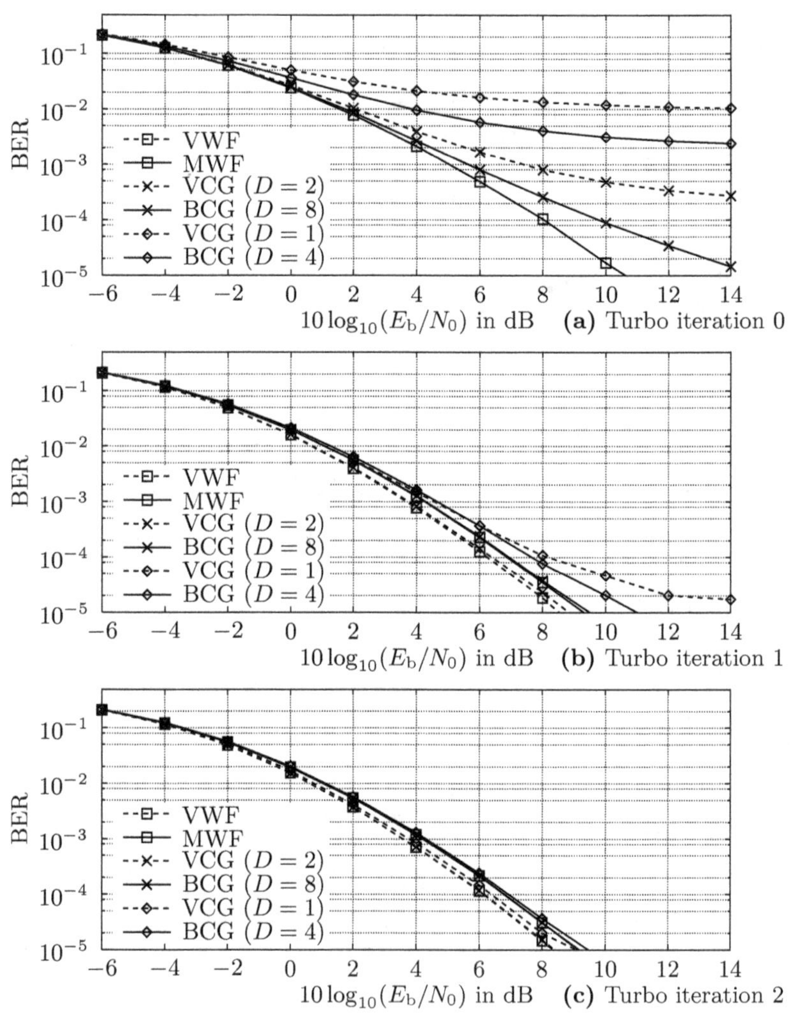

Fig. 6.16. BER performance of vector and matrix equalizers in a multiuser system with random channel of an exponential power delay profile ($K = \chi = 4$, $R = 2$, $F = 25$)

Finally, the remaining simulation results investigate the BER performance of the proposed equalizers if the channel is no longer perfectly known to the receiver. As in Subsection 6.2.4, we consider LS channel estimation (cf., e. g., [188, 53]) due to its computational efficiency.[11]

In Figure 6.17 where the channel has been estimated based on 15 training symbols, the BER performances of the MWF and its reduced-rank approximations are similar to the ones where the receiver has perfect CSI available except from the fact that they are shifted by approximately 3 dB to the right. The further difference is that errors in the channel estimates seem to slow down the convergence of the BCG to the full-rank solution. Whereas after two Turbo iterations, the rank $D = 4$ BCG is already close to the MWF in case of perfect CSI, the rank $D = 4$ BCG has a higher BER than the MWF if the channel is estimated (cf. Figure 6.17(c)).

Figure 6.18 shows the BER performance of a receiver without any Turbo feedback if the channel is estimated with only 5 training symbols which results in severe estimation errors. Thus, the BER compared to the case where 15 training symbols have been used (see Figure 6.17(a)) degrades drastically. However, in this case, we see the regularizing effect of the BCG algorithm as discussed in Subsection 3.4.3, i. e., the BCG algorithm with $D = 8$ achieves amazingly a smaller BER than the full-rank MWF. Note that in Turbo receivers and for better channel estimations as presented in Figure 6.17, the intrinsic regularization of the BCG algorithm is no longer visible. In such cases, it remains to introduce additional regularization via, e. g., *diagonal loading* (see Equation A.24 of Appendix A.4), which can further improve the performance as was shown in [119, 43] for systems without any coding and Turbo feedback.

---

[11] Note that the LS method also requires the solution of a system of linear equations where the system matrix consists of the training symbols. Thus, if the training symbols do not change, the inverse of the system matrix can be computed off-line and the channel estimation reduces to a computationally cheap matrix-vector multiplication. Contrary to the LS method, linear *MMSE channel estimation* requires the solution of a system of linear equations each time when the statistics of the interference and/or noise is changing. Instead of using LS channel estimation for a low-complexity solution, one could also apply the proposed reduced-complexity methods to the linear MMSE channel estimator [63].

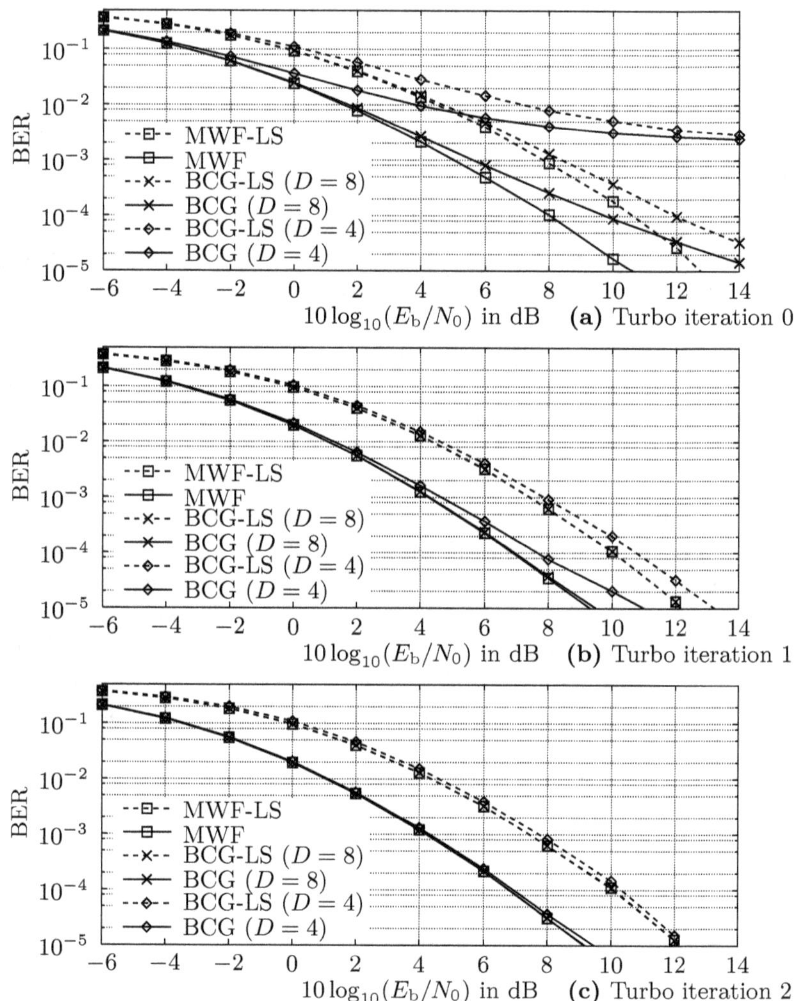

Fig. 6.17. BER performance of different equalizers in a multiuser system with random channel of an exponential power delay profile ($K = \chi = 4$, $R = 2$, $F = 25$) assuming perfect or estimated CSI (15 pilot symbols)

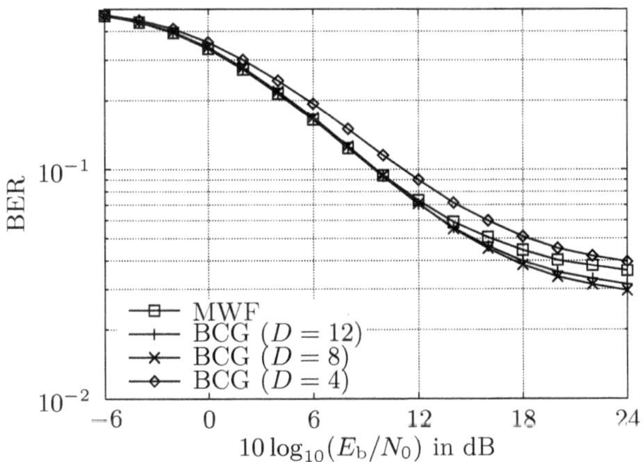

Fig. 6.18. BER performance of different equalizers in a multiuser system with random channel of an exponential power delay profile ($K = \chi = 4$, $R = 2$, $F - 25$) assuming estimated CSI (5 pilot symbols)

7
Conclusions

The first part of this book considered the linear estimation of multivariate non-zero mean and non-stationary random sequences. The design of the *Mean Square Error* (MSE) optimal *Matrix Wiener Filter* (MWF) lead to the Wiener–Hopf equation, i.e., a system of linear equations with multiple right-hand sides. Due to the non-stationarity of the signals, the MWF is time-variant, i.e., it has to be recomputed at each time index. Thus, low-complexity solutions of the MWF are of great value.

Besides solving the Wiener–Hopf equation based on the *Cholesky method*, we also suggested a reduced-complexity algorithm which exploits the time-dependent submatrix structure of the auto-covariance matrix of the observation, i.e., the system matrix of the Wiener–Hopf equation. Moreover, we discussed iterative *block Krylov methods* which are popular procedures in numerical linear algebra to solve very large and sparse systems of linear equations. A detailed *convergence analysis* of the *Block Conjugate Gradient* (BCG) *algorithm* as one of the considered block Krylov methods has shown that the BCG method converges linearly to the optimum solution.

Another strategy to reduce computational complexity is given by reduced-rank signal processing. Contrary to the *Principal Component* (PC) and *Cross-Spectral* (CS) *method*, the *MultiStage MWF* (MSMWF) does not approximate the MWF in an eigensubspace of the auto-covariance matrix of the observation. This book presented the derivation of the MSMWF in its most general form leading to the insights that first, the typical structure of the MSMWF only occurs if it is initialized with the cross-covariance matrix, i.e., the right-hand side matrix of the Wiener–Hopf equation, and second, the performance of the MSMWF strongly depends on the choice of the *blocking matrices*. Moreover, this book revealed the restrictions which are necessary such that the MSMWF is equal to one of the block Krylov methods. This relationship has been exploited to get computationally efficient Krylov subspace based implementations of the MSMWF with a more flexible rank selection than recently published versions. The detailed investigation of the number of *FLoating point OPerations* (FLOPs) required to get the different low-complexity solutions showed that the new MSMWF implementations beat the already very efficient Cholesky based implementation of the MWF if the rank is chosen adequately small.

The second part of this book included the application of the different MWF implementations to a *iterative* or *Turbo receiver* of a coded *Direct Sequence Code Division Multiple-Access* (DS-CDMA) *system*. The exploitation of the *a priori information* obtained from the decoder was the reason for the non-zero mean and non-stationary signal model in the first part of this book.

The evaluation of the *Bit Error Rate* (BER) based on *Monte Carlo* simulations showed that the rank of the MSMWF can be reduced drastically without changing its performance leading to a computationally efficient reduced-rank MWF implementation. Moreover, there exists a trade-off between number of Turbo iterations and the rank of the equalizer, i. e., the higher the number of iterations the smaller the required rank. We compared also the BER performances of the time-variant equalizers with those of the time-invariant approximations thereof which revealed that their BER is almost the same although the time-invariant approximations are computationally much cheaper. Thus, it is recommended to use time-invariant equalizer approximations if one is interested in low-complexity receivers. Besides, we simulated the BER of *Vector Wiener Filters* (VWFs), i. e., vector versions of the MWF which estimate the signal of each user separately. Whereas the VWF and its derivates achieve a smaller BER in systems with many Turbo iterations, the MWF is preferable in systems where the number of iterations is small. Finally, the BERs are discussed in the case where the channel is no longer perfectly known at the receiver but estimated using the *Least Squares* (LS) *method*. If the channel estimation is based on too few training symbols, we observed the regularizing property of the BCG algorithm which we analyzed by means of *filter factors* in the first part of this book. The regularizing effect of the BCG method has the consequence that a smaller rank than the dimension of the observation vector leads to a better performance.

Besides the BER performance analysis, we investigated the MSE performance of the different reduced-rank MWFs based on *EXtrinsic Information Transfer* (EXIT) *charts*. Here, we calculated the EXIT characteristics of the proposed equalizers in a semianalytical way without the simulated transmission of bits. Comparing the MSE performance of the MSMWF with the one of the eigensubspace methods showed that neither the PC nor the MSE optimal CS method are suited for the application to the given communication system. Finally, the introduction of so-called *complexity–performance charts* revealed that it is better to use the MSMWF with a very small rank instead of reducing the filter order if one is interested in low-complexity receivers.

A

Mathematical Basics

A.1 Inverses of Structured Matrices

This subsection presents the inverses of structured matrices which occur frequently in this book. In the following, the matrices $A, D \in \mathbb{C}^{M \times M}$ and $C, F \in \mathbb{C}^{N \times N}$ are Hermitian and invertible, and $B \in \mathbb{C}^{N \times M}$, $N, M \in \mathbb{N}$.

A.1.1 Inverse of a Schur Complement

If the matrix F is the *Schur complement* (see, e.g., [210]) of C, i.e.,

$$F = C - B A^{-1} B^{\mathrm{H}}, \tag{A.1}$$

where the matrices A, B, and C are partitions of a larger matrix (cf. Equation A.4), its inverse computes as

$$F^{-1} = C^{-1} + C^{-1} B D^{-1} B^{\mathrm{H}} C^{-1}, \tag{A.2}$$

with the matrix

$$D = A - B^{\mathrm{H}} C^{-1} B, \tag{A.3}$$

being the Schur complement of A. In [210], the inversion of F is presented as the *matrix inversion lemma*.

A.1.2 Inverse of a Partitioned Matrix

If the Hermitian and non-singular matrix $X \in \mathbb{C}^{(M+N) \times (M+N)}$ can be partitioned according to

$$X = \begin{bmatrix} A & B^{\mathrm{H}} \\ B & C \end{bmatrix}, \tag{A.4}$$

its inverse computes as

$$
\begin{aligned}
\boldsymbol{X}^{-1} &= \begin{bmatrix} \boldsymbol{D}^{-1} & -\boldsymbol{A}^{-1}\boldsymbol{B}^{\mathrm{H}}\boldsymbol{F}^{-1} \\ -\boldsymbol{F}^{-1}\boldsymbol{B}\boldsymbol{A}^{-1} & \boldsymbol{F}^{-1} \end{bmatrix} \\
&= \begin{bmatrix} \boldsymbol{D}^{-1} & -\boldsymbol{D}^{-1}\boldsymbol{B}^{\mathrm{H}}\boldsymbol{C}^{-1} \\ -\boldsymbol{C}^{-1}\boldsymbol{B}\boldsymbol{D}^{-1} & \boldsymbol{F}^{-1} \end{bmatrix},
\end{aligned}
\tag{A.5}
$$

with the Schur complements $\boldsymbol{D}$ and $\boldsymbol{F}$ as given in Equations (A.3) and (A.1), respectively. The first line of Equation (A.5) is the well-known *inversion lemma for partitioned matrices* (see, e. g., [210]). Note that the second equality of Equation (A.5) holds if

$$
\boldsymbol{A}^{-1}\boldsymbol{B}^{\mathrm{H}}\boldsymbol{F}^{-1} = \boldsymbol{D}^{-1}\boldsymbol{B}^{\mathrm{H}}\boldsymbol{C}^{-1},
\tag{A.6}
$$

which can be easily shown by using Equation (A.2) and the inverse

$$
\boldsymbol{D}^{-1} = \boldsymbol{A}^{-1} + \boldsymbol{A}^{-1}\boldsymbol{B}^{\mathrm{H}}\boldsymbol{F}^{-1}\boldsymbol{B}\boldsymbol{A}^{-1},
\tag{A.7}
$$

of the Schur complement $\boldsymbol{D}$ of $\boldsymbol{A}$.

Combining Equation (A.7) with Equation (A.5) yields the following alternative representation of the inverse:

$$
\boldsymbol{X}^{-1} = \begin{bmatrix} \boldsymbol{A}^{-1} & \boldsymbol{0}_{M\times N} \\ \boldsymbol{0}_{N\times M} & \boldsymbol{0}_{N\times N} \end{bmatrix} + \begin{bmatrix} \boldsymbol{A}^{-1}\boldsymbol{B}^{\mathrm{H}} \\ -\boldsymbol{I}_N \end{bmatrix} \boldsymbol{F}^{-1} \begin{bmatrix} \boldsymbol{B}\boldsymbol{A}^{-1} & -\boldsymbol{I}_N \end{bmatrix}.
\tag{A.8}
$$

A.2 Matrix Norms

Let $\boldsymbol{X}, \boldsymbol{Y} \in \mathbb{C}^{N\times M}$, $N, M \in \mathbb{N}$, and $x \in \mathbb{C}$. Then, $\|\cdot\|$ is a *matrix norm* if the following conditions are fulfilled (e. g., [237]):

1. $\|\boldsymbol{X}\| \geq 0$, and $\|\boldsymbol{X}\| = 0$ only if $\boldsymbol{X} = \boldsymbol{0}_{N\times M}$,
2. $\|\boldsymbol{X} + \boldsymbol{Y}\| \leq \|\boldsymbol{X}\| + \|\boldsymbol{Y}\|$, and
3. $\|x\boldsymbol{X}\| = |x|\,\|\boldsymbol{X}\|$.

The next three subsections review some examples of matrix norms which are of great importance for this book.

A.2.1 Hilbert–Schmidt or Frobenius norm

The most important matrix norm is the *Hilbert–Schmidt* or *Frobenius norm* defined as

$$
\|\boldsymbol{X}\|_{\mathrm{F}} = \sqrt{\sum_{n=0}^{N-1}\sum_{m=0}^{M-1} |x_{n,m}|^2} = \sqrt{\mathrm{tr}\,\{\boldsymbol{X}^{\mathrm{H}}\boldsymbol{X}\}},
\tag{A.9}
$$

where $x_{n,m}$ is the element in the $(n+1)$th row and $(m+1)$th column of $\boldsymbol{X}$ and $\mathrm{tr}\{\cdot\}$ denotes the trace of a matrix. If $M = 1$, i. e., the matrix $\boldsymbol{X} \in \mathbb{C}^{N\times M}$

reduces to the vector $\boldsymbol{x} \in \mathbb{C}^N$, the Frobenius norm $\|\boldsymbol{x}\|_F$ is equal to the *Euclidean vector norm* or *vector 2-norm* $\|\boldsymbol{x}\|_2$, i.e.,

$$\|\boldsymbol{x}\|_2 = \sqrt{\sum_{n=0}^{N-1} |x_n|^2} = \sqrt{\boldsymbol{x}^H \boldsymbol{x}} = \|\boldsymbol{x}\|_F, \qquad (A.10)$$

where x_n denotes the $(n+1)$th entry in $\boldsymbol{x}$.

A.2.2 A-norm

If $\boldsymbol{A} \in \mathbb{C}^{N \times N}$ is Hermitian and positive definite, we define the $\boldsymbol{A}$-norm of the matrix $\boldsymbol{X} \in \mathbb{C}^{N \times M}$ as

$$\|\boldsymbol{X}\|_{\boldsymbol{A}} = \sqrt{\operatorname{tr}\{\boldsymbol{X}^H \boldsymbol{A} \boldsymbol{X}\}}. \qquad (A.11)$$

If $\boldsymbol{A} = \boldsymbol{I}_N$, the $\boldsymbol{A}$-norm and the Frobenius norm are identical, i.e., $\|\boldsymbol{X}\|_{\boldsymbol{I}_N} = \|\boldsymbol{X}\|_F$. Note that the $\boldsymbol{A}$-norm is not defined for arbitrary matrices $\boldsymbol{A}$.

A.2.3 2-norm

Compared to the Frobenius or $\boldsymbol{A}$-norm which are *general matrix norms*, the 2-*norm* of a matrix is an *induced matrix norm*, i.e., it is defined with respect to the behavior of the matrix as an operator between two normed vector spaces. The 2-norm of $\boldsymbol{X}$ reads as

$$\|\boldsymbol{X}\|_2 = \sup_{\boldsymbol{a} \neq \boldsymbol{0}_M} \frac{\|\boldsymbol{X}\boldsymbol{a}\|_2}{\|\boldsymbol{a}\|_2} = \sup_{\substack{\boldsymbol{a} \in \mathbb{C}^M \\ \|\boldsymbol{a}\|_2 = 1}} \|\boldsymbol{X}\boldsymbol{a}\|_2, \qquad (A.12)$$

and is equal to the largest singular value $\sigma_{\max}(\boldsymbol{X})$ of $\boldsymbol{X}$, i.e., $\|\boldsymbol{X}\|_2 = \sigma_{\max}(\boldsymbol{X})$.

With the 2-norm, one can express the *condition number* $\kappa(\boldsymbol{X})$ of a matrix $\boldsymbol{X}$ as follows:

$$\kappa(\boldsymbol{X}) = \frac{\sigma_{\max}(\boldsymbol{X})}{\sigma_{\min}(\boldsymbol{X})} = \|\boldsymbol{X}\|_2 \|\boldsymbol{X}^\dagger\|_2, \qquad (A.13)$$

where $\sigma_{\min}(\boldsymbol{X})$ denotes the smallest singular value of $\boldsymbol{X}$ and $\boldsymbol{X}^\dagger = (\boldsymbol{X}^H \boldsymbol{X})^{-1} \boldsymbol{X}^H$ is the left-hand side *Moore–Penrose pseudoinverse* [167, 168, 184, 97] of $\boldsymbol{X}$. If $\boldsymbol{X}$ is square and non-singular, $\kappa(\boldsymbol{X}) = \|\boldsymbol{X}\|_2 \|\boldsymbol{X}^{-1}\|_2$.

A.3 Chebyshev Polynomials

The *Chebyshev polynomial* $T^{(n)}(x)$ of the first kind and degree $n \in \mathbb{N}_0$ [28] is defined as

$$T^{(n)}(x) = \frac{1}{2}\left(\left(x + (x^2 - 1)^{\frac{1}{2}}\right)^n + \left(x - (x^2 - 1)^{\frac{1}{2}}\right)^n\right) \in \mathbb{R}, \quad x \in \mathbb{R}. \tag{A.14}$$

With the inverse cosine defined as [1]

$$\arccos(x) = -\mathrm{j}\,\log\left(x + (x^2 - 1)^{\frac{1}{2}}\right) \in \mathbb{C}, \tag{A.15}$$

and due to the fact that $(x + (x^2 - 1)^{1/2})(x - (x^2 - 1)^{1/2}) = 1$, Equation (A.14) can be rewritten as

$$T^{(n)}(x) = \frac{1}{2}\left(\exp\left(\mathrm{j}\,n\arccos(x)\right)) + \exp\left(-\mathrm{j}\,n\arccos(x)\right)\right). \tag{A.16}$$

Finally, with *Euler's formula* $\exp(z) = \cos(z) + \mathrm{j}\sin(z)$, the cosine can be replaced by $\cos(z) = (\exp(\mathrm{j}\,z) + \exp(-\mathrm{j}\,z))/2$, and Equation (A.16) reads as

$$T^{(n)}(x) = \cos\left(n\arccos(x)\right) \in \mathbb{R}, \quad x \in \mathbb{R}, \tag{A.17}$$

which is an alternative expression of the Chebyshev polynomial of the first kind and degree n. A third representation of the Chebyshev polynomial $T^{(n)}(x)$ is given by the recursion formulas (e. g., [1])

$$T^{(n)}(x) = 2xT^{(n-1)}(x) - T^{(n-2)}(x), \quad n \geq 2, \tag{A.18}$$

$$T^{(2n)}(x) = 2T^{(n),2}(x) - 1, \quad n \geq 0, \tag{A.19}$$

initialized with $T^{(0)}(x) = 1$ and $T^{(1)}(x) = x$. Figure A.1 depicts exemplarily the Chebyshev polynomials $T^{(n)}(x)$ of degree $n \in \{2, 3, 8\}$.

It follows the most important properties of Chebyshev polynomials of the first kind. From Equation (A.17), we see that $|T^{(n)}(x)| \leq 1$ for $x \in [-1, 1]$ and $T^{(n)}(0) = 1$. Besides, the Chebyshev polynomial $T^{(n)}(x)$ is even or odd, if n is even or odd, respectively (cf. Equation A.14). Finally, it can be easily verified that

$$T^{(n)}(x_i) = (-1)^i \quad \text{with} \quad x_i = \cos\left(\frac{i\pi}{n}\right), \quad i \in \{0, 1, \ldots, n\}, \tag{A.20}$$

are the $n + 1$ local extrema of the Chebyshev polynomial $T^{(n)}(x)$ if x is restricted to the region $[-1, 1]$. Note that $T^{(n)}(-1) = (-1)^n$ and $T^{(n)}(1) = 1$ are no longer local extrema of $T^{(n)}(x)$ if $x \in \mathbb{R}$ (cf. also Figure A.1). Since a polynomial in $\mathbb{R}$ of degree n has at most $n - 1$ extrema and due to the fact that all of the extrema of $T^{(n)}(x)$ lie in $[-1, 1]$ and $|T^{(n)}(\pm 1)| = 1$, the values $|T^{(n)}(x)|$ for x outside of $[-1, 1]$ are larger than one, i. e., $|T^{(n)}(x)| > 1$ for $|x| > 1$.

A.4 Regularization

Given the operator $A : \mathcal{X} \to \mathcal{B}, x \mapsto b$, between the abstract vector spaces $\mathcal{X}$ and $\mathcal{B}$, and an output vector b, the *inverse problem* (e. g., [110]) is to find the

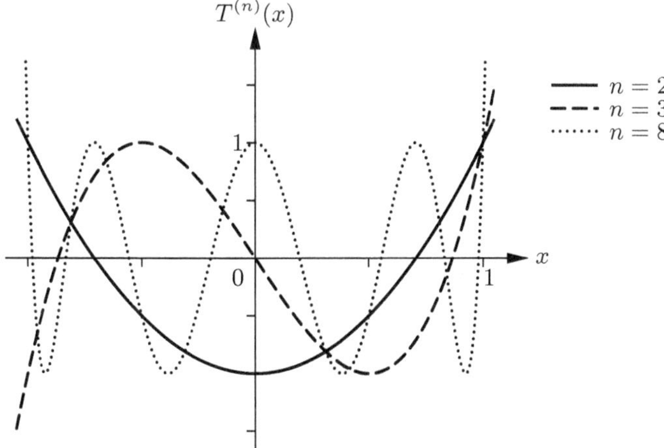

Fig. A.1. Chebyshev polynomials $T^{(n)}(x)$ of the first kind and degree $n \in \{2, 3, 8\}$

corresponding input vector x such that the application of A to x results in b. Although the operator can be either linear or non-linear, we restrict ourselves in the following to the linear case which is the focus of this book. One example for a linear inverse problem is to solve the *Fredholm integral equation of the first kind* (see, e. g., [99]), i. e.,

$$\int_0^1 A(s,t)x(t)\,\mathrm{d}t = b(s), \quad 0 \le s \le 1, \tag{A.21}$$

with respect to the unknown function $x : [0,1] \to \mathbb{C}, t \mapsto x(t)$. Here, the square integrable kernel function $A : [0,1] \times [0,1] \to \mathbb{C}, (s,t) \mapsto A(s,t)$, and the right-hand side function $b : [0,1] \to \mathbb{C}, s \mapsto b(s)$, are known functions. Another example of a linear inverse problem where the vector spaces have the finite dimension NM, is solving the system

$$\sum_{m=0}^{N-1} a_{n,m}\bar{\boldsymbol{x}}_m^{\mathrm{T}} = \bar{\boldsymbol{b}}_n^{\mathrm{T}}, \quad n \in \{0,1,\ldots,N-1\}, \quad \text{or} \quad \boldsymbol{A}\boldsymbol{X} = \boldsymbol{B}, \tag{A.22}$$

of linear equations with multiple right-hand sides with respect to the unknown matrix $\boldsymbol{X} = [\bar{\boldsymbol{x}}_0, \bar{\boldsymbol{x}}_1, \ldots, \bar{\boldsymbol{x}}_{N-1}]^{\mathrm{T}} \in \mathbb{C}^{N \times M}$, $N, M \in \mathbb{N}$. The transpose of the vector $\bar{\boldsymbol{x}}_m \in \mathbb{C}^M$ denotes the $(m+1)$th row of the matrix $\boldsymbol{X}$. Again, the system matrix $\boldsymbol{A} \in \mathbb{C}^{N \times N}$ whose element in the $(n+1)$th row and $(m+1)$th column is denoted as $a_{n,m} = \boldsymbol{e}_{n+1}^{\mathrm{T}}\boldsymbol{A}\boldsymbol{e}_{m+1}$, and the right-hand side matrix $\boldsymbol{B} = [\bar{\boldsymbol{b}}_0, \bar{\boldsymbol{b}}_1, \ldots, \bar{\boldsymbol{b}}_{N-1}]^{\mathrm{T}} \in \mathbb{C}^{N \times M}$ are known. In the sequel, we restrict ourselves to the system of linear equations as an example of an inverse problem. However, the following derivation can be easily extended to the general inverse problem defined above.

Solving an inverse problem can be challenging if the problem is either *ill-posed* or *ill-conditioned*. Hadamard [104] defined a problem as ill-posed if no unique solution exists or if it is not a continuous function in the right-hand side, i. e., an arbitrarily small error in B can cause an arbitrarily large error in the solution X. A problem is denoted as ill-conditioned if the condition number of the system matrix A (cf. Appendix A.2) is very large, i. e., a small error in B causes a large error in the solution X. Especially, in floating or fixed point arithmetic, ill-conditioned problems lead to numerical instability.

The method of *regularization* (e. g., [110]) introduces additional assumptions about the solution like, e. g., smoothness or a bounded norm, in order to handle ill-posed or ill-conditioned problems. In the remainder of this section, we discuss *Tikhonov regularization* [234, 235] in order to illustrate the basic idea of regularizing the solution of an inverse problem. With the system of linear equations defined in Equation (A.22), the Tikhonov regularized solution $X_{\text{reg}} \in \mathbb{C}^{N \times M}$ can be found via the following optimization:

$$X_{\text{reg}} = \underset{X \in \mathbb{C}^{N \times M}}{\operatorname{argmin}} \left(\|AX - B\|_{\text{F}}^2 + \alpha \|X\|_{\text{F}}^2 \right). \tag{A.23}$$

Here, $\| \cdot \|_{\text{F}}$ denotes the Frobenius norm as defined in Appendix A.2, and $\alpha \in \mathbb{R}_{0,+}$ is the so-called *Tikhonov factor*. Finally, the regularized solution computes as

$$X_{\text{reg}} = \left(A^{\text{H}} A + \alpha I_N \right)^{-1} A^{\text{H}} B. \tag{A.24}$$

It remains to choose the factor α properly. Note that if we set $\alpha = 0$, the regularized solution in Equation (A.24) reduces to the *Least Squares* (LS) *solution*. Besides heuristic choices for α, it can be derived via a *Bayesian approach* if the solution and the right-hand side are assumed to be multivariate random variables (see, e. g., [235]).

The *filter factors* are a well-known tool to characterize regularization. If $A = U \Lambda U^{\text{H}}$ denotes the *eigen decomposition* (see, e. g., [131]) of the system matrix (cf. Equation 3.74), the exact solution of Equation (A.22) can be written as

$$X = A^{-1} B = U \Lambda^{-1} U^{\text{H}} B. \tag{A.25}$$

Then, the regularized solution is defined as

$$X_{\text{reg}} = U F \Lambda^{-1} U^{\text{H}} B, \tag{A.26}$$

where the diagonal matrix $F \in \mathbb{R}_{+}^{N \times N}$, in the following denoted as the *filter factor matrix*, comprises the filter factors. The diagonal elements in F attenuate the inverse eigenvalues in Λ^{-1} in order to suppress the eigenmodes which have the highest influence on the error in the solution. In the case of Tikhonov regularization, we get (cf. Equation A.24)

$$X_{\text{reg}} = U \left(\Lambda^2 + \alpha I_N \right)^{-1} \Lambda U^{\text{H}} B, \tag{A.27}$$

leading to the filter factor matrix

$$F = (\Lambda^2 + \alpha I_N)^{-1} \Lambda^2, \quad \text{with} \quad f_i = e_{i+1}^{\text{T}} F e_{i+1} = \frac{\lambda_i^2}{\lambda_i^2 + \alpha} \in \mathbb{R}_+, \tag{A.28}$$

being the Tikhonov filter factors. We observe that

$$\lim_{\lambda_i \to 0} f_i = \lim_{\lambda_i} \frac{\lambda_i^2}{\lambda_i^2 + \alpha} = 0, \tag{A.29}$$

i.e., the filter factors f_i corresponding to small eigenvalues are very small. If we remember that especially errors in the small eigenvalues cause a large error in the solution due to their inversion in Equation (A.25), we see that the suppression of the eigenmodes corresponding to small eigenvalues, by the Tikhonov filter factors decreases the error in the regularized solution X_{reg}.

A.5 Vector Valued Function of a Matrix and Kronecker Product

The *vector valued function* of the matrix $A \in \mathbb{C}^{N \times M}$, $N, M \in \mathbb{N}$, in the following denoted as the vec-*operation*, is defined as (e.g., [17])

$$\text{vec}\{A\} = \begin{bmatrix} a_0 \\ a_1 \\ \vdots \\ a_{M-1} \end{bmatrix} \in \mathbb{C}^{NM}, \tag{A.30}$$

where the vector $a_m = A e_{m+1} \in \mathbb{C}^N$, $m \in \{0, 1, \ldots, M-1\}$, is the $(m+1)$th column of A.

The *Kronecker product* between the matrix $A \in \mathbb{C}^{N \times M}$ and the matrix $B \in \mathbb{C}^{Q \times P}$, $Q, P \in \mathbb{N}$, is defined as (e.g., [17])

$$A \otimes B = \begin{bmatrix} a_{0,0} B & a_{0,1} B & \cdots & a_{0,M-1} B \\ a_{1,0} B & a_{1,1} B & \cdots & a_{1,M-1} B \\ \vdots & \vdots & \ddots & \vdots \\ a_{N-1,0} B & a_{N-1,1} B & \cdots & a_{N-1,M-1} B \end{bmatrix} \in \mathbb{C}^{NQ \times MP}, \tag{A.31}$$

where the scalar $a_{n,m} \in \mathbb{C}$, $n \in \{0, 1, \ldots, N-1\}$, $m \in \{0, 1, \ldots, M-1\}$, denotes the element in the $(n+1)$th row and $(m+1)$th column of A, i.e., $a_{n,m} = e_{n+1}^{\text{T}} A e_{m+1}$.

It follows a summary of the most important properties of the vec-operation and the Kronecker product which are used in this book. A detailed derivation of the formulas can be found in [17]. If $\boldsymbol{A} \in \mathbb{C}^{N \times M}$, $\boldsymbol{B} \in \mathbb{C}^{Q \times P}$, $\boldsymbol{C} \in \mathbb{C}^{M \times S}$, $\boldsymbol{D} \in \mathbb{C}^{P \times R}$, $\boldsymbol{E} \in \mathbb{C}^{S \times R}$, $\boldsymbol{F} \in \mathbb{C}^{N \times N}$, and $\boldsymbol{G} \in \mathbb{C}^{M \times M}$, $S, R \in \mathbb{N}$, with the latter two matrices being non-singular, i. e., the inverse matrices exist, it holds:

$$(\boldsymbol{A} \otimes \boldsymbol{B})\,(\boldsymbol{C} \otimes \boldsymbol{D}) = \boldsymbol{A}\boldsymbol{C} \otimes \boldsymbol{B}\boldsymbol{D} \in \mathbb{C}^{NQ \times SR}, \tag{A.32}$$

$$\operatorname{vec}\{\boldsymbol{A}\boldsymbol{C}\boldsymbol{E}\} = \left(\boldsymbol{E}^{\mathrm{T}} \otimes \boldsymbol{A}\right) \operatorname{vec}\{\boldsymbol{C}\} \in \mathbb{C}^{NR}, \tag{A.33}$$

$$(\boldsymbol{F} \otimes \boldsymbol{G})^{-1} = \boldsymbol{F}^{-1} \otimes \boldsymbol{G}^{-1} \in \mathbb{C}^{NM \times NM}. \tag{A.34}$$

A.6 Square Root Matrices

Let $\boldsymbol{A} \in \mathbb{C}^{N \times N}$, $N \in \mathbb{N}$, be a Hermitian and positive semidefinite matrix with rank R. Then, there exists the *eigen decomposition* (see, e. g., [131])

$$\boldsymbol{A} = \sum_{i=0}^{R-1} \lambda_i \boldsymbol{u}_i \boldsymbol{u}_i^{\mathrm{H}}, \tag{A.35}$$

with the *eigenvalues* $\lambda_i \in \mathbb{R}_+$, $i \in \{0, 1, \ldots, R-1\}$, and the corresponding orthonormal *eigenvectors* $\boldsymbol{u}_i \in \mathbb{C}^N$, i. e., $\boldsymbol{u}_i^{\mathrm{H}} \boldsymbol{u}_j = \delta_{i,j}$, $i, j \in \{0, 1, \ldots, R-1\}$. Here, $\delta_{i,j}$ denotes the *Kronecker delta* being one only for $i = j$ and zero otherwise. In the following, we review two different definitions of the *square root matrix* of $\boldsymbol{A}$. Note that we assume for all definitions that the square root matrix is Hermitian.

The *general square root matrices* $\sqrt{\boldsymbol{A}} \in \mathbb{C}^{N \times N}$ of $\boldsymbol{A}$ with $\sqrt{\boldsymbol{A}}\sqrt{\boldsymbol{A}} = (\sqrt{\boldsymbol{A}})^2 = \boldsymbol{A}$ are defined as

$$\sqrt{\boldsymbol{A}} := \sum_{i=0}^{R-1} \pm\sqrt{\lambda_i}\,\boldsymbol{u}_i \boldsymbol{u}_i^{\mathrm{H}}. \tag{A.36}$$

Due to the fact that the signs of the square roots $\sqrt{\lambda_i}$ can be chosen arbitrarily without changing the product $(\sqrt{\boldsymbol{A}})^2 = \boldsymbol{A}$, there exists 2^R different general square root matrices.

If we are interested in a unique definition of the square root matrix of $\boldsymbol{A}$, we can restrict it to be not only Hermitian but also positive semidefinite (see, e. g., [131]), i. e., we choose the one of the general square root matrices of $\boldsymbol{A}$ where the eigenvalues are non-negative. The unique *positive semidefinite square root matrix* reads as

$$\sqrt[\oplus]{\boldsymbol{A}} := \sum_{i=0}^{R-1} \sqrt{\lambda_i}\,\boldsymbol{u}_i \boldsymbol{u}_i^{\mathrm{H}} \in \mathbb{C}^{N \times N}. \tag{A.37}$$

A.7 Binary Galois Field

The set $\mathbb{F}$ together with the operations '$\oplus$' and '$\diamond$', in the following denoted as $(\mathbb{F}, \oplus, \diamond)$, is a *field* if $a, b, c \in \mathbb{F}$ satisfy the following conditions (e.g., [118]):

1. *Closure of $\mathbb{F}$ under '$\oplus$' and '$\diamond$'*:
   $a \oplus b \in \mathbb{F}$ and $a \diamond b \in \mathbb{F}$.
2. *Associative law of both operations*:
   $a \oplus (b \oplus c) = (a \oplus b) \oplus c$ and $a \diamond (b \diamond c) = (a \diamond b) \diamond c$.
3. *Commutative law of both operations*:
   $a \oplus b = b \oplus a$ and $a \diamond b = b \diamond a$.
4. *Distributive law of '$\diamond$' over '$\oplus$'*:
   $a \diamond (b \oplus c) = (a \diamond b) \oplus (a \diamond c)$.
5. *Existence of an additive and a multiplicative identity*:
   There exists the *zero element* or *additive identity* $n \in \mathbb{F}$ such that $a \oplus n = a$, and the *unit element* or *multiplicative identity* $e \in \mathbb{F}$ such that $a \diamond e = a$.
6. *Existence of additive and multiplicative inverses*:
   For every $a \in \mathbb{F}$, there exists an *additive inverse* $\bar{a}$ such that $a \oplus \bar{a} = n$ and if $a \neq n$, there exists a *multiplicative inverse* $\underline{a}$ such that $a \diamond \underline{a} = e$.

The most famous example of a field is $(\mathbb{R}, +, \cdot)$ with the common addition '$+$', the common multiplication '$\cdot$', the zero element 0, the unit element 1, the additive inverse $-a$ of $a \in \mathbb{R}$, and the multiplicative inverse $1/a$ of $a \in \mathbb{R}\backslash\{0\}$.

The field $(\mathbb{F}, \oplus, \diamond)$ with a finite set $\mathbb{F}$ is called *Galois field* (e.g., [118]). If q is the cardinality of $\mathbb{F}$, the respective Galois field is denoted as $GF(q)$. For example, the field $(\{0, 1\}, \boxplus, \boxdot)$, i.e., the set $\{0, 1\}$ together with the modulo 2 addition '$\boxplus$' and the modulo 2 multiplication '$\boxdot$' as defined in Table A.1, is the *binary Galois field* $GF(2)$ (cf., e.g., [163, 118]) which plays a fundamental role in digital communications (see Chapter 5).

Table A.1. Modulo 2 addition '$\boxplus$' and modulo 2 multiplication '$\boxdot$'

$\boxplus$	0	1		$\boxdot$	0	1
0	0	1		0	0	0
1	1	0		1	0	1

Note that in $GF(2)$, 0 is the zero element since $a \boxplus 0 = a$, $a \in \{0, 1\}$, 1 the unit element since $a \boxdot 1 = a$, 0 the additive inverse of 0 since $0 \boxplus 0 = 0$, 1 the additive inverse of 1 since $1 \boxplus 1 = 0$, 1 the multiplicative inverse of 1 since $1 \boxdot 1 = 1$. Recall that the zero element has no multiplicative inverse. Clearly, the two elements of the binary Galois field $GF(2)$ could be also represented by other symbols than '0' and '1' if the operations in Table A.1 are defined adequately.

B

Derivations and Proofs

B.1 Inversion of Hermitian and Positive Definite Matrices

The procedure to invert a Hermitian and positive definite matrix $\boldsymbol{A} \in \mathbb{C}^{N \times N}$, $N \in \mathbb{N}$, is as follows (see, e. g., [218, 136]):

1. *Cholesky Factorization* (CF):
   Compute the lower triangular matrix $\boldsymbol{L} \in \mathbb{C}^{N \times N}$ such that $\boldsymbol{A} = \boldsymbol{L}\boldsymbol{L}^{\mathrm{H}}$.
2. *Lower Triangular Inversion* (LTI):
   Compute the inverse $\boldsymbol{L}^{-1} \in \mathbb{C}^{N \times N}$ of the lower triangular matrix $\boldsymbol{L}$.
3. *Lower Triangular Product* (LTP):
   Construct the inverse $\boldsymbol{A}^{-1}$ via the matrix-matrix product $\boldsymbol{A}^{-1} = \boldsymbol{L}^{-1,\mathrm{H}}\boldsymbol{L}^{-1}$.

First, the CF is given by Algorithm 2.1. which has a computational complexity of (cf. Equation 2.23)

$$\zeta_{\mathrm{CF}}(N) = \frac{1}{3}N^3 + \frac{1}{2}N^2 + \frac{1}{6}N \quad \text{FLOPs.} \tag{B.1}$$

Second, in order to compute the inverse of the lower triangular matrix $\boldsymbol{L}$, we have to solve the system $\boldsymbol{L}\boldsymbol{L}^{-1} = \boldsymbol{I}_N$ of linear equations according to the lower triangular matrix $\boldsymbol{L}^{-1}$. For $i \in \{0, 1, \ldots, N-1\}$ and $j \in \{0, 1, \ldots, i\}$, the element in the $(i+1)$th row and $(j+1)$th column of $\boldsymbol{L}\boldsymbol{L}^{-1}$ can be written as

$$[\boldsymbol{L}]_{i,:} \left[\boldsymbol{L}^{-1}\right]_{:,j} = [\boldsymbol{L}]_{i,j:i} \left[\boldsymbol{L}^{-1}\right]_{j:i,j} = \delta_{i,j}, \tag{B.2}$$

with the *Kronecker delta* $\delta_{i,j}$ being one only for $i = j$ and zero otherwise, and the notation for submatrices defined in Subsection 2.2.1 and Section 1.2, respectively. In Equation (B.2), we exploit the fact that the non-zero entries of the $(i+1)$th row of $\boldsymbol{L}$ and the $(j+1)$th column of $\boldsymbol{L}^{-1}$ whose product yields the element of the $(i+1)$th row and $(j+1)$th column of $\boldsymbol{L}\boldsymbol{L}^{-1}$, overlap only from the $(j+1)$th element to the $(i+1)$th element ($j \leq i$). All elements of the $(i+1)$th row of L with an index higher than i and all elements of the $(j+1)$th column of L^{-1} with an index smaller than j are zero due to the lower triangular structure of L and L^{-1}. Remember that the inverse of a lower triangular matrix is again lower triangular. If we extract the term $[L]_{i,i}[L^{-1}]_{i,j}$ from the sum given by Equation (B.2), and resolve the resulting equation according to $[L^{-1}]_{i,j}$, we get

$$[L^{-1}]_{i,j} = \begin{cases} \dfrac{1}{[L]_{i,i}}, & j = i, \\[2ex] -\dfrac{[L]_{i,j:i-1}\,[L^{-1}]_{j:i-1,j}}{[L]_{i,i}}, & j < i,\ i \in \mathbb{N}. \end{cases} \tag{B.3}$$

Algorithm B.1. computes the inverse of the lower triangular matrix L based on Equation (B.3) where the number in angle brackets denotes the number of FLOPs required to perform the respective line of the algorithm. Recall from Subsection 2.2.1 that the matrix-matrix product of an $n \times m$ matrix by an $m \times p$ matrix requires $np(2m-1)$ FLOPs.

Algorithm B.1. Lower Triangular Inversion (LTI)

>   **for** $i = 0, 1, \ldots, N-1$ **do**
> 2: $L^{-1} \leftarrow \mathbf{0}_{N \times N}$
> $[L^{-1}]_{i,i} \leftarrow 1/[L]_{i,i}$ $\langle 1 \rangle$
> 4:   **for** $j = 0, 1, \ldots, i-1 \ \wedge\ i \in \mathbb{N}$ **do**
> $[L^{-1}]_{i,j} \leftarrow -[L]_{i,j:i-1}[L^{-1}]_{j:i-1,j}/[L]_{i,i}$ $\langle 2(i-j) \rangle$
> 6:   **end for**
>   **end for**

The total number of FLOPs needed for LTI performed by Algorithm B.1., calculates as

$$\zeta_{\mathrm{LTI}}(N) = \sum_{\substack{i=0 \\ i \in \mathbb{N}}}^{N-1} \left(1 + \sum_{j=0}^{i-1} 2(i-j) \right) = \frac{1}{3}N^3 + \frac{2}{3}N. \tag{B.4}$$

Third, it remains to compute the matrix-matrix product $L^{-1,\mathrm{H}}L^{-1}$ which is the wanted inverse A^{-1}. The element in the $(i+1)$th row and $(j+1)$th column of A^{-1} can be written as

$$[A^{-1}]_{i,j} = [L^{-1,\mathrm{H}}]_{i,:}\,[L^{-1}]_{:,j} = [L^{-1}]_{j:N-1,i}^{\mathrm{H}}\,[L^{-1}]_{j:N-1,j}, \tag{B.5}$$

if $i \in \{0, 1, \ldots, N-1\}$ and $j \in \{i, i+1, \ldots, N-1\}$. In Equation (B.5), we used the lower triangular property of L^{-1}, i.e., for each column with index i or j, the elements with an index smaller than i or j, respectively, are zero. Since $j \geq i$, the index j determines the first non-zero addend in the sum given by Equation (B.5). If we exploit additionally the Hermitian structure of A^{-1},

Algorithm B.2. Lower Triangular Product (LTP)

 for $i = 0, 1, \ldots, N - 1$ **do**
2: **for** $j = i, i+1, \ldots, N - 1$ **do**
   $[\boldsymbol{A}^{-1}]_{i,j} \leftarrow [\boldsymbol{L}^{-1}]^{\mathrm{H}}_{j:N-1,i}[\boldsymbol{L}^{-1}]_{j:N-1,j}$ $\langle 2(N - j) - 1\rangle$
4:   **if** $j \neq i$ **then**
    $[\boldsymbol{A}^{-1}]_{j,i} = [\boldsymbol{A}^{-1}]^*_{i,j}$
6:   **end if**
  end for
8: **end for**

we get Algorithm B.2.. Note that the inverse operation does not change the Hermitian property.

The number of FLOPs required to perform the LTP according to Algorithm B.2., is

$$\zeta_{\mathrm{LTP}}(N) = \sum_{i=0}^{N-1}\sum_{j=i}^{N-1}(2(N - j) - 1) = \frac{1}{3}N^3 + \frac{1}{2}N^2 + \frac{1}{6}N, \qquad (\mathrm{B.6})$$

which is the same computational complexity as the one of the CF. Finally, the number of FLOPs needed to invert a Hermitian and positive definite $N \times N$ matrix (*Hermitian Inversion*, HI) is given by the sum of FLOPs in Equations (B.1), (B.4), and (B.6), i. e.,

$$\left| \begin{aligned} \zeta_{\mathrm{HI}}(N) &= \zeta_{\mathrm{CF}}(N) + \zeta_{\mathrm{LTI}}(N) + \zeta_{\mathrm{LTP}}(N) \\ &= N^3 + N^2 + N. \end{aligned} \right| \qquad (\mathrm{B.7})$$

With this knowledge, it is interesting to discuss the question whether it is computational cheaper to solve systems of linear equations, i. e., $\boldsymbol{AX} = \boldsymbol{B} \in \mathbb{C}^{N \times M}$, $N, M \in \mathbb{N}$, based on the Cholesky factorization as described in Subsection 2.2.1, or via calculating $\boldsymbol{X} = \boldsymbol{A}^{-1}\boldsymbol{B} \in \mathbb{C}^{N \times M}$ with the explicit inversion of $\boldsymbol{A}$ as presented above. Whereas the Cholesky based method requires $\zeta_{\mathrm{MWF}}(N, M)$ FLOPs as given in Equation (2.28), the HI based approach has a total computational complexity of $\zeta_{\mathrm{HI}}(N) + MN(2N - 1)$ FLOPs where the additional term arises from the required multiplication of $\boldsymbol{A}^{-1} \in \mathbb{C}^{N \times N}$ by $\boldsymbol{B} \in \mathbb{C}^{N \times M}$. With some straightforward manipulations, it can be easily shown that

$$\zeta_{\mathrm{HI}}(N) + MN(2N - 1) > \zeta_{\mathrm{MWF}}(N, M) \quad \leftrightarrow \quad M < \frac{2}{3}N^2 + \frac{1}{2}N + \frac{5}{6}. \quad (\mathrm{B.8})$$

Thus and due the fact that $N < 2N^2/3 + N/2 + 5/6$ for all $N \in \mathbb{N}$, the Cholesky based algorithm is computationally cheaper than the HI based method if $M \leq N$. Further, for values of M which fulfill $N < M < 2N^2/3 + N/2 + 5/6$, it remains computationally less expensive. However, in the case where $M > 2N^2/3 + N/2 + 5/6$ or if the inverse of the system matrix $\boldsymbol{A}$ is explicitly needed, it is recommended to use the HI based approach to solve systems of linear equations.

B.2 Real-Valued Coefficients of Krylov Subspace Polynomials

Consider the approximation of the solution $\boldsymbol{x} \in \mathbb{C}^N$ of the system of linear equations $\boldsymbol{A}\boldsymbol{x} = \boldsymbol{b} \in \mathbb{C}^N$ where $\boldsymbol{A} \in \mathbb{C}^{N \times N}$ is Hermitian and positive definite, in the D-dimensional Krylov subspace

$$\mathcal{K}^{(D)}\left(\boldsymbol{A}, \boldsymbol{b}\right) = \operatorname{range}\left\{\left[\boldsymbol{b} \; \boldsymbol{A}\boldsymbol{b} \cdots \boldsymbol{A}^{D-1}\boldsymbol{b}\right]\right\}. \tag{B.9}$$

In other words, the solution $\boldsymbol{x}$ is approximated by the polynomial

$$\boldsymbol{x} \approx \boldsymbol{x}_D = \boldsymbol{\Psi}^{(D-1)}(\boldsymbol{A})\boldsymbol{b} = \sum_{\ell=0}^{D-1} \psi_\ell \boldsymbol{A}^\ell \boldsymbol{b} \in \mathbb{C}^N. \tag{B.10}$$

The polynomial coefficients ψ_ℓ are chosen to minimize the squared $\boldsymbol{A}$-norm of the error vector $\boldsymbol{e}'_D = \boldsymbol{x} - \boldsymbol{x}'_D$ with $\boldsymbol{x}'_D = \sum_{\ell=0}^{D-1} \psi'_\ell \boldsymbol{A}^\ell \boldsymbol{b}$ (cf. Section 3.4 and Appendix A.2), i.e.,

$$(\psi_0, \psi_1, \ldots, \psi_{D-1}) = \operatorname*{argmin}_{\left(\psi'_0, \psi'_1, \ldots, \psi'_{D-1}\right)} \|\boldsymbol{e}'_D\|_{\boldsymbol{A}}^2. \tag{B.11}$$

Setting the partial derivatives of

$$\begin{aligned}
\|\boldsymbol{e}_D\|_{\boldsymbol{A}}^2 = {}&\boldsymbol{x}^{\mathrm{H}}\boldsymbol{x} - \boldsymbol{x}^{\mathrm{H}}\sum_{\ell=0}^{D-1} \psi'_\ell \boldsymbol{A}^\ell \boldsymbol{b} - \sum_{\ell=0}^{D-1} \psi'^{*}_\ell \boldsymbol{b}^{\mathrm{H}}\boldsymbol{A}^\ell \boldsymbol{x} \\
&+ \sum_{\ell=0}^{D-1}\sum_{m=0}^{D-1} \psi'^{*}_\ell \boldsymbol{b}^{\mathrm{H}}\boldsymbol{A}^{\ell+m}\boldsymbol{b}\psi'_m,
\end{aligned} \tag{B.12}$$

with respect to ψ'^{*}_ℓ, $\ell \in \{0, 1, \ldots, D-1\}$, equal to zero at $\psi'_m = \psi_m$, yields the set of linear equations:

$$\sum_{m=0}^{D-1} \boldsymbol{b}^{\mathrm{H}}\boldsymbol{A}^{\ell+m}\boldsymbol{b}\psi_m = \boldsymbol{b}^{\mathrm{H}}\boldsymbol{A}^\ell \boldsymbol{x}, \quad \ell \in \{0, 1, \ldots, D-1\}, \tag{B.13}$$

or

$$\sum_{m=0}^{D-1} \boldsymbol{b}^{\mathrm{H}}\boldsymbol{A}^{\ell+m}\boldsymbol{b}\psi_m = \boldsymbol{b}^{\mathrm{H}}\boldsymbol{A}^{\ell-1}\boldsymbol{b}, \quad \ell \in \{0, 1, \ldots, D-1\}, \tag{B.14}$$

if we remember that $\boldsymbol{b} = \boldsymbol{A}\boldsymbol{x}$. Due to the fact that $\boldsymbol{A}$ is Hermitian and positive definite, the matrices $\boldsymbol{A}^i$, $i \in \mathbb{N}_0$, are also Hermitian and positive definite, and the terms $\boldsymbol{b}^{\mathrm{H}}\boldsymbol{A}^i\boldsymbol{b}$ are real-valued and positive. Thus, all elements of the linear system in Equation (B.14) are real-valued and positive and finally, the optimal coefficients $\psi_\ell \in \mathbb{R}$ for all $\ell \in \{0, 1, \ldots, D-1\}$ (cf. also [67]).

In the limiting case where $D = N$, the solution $\boldsymbol{x}$ is exactly represented by the optimal linear combination of the Krylov basis vectors using only real-valued coefficients, i.e., $\boldsymbol{x} = \boldsymbol{x}_N$. However, we are not operating in a real-valued vector space because we still need complex-valued coefficients in order to reach any $\boldsymbol{x}' \in \mathbb{C}^N$ with the linear combination of the Krylov basis vectors $\boldsymbol{b}$, $\boldsymbol{Ab}$, $\ldots$, and $\boldsymbol{A}^{N-1}\boldsymbol{b}$. Especially, if $\boldsymbol{Ax}' \neq \boldsymbol{b}$, minimizing the squared $\boldsymbol{A}$-norm of the error between $\boldsymbol{x}'_N$ and $\boldsymbol{x}'$ leads to Equation (B.13) where $\boldsymbol{x}$ is replaced by $\boldsymbol{x}'$. In this case, Equation (B.13) cannot be simplified to Equation (B.14) since $\boldsymbol{b} \neq \boldsymbol{Ax}'$, and the right-hand sides are no longer real-valued. Thus, the optimal coefficients ψ_ℓ are complex-valued for all $\ell \in \{0, 1, \ldots, D - 1\}$. Only the optimal solution $\boldsymbol{x}$ of the system of linear equations $\boldsymbol{Ax} = \boldsymbol{b}$ can be represented exactly by the polynomial $\Psi^{(N-1)}(\boldsymbol{A})\boldsymbol{b}$ with real-valued coefficients because of the matching between $\boldsymbol{x}$ and the special choice of the Krylov basis vectors according to the system matrix $\boldsymbol{A}$ and the right-hand side $\boldsymbol{b}$.

If the Hermitian and positive definite system of linear equations has multiple right-hand sides, the matrix coefficients have no special structure, i.e., they are in general neither Hermitian nor are their elements real-valued.

B.3 QR Factorization

The *QR factorization* [94, 237] of the matrix $\boldsymbol{A} \in \mathbb{C}^{N \times M}$, $N, M \in \mathbb{N}$, $N \geq M$, is defined as the decomposition $\boldsymbol{A} = \boldsymbol{QR}$ where $\boldsymbol{Q} \in \mathbb{C}^{N \times M}$ is a matrix with mutually orthonormal columns, i.e., $\boldsymbol{Q}^{\mathrm{H}}\boldsymbol{Q} = \boldsymbol{I}_M$, and $\boldsymbol{R} \in \mathbb{C}^{M \times M}$ is an upper triangular matrix. To access the factors $\boldsymbol{Q}$ and $\boldsymbol{R}$ of the QR factorization of $\boldsymbol{A}$, we define the operator 'QR$(\cdot)$' such that $(\boldsymbol{Q}, \boldsymbol{R}) = \mathrm{QR}(\boldsymbol{A})$.

The derivation of the QR factorization is very similar to the one of the block Arnoldi algorithm in Section 3.2. However, instead of orthogonalizing the basis of the Krylov subspace, the QR factorization orthogonalizes the columns of the matrix $\boldsymbol{A}$. If we recall the *modified Gram–Schmidt procedure* [94, 237] of Equation (3.15) and apply it to orthogonalize the $(j + 1)$th column of $\boldsymbol{A}$, i.e., $[\boldsymbol{A}]_{:,j}$, $j \in \{0, 1, \ldots, M - 1\}$, with respect to the already orthonormal columns of $[\boldsymbol{Q}]_{:,0:j-1}$, we get

$$[\boldsymbol{Q}]_{:,j}[\boldsymbol{R}]_{j,j} = \left(\prod_{\substack{i=0 \\ j \in \mathbb{N}}}^{j-1} \boldsymbol{P}_{\perp[\boldsymbol{Q}]_{:,i}} \right) [\boldsymbol{A}]_{:,j}, \tag{B.15}$$

where $\boldsymbol{P}_{\perp[\boldsymbol{Q}]_{:,i}} = \boldsymbol{I}_N + [\boldsymbol{Q}]_{:,i}[\boldsymbol{Q}]_{:,i}^{\mathrm{H}} \in \mathbb{C}^{N \times N}$ is a projector onto the subspace which is orthogonal to the $(i+1)$th column of $\boldsymbol{Q}$, i.e., $[\boldsymbol{Q}]_{:,i}$, $i \in \{0, 1, \ldots, j-1\}$, $j \in \mathbb{N}$. Again, we used the notation for submatrices as defined in Subsection 2.2.1 and Section 1.2, respectively. Due to the fact that each column of $\boldsymbol{A}$ is only orthogonalized with respect to the j already computed orthonormal vectors, the entries of the lower triangular part of $\boldsymbol{R} = \boldsymbol{Q}^{\mathrm{H}}\boldsymbol{A}$ are equal to zero.

Algorithm B.3. performs the QR factorization based on the modified Gram–Schmidt procedure. Immediately after $[\boldsymbol{Q}]_{:,j}$ is known, the projector $\boldsymbol{P}_{\perp[\boldsymbol{Q}]_{:,j}} = \boldsymbol{I}_N + [\boldsymbol{Q}]_{:,j}[\boldsymbol{Q}]_{:,j}^{\mathrm{H}}$ is applied to each column of $[\boldsymbol{A}]_{:,j+1:M-1}$ (cf. inner **for**-loop of Algorithm B.3.). Here, the columns of $\boldsymbol{A}$ which are no longer needed in the subsequent iterations are overwritten to save storage. We refer to [94, 237] for a detailed derivation of Algorithm B.3..

Algorithm B.3. QR factorization

	for $j = 0, 1, \ldots, M-1$ **do**	
2:	$\quad [\boldsymbol{R}]_{j,j} \leftarrow \|[\boldsymbol{A}]_{:,j}\|_2$	$\langle 2N \rangle$
	$\quad [\boldsymbol{Q}]_{:,j} \leftarrow [\boldsymbol{A}]_{:,j}/[\boldsymbol{R}]_{j,j}$	$\langle N \rangle$
4:	$\quad$ **for** $i = j+1, j+2, \ldots, M-1 \wedge j < M-1$ **do**	
	$\quad\quad [\boldsymbol{R}]_{j,i} \leftarrow [\boldsymbol{Q}]_{:,j}^{\mathrm{H}}[\boldsymbol{A}]_{:,i}$	$\langle 2N - 1 \rangle$
6:	$\quad\quad [\boldsymbol{A}]_{:,i} \leftarrow [\boldsymbol{A}]_{:,i} - [\boldsymbol{Q}]_{:,j}[\boldsymbol{R}]_{j,i}$	$\langle 2N \rangle$
	$\quad$ **end for**	
8:	**end for**	

In each line of Algorithm B.3., the number in angle brackets denotes the number of FLOPs required to perform the corresponding line of the algorithm. Remember from Section 2.2 that the scalar product of two n-dimensional vectors needs $2n - 1$ FLOPs. Finally, the total number of FLOPs which are necessary to perform the QR factorization based on the modified Gram–Schmidt procedure according to Algorithm B.3., computes as

$$
\zeta_{\mathrm{QR}}(N, M) = \sum_{j=0}^{M-1} \left(3N + \sum_{\substack{i=j+1 \\ j<M-1}}^{M-1} (4N - 1) \right) \tag{B.16}
$$

$$
= (2M + 1)MN - \frac{1}{2}(M - 1)M.
$$

B.4 Proof of Proposition 3.3

The columns of $\boldsymbol{Q}$, viz., the vectors $\boldsymbol{q}_j \in \mathbb{C}^N$, $j \in \{0, 1, \ldots, D-1\}$, are orthonormal due to the Gram–Schmidt procedure. If D is an integer multiple of M, i.e., $\mu = M$, Proposition 3.3 is true because the block Lanczos–Ruhe algorithm is equal to the block Lanczos procedure by construction and $\mathcal{K}_{\mathrm{Ruhe}}^{(D)}(\boldsymbol{A}, \boldsymbol{q}_0, \boldsymbol{q}_1, \ldots, \boldsymbol{q}_{M-1}) = \mathcal{K}^{(D)}(\boldsymbol{A}, \boldsymbol{Q}_0)$. It remains to prove Proposition 3.3 for $D \neq dM$, i.e., the vectors $\boldsymbol{q}_j$, $j \in \{0, 1, \ldots, D-1\}$, are given by the linear combination

$$q_j = \Psi_{0,j}^{(\iota(j)-1)}(A)q_0 + \Psi_{1,j}^{(\iota(j)-1)}(A)q_1 + \ldots + \Psi_{\varrho(j)-1,j}^{(\iota(j)-1)}(A)q_{\varrho(j)-1}$$

$$+ \Psi_{\varrho(j),j}^{(\iota(j)-2)}(A)q_{\varrho(j)} + \Psi_{\varrho(j)+1,j}^{(\iota(j)-2)}(A)q_{\varrho(j)+1} + \ldots + \Psi_{M-1,j}^{(\iota(j)-2)}(A)q_{M-1}$$

$$= \sum_{k=0}^{\varrho(j)-1} \Psi_{k,j}^{(\iota(j)-1)}(A)q_k + \sum_{\substack{k=\varrho(j) \\ \varrho(j)<M}}^{M-1} \Psi_{k,j}^{(\iota(j)-2)}(A)q_k \in \mathbb{C}^N, \tag{B.17}$$

where the polynomials $\Psi_{k,j}^{(\iota(j)-1)}(A)$, $k \in \{0,1,\ldots,\varrho(j)-1\}$, of degree $\iota(j)-1$ and the polynomials $\Psi_{k,j}^{(\iota(j)-2)}(A)$, $k \in \{\varrho(j),\varrho(j)+1,\ldots,M-1\}$, of degree $\iota(j)-2$ are defined analogous to Equation (3.3).[1] The graphs of the functions $\iota(j) = \lceil(j+1)/M\rceil \in \{1,2,\ldots,d\}$, $d = \lceil D/M\rceil$, which is one plus the maximal polynomial degree, and $\varrho(j) = j + 1 - (\iota(j)-1)M \in \{1,2,\ldots,M\}$ which is the number of polynomials with maximal degree $\iota(j)-1$, are depicted in Figure B.1.

We prove by induction on j. For $j \in \{0,1,\ldots,M-1\}$, $q_0 = \Psi_{0,0}^{(0)}(A)q_0$ with $\Psi_{0,0}^{(0)}(A) := I_N$, $q_1 = \Psi_{0,1}^{(0)}(A)q_0 + \Psi_{1,1}^{(0)}(A)q_1$ with $\Psi_{0,1}^{(0)}(A) := 0_{N\times N}$ and $\Psi_{1,1}^{(0)}(A) := I_N$, $\ldots$, and $q_{M-1} = \sum_{k=0}^{M-1} \Psi_{k,M-1}^{(0)}(A)q_k$ with $\Psi_{k,M-1}^{(0)}(A) := 0_{N\times N}$ for $k < M-1$ and $\Psi_{M-1,M-1}^{(0)}(A) := I_N$. Assume that Equation (B.17) is true for all $M-1 \leq i \leq j$, $j \in \{M-1,M,\ldots,D-2\}$.[2] Then, we have to show that it is also true for $i = j+1$. Use Equation (3.25) and replace m by $j-M+1$ to get

$$q_{j+1}h_{j+1,j-M+1} = Aq_{j-M+1} - \sum_{\substack{i=j-2M+1 \\ i\in\mathbb{N}_0}}^{j} q_i h_{i,j-M+1}. \tag{B.18}$$

With the relations $\iota(j-M+1) = \iota(j+1)-1$ and $\varrho(j-M+1) = \varrho(j+1)$,[3] Equation (B.18) may be rewritten as

---

$$\boldsymbol{q}_{j+1} h_{j+1,j-M+1} = \boldsymbol{A} \left(\sum_{k=0}^{\varrho(j+1)-1} \Psi_{k,j-M+1}^{(\iota(j+1)-2)}(\boldsymbol{A})\boldsymbol{q}_k + \sum_{\substack{k=\varrho(j+1) \\ \varrho(j+1)<M}}^{M-1} \Psi_{k,j-M+1}^{(\iota(j+1)-3)}(\boldsymbol{A})\boldsymbol{q}_k \right.$$

$$- \sum_{\substack{i=j-2M+1 \\ i\in\mathbb{N}_0}}^{j} h_{i,j-M+1} \left(\sum_{k=0}^{\varrho(i)-1} \Psi_{k,i}^{(\iota(i)-1)}(\boldsymbol{A})\boldsymbol{q}_k \right.$$

$$\left. \left. + \sum_{\substack{k=\varrho(i) \\ \varrho(i)<M}}^{M-1} \Psi_{k,i}^{(\iota(i)-2)}(\boldsymbol{A})\boldsymbol{q}_k \right) \right) , \quad (B.19)$$

if we recall the polynomial representation of $\boldsymbol{q}_j$ in Equation (B.17). First consider the case where $j = (\ell+1)M-1$, $\ell \in \{0,1,\dots,d-2\}$, i.e., $\iota(j) = \ell+1$, $\iota(j+1) = \iota(j)+1 = \ell+2$, $\varrho(j) = M$, and $\varrho(j+1) = 1$ (cf. Figure B.1). Hence,

$$\boldsymbol{q}_{j+1} = \underbrace{h_{j+1,j-M+1}^{-1} \boldsymbol{A}\Psi_{0,j-M+1}^{(\iota(j+1)-2)}(\boldsymbol{A})\boldsymbol{q}_0}_{k=0,\ \text{degree}\ \iota(j+1)-1} + \underbrace{h_{j+1,j-M+1}^{-1} \sum_{k=1}^{M-1} \boldsymbol{A}\Psi_{k,j-M+1}^{(\iota(j+1)-3)}(\boldsymbol{A})\boldsymbol{q}_k}_{k\in\{1,2,\dots,M-1\},\ \text{maximal degree}\ \iota(j+1)-2}$$

$$\underbrace{- \sum_{\substack{i=j-2M+1 \\ i\in\mathbb{N}_0}}^{j} \frac{h_{i,j-M+1}}{h_{j+1,j-M+1}} \left(\sum_{k=0}^{\varrho(i)-1} \Psi_{k,i}^{(\iota(i)-1)}(\boldsymbol{A})\boldsymbol{q}_k + \sum_{\substack{k=\varrho(i) \\ \varrho(i)<M}}^{M-1} \Psi_{k,i}^{(\iota(i)-2)}(\boldsymbol{A})\boldsymbol{q}_k \right)}_{k\in\{0,1,\dots,M-1\},\ \text{maximal degree}\ \iota(j)-1 = \iota(j+1)-2}$$

$$= \Psi_{0,j+1}^{(\iota(j+1)-1)}(\boldsymbol{A})\boldsymbol{q}_0 + \sum_{k=1}^{M-1} \Psi_{k,j+1}^{(\iota(j+1)-2)}(\boldsymbol{A})\boldsymbol{q}_k, \quad (B.20)$$

which confirms Equation (B.17) for the special case of $j = (\ell+1)M-1$. In Equation (B.20), only the terms with $k = 0$ have a maximal degree of $\iota(j+1)-1$ which justifies the last equality. It remains to find the polynomial representation of $\boldsymbol{q}_{j+1}$ for $j \neq (\ell+1)M-1$, $\ell \in \{0,1,\dots,d-2\}$. In this case, $\iota(j+1) = \iota(j)$ and $\varrho(j+1) = \varrho(j)+1$ (cf. Figure B.1), and Equation (B.19) reads as

$$
\boldsymbol{q}_{j+1}h_{j+1,j-M+1} = \underbrace{\sum_{k=0}^{\varrho(j)-1}\boldsymbol{A}\Psi_{k,j-M+1}^{(\iota(j+1)-2)}(\boldsymbol{A})\boldsymbol{q}_k - \sum_{i=(\iota(j)-1)M}^{j} h_{i,j-M+1}\sum_{k=0}^{\varrho(i)-1}\Psi_{k,i}^{(\iota(i)-1)}(\boldsymbol{A})\boldsymbol{q}_k}_{k\in\{0,1,\dots,\varrho(j)-1\},\ \text{degree } \iota(j)-1 = \iota(j+1)-1}
$$

$$
+ \underbrace{\boldsymbol{A}\Psi_{\varrho(j),j-M+1}^{(\iota(j+1)-2)}(\boldsymbol{A})\boldsymbol{q}_{\varrho(j)}}_{\substack{k\,=\,\varrho(j),\ \varrho(j)\,<\\ M,\\ \text{degree } \iota(j+1)-1}} + \underbrace{\sum_{\substack{k=\varrho(j)+1\\ \varrho(j)<M-1}}^{M-1}\boldsymbol{A}\Psi_{k,j-M+1}^{(\iota(j+1)-3)}(\boldsymbol{A})\boldsymbol{q}_k}_{\substack{k\in\{\varrho(j)+1,\varrho(j)+2,\dots,M-\\ 1\},\ \varrho(j)<M-1,\ \text{degree } \iota(j+\\ 1)-2}}
$$

$$
- \underbrace{\sum_{\substack{i=j-2M+1\\ i\in\mathbb{N}_0}}^{(\iota(j)-1)M-1} h_{i,j-M+1}\sum_{k=0}^{\varrho(i)-1}\Psi_{k,i}^{(\iota(i)-1)}(\boldsymbol{A})\boldsymbol{q}_k}_{k\in\{0,1,\dots,M-1\},\ \text{maximal degree } \iota(j)-2 = \iota(j+1)-2}
$$

$$
- \underbrace{\sum_{\substack{i=j-2M+1\\ i\in\mathbb{N}_0}}^{j} h_{i,j-M+1}\sum_{\substack{k=\varrho(i)\\ \varrho(i)<M}}^{M-1}\Psi_{k,i}^{(\iota(i)-2)}(\boldsymbol{A})\boldsymbol{q}_k}_{k\in\{0,1,\dots,M-1\},\ \text{maximal degree } \iota(j)-2 = \iota(j+1)-2}. \tag{B.21}
$$

Since the degree of the polynomial with $k = \varrho(j) = \varrho(j+1)-1$ is increased by one, only the terms with $k \in \{0,1,\dots,\varrho(j+1)-1\}$ have the maximal degree of $\iota(j+1)-1$, and we can write

$$
\boldsymbol{q}_{j+1} = \sum_{k=0}^{\varrho(j+1)-1}\Psi_{k,j+1}^{(\iota(j+1)-1)}(\boldsymbol{A})\boldsymbol{q}_k + \sum_{\substack{k=\varrho(j+1)\\ \varrho(j+1)<M}}^{M-1}\Psi_{k,j+1}^{(\iota(j+1)-2)}(\boldsymbol{A})\boldsymbol{q}_k, \tag{B.22}
$$

which completes the proof.

B.5 Correlated Subspace

Consider the observation of the unknown vector random sequence $\boldsymbol{x}[n] \in \mathbb{C}^M$, $M \in \mathbb{N}$, with the mean $\boldsymbol{m}_x \in \mathbb{C}^M$ via the vector random sequence $\boldsymbol{y}[n] \in \mathbb{C}^N$, $N \in \mathbb{N}$, $N > M$, with the mean $\boldsymbol{m}_y \in \mathbb{C}^N$ (cf. Section 2.1). For simplicity, we assume in the following considerations that the random sequences are stationary. However, the results can be easily generalized to the non-stationary case by introducing a time-dependency as in Section 2.1. The *normalized matrix matched filter* [88] or *normalized matrix correlator* $\boldsymbol{M} \in \mathbb{C}^{N\times M}$ with orthonormal columns is defined to maximize the *correlation measure* $\mathrm{Re}\{\mathrm{tr}\{\boldsymbol{C}_{\hat{x},x}\}\}$, i.e., the real part of the sum of cross-covariances between each element of

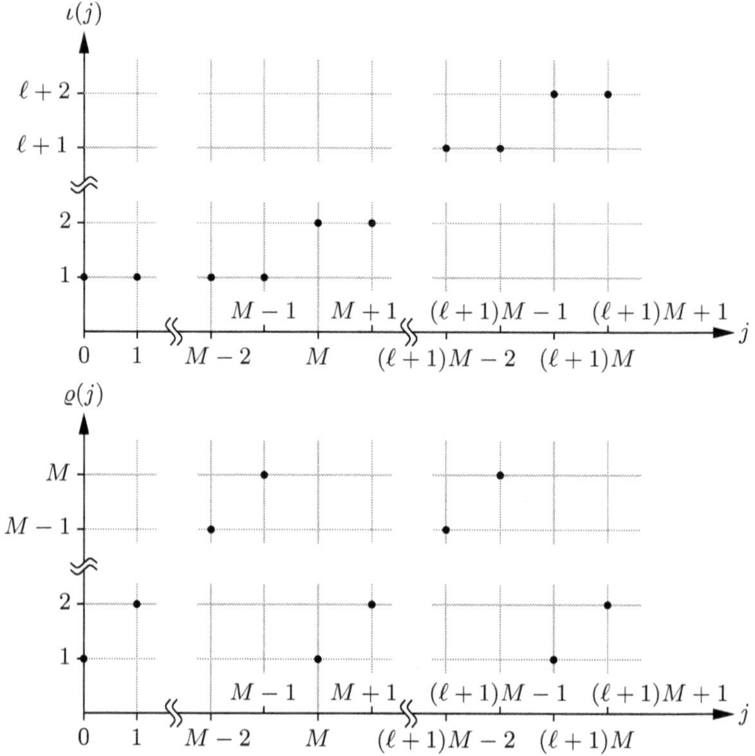

Fig. B.1. Graphs of functions $\iota(j)$, i.e., one plus the maximal polynomial degree, and the number $\varrho(j)$ of polynomials with maximal degree $\iota(j) - 1$

$\boldsymbol{x}[n]$ and the corresponding element of its output $\hat{\boldsymbol{x}}[n] = \boldsymbol{M}^{\mathrm{H}}\boldsymbol{y}[n] \in \mathbb{C}^{M}$.[4] The cross-covariance matrix between $\hat{\boldsymbol{x}}[n]$ and $\boldsymbol{x}[n]$ reads as $\boldsymbol{C}_{\hat{x},x} = \boldsymbol{M}^{\mathrm{H}}\boldsymbol{C}_{y,x} = \boldsymbol{M}^{\mathrm{H}}(\boldsymbol{R}_{y,x} - \boldsymbol{m}_{y}\boldsymbol{m}_{x}^{\mathrm{H}})$.

In the sequel, we are interested in the subspace $\mathcal{M} \in G(N, M)$ spanned by the columns of $\boldsymbol{M} \in \mathbb{C}^{N \times M}$ where $G(N, M)$ denotes the *Grassmann manifold* (e.g., [13]), i.e., the set of all M-dimensional subspaces in $\mathbb{C}^{N}$. We denote $\mathcal{M}$ as the *correlated subspace* because the projection of the observed vector random sequence $\boldsymbol{y}[n]$ on $\mathcal{M}$ still implies the signal which is maximal correlated to $\boldsymbol{x}[n]$, i.e., which maximizes $\mathrm{Re}\{\mathrm{tr}\{\boldsymbol{C}_{\hat{x},x}\}\}$. If we define the *subspace correlation measure* $\rho(\mathcal{M}') \in \mathbb{R}$ of the subspace $\mathcal{M}' \in G(N, M)$ as

---

[4] Note that we choose the real part instead of the absolute value of the sum of cross-covariances to ensure that the phase of the solution is unique. This choice forces the output $\hat{\boldsymbol{x}}[n]$ of the normalized matrix correlator to be in-phase with $\boldsymbol{x}[n]$ motivated by the trivial case where $\hat{\boldsymbol{x}}[n] = \boldsymbol{x}[n]$.

$$\rho\left(\mathcal{M}'\right) = \max_{\boldsymbol{\Psi}'\in\mathbb{C}^{M\times M}} \mathrm{Re}\left\{\mathrm{tr}\left\{\boldsymbol{\Psi}'^{,\mathrm{H}}\boldsymbol{M}'^{,\mathrm{H}}\boldsymbol{C}_{\mathbf{y},\mathbf{x}}\right\}\right\}$$
$$\text{s.t.}\quad \mathrm{range}\left\{\boldsymbol{M}'\right\} = \mathcal{M}' \text{ and } \boldsymbol{\Psi}'^{,\mathrm{H}}\boldsymbol{M}'^{,\mathrm{H}}\boldsymbol{M}'\boldsymbol{\Psi}' = \boldsymbol{I}_M, \tag{B.23}$$

where $\boldsymbol{M}' \in \mathbb{C}^{N\times M}$ is an arbitrary basis of the M-dimensional subspace $\mathcal{M}'$ and $\boldsymbol{\Psi}' \in \mathbb{C}^{M\times M}$ is chosen to maximize the cost function as well as to ensure that the columns of the matrix $\boldsymbol{M}'\boldsymbol{\Psi}' \in \mathbb{C}^{N\times M}$ are orthonormal to each other, the correlated subspace $\mathcal{M}$ is the subspace which maximizes $\rho(\mathcal{M}')$ over all subspaces $\mathcal{M}' \in G(N, M)$, i.e.,

$$\mathcal{M} = \underset{\mathcal{M}'\in G(N,M)}{\mathrm{argmax}}\ \rho\left(\mathcal{M}'\right). \tag{B.24}$$

Before solving the optimization in Equation (B.24), we simplify the subspace correlation measure $\rho(\mathcal{M}')$ given by Equation (B.23). Partially differentiating the real-valued Lagrangian function

$$L\left(\boldsymbol{\Psi}', \boldsymbol{\Lambda}\right) - \mathrm{Re}\left\{\mathrm{tr}\left\{\boldsymbol{\Psi}'^{,\mathrm{H}}\boldsymbol{M}'^{,\mathrm{H}}\boldsymbol{C}_{\mathbf{y},\mathbf{x}} - \left(\boldsymbol{\Psi}'^{,\mathrm{H}}\boldsymbol{M}'^{,\mathrm{H}}\boldsymbol{M}'\boldsymbol{\Psi}' - \boldsymbol{I}_M\right)\boldsymbol{\Lambda}\right\}\right\}, \tag{B.25}$$

$\boldsymbol{\Lambda} \in \mathbb{C}^{M\times M}$, with respect to $\boldsymbol{\Psi}'^{,*}$ and setting the result equal to $\boldsymbol{0}_{M\times M}$ for $\boldsymbol{\Psi}' = \boldsymbol{\Psi}$, yields

$$\left.\frac{\partial L\left(\boldsymbol{\Psi}', \boldsymbol{\Lambda}\right)}{\partial \boldsymbol{\Psi}'^{,*}}\right|_{\boldsymbol{\Psi}'=\boldsymbol{\Psi}} = \frac{1}{2}\left(\boldsymbol{M}'^{,\mathrm{H}}\boldsymbol{C}_{\mathbf{y},\mathbf{x}} - \boldsymbol{M}'^{,\mathrm{H}}\boldsymbol{M}'\boldsymbol{\Psi}\left(\boldsymbol{\Lambda} + \boldsymbol{\Lambda}^{\mathrm{H}}\right)\right) \overset{!}{=} \boldsymbol{0}_{M\times M}. \tag{B.26}$$

With the Hermitian matrix $\tilde{\boldsymbol{\Lambda}} := \boldsymbol{\Lambda} + \boldsymbol{\Lambda}^{\mathrm{H}} \in \mathbb{C}^{M\times M}$, we get the solution

$$\boldsymbol{\Psi}\tilde{\boldsymbol{\Lambda}} = \left(\boldsymbol{M}'^{,\mathrm{H}}\boldsymbol{M}'\right)^{-1}\boldsymbol{M}'^{,\mathrm{H}}\boldsymbol{C}_{\mathbf{y},\mathbf{x}}. \tag{B.27}$$

Note that $\tilde{\boldsymbol{\Lambda}}$ is not necessarily invertible. It remains to choose $\tilde{\boldsymbol{\Lambda}}$ such that $\boldsymbol{\Psi}^{\mathrm{H}}\boldsymbol{M}'^{,\mathrm{H}}\boldsymbol{M}'\boldsymbol{\Psi} = \boldsymbol{I}_M$. Multiplying the latter equation with $\tilde{\boldsymbol{\Lambda}}$ on the left-hand and right-hand side, and replacing $\boldsymbol{\Psi}\tilde{\boldsymbol{\Lambda}}$ based on Equation (B.27) yields

$$\tilde{\boldsymbol{\Lambda}}^2 - \boldsymbol{C}_{\mathbf{y},\mathbf{x}}^{\mathrm{H}}\boldsymbol{M}'\left(\boldsymbol{M}'^{,\mathrm{H}}\boldsymbol{M}'\right)^{-1}\boldsymbol{M}'^{,\mathrm{H}}\boldsymbol{C}_{\mathbf{y},\mathbf{x}} \tag{B.28}$$

With the Hermitian general square root matrix as defined in Appendix A.6 and the projector $\boldsymbol{P}_{\mathcal{M}'} = \boldsymbol{M}'(\boldsymbol{M}'^{,\mathrm{H}}\boldsymbol{M}')^{-1}\boldsymbol{M}'^{,\mathrm{H}}$ projecting onto the subspace $\mathcal{M}'$, i.e., the subspace spanned by the columns of $\boldsymbol{M}'$, we get

$$\tilde{\boldsymbol{\Lambda}} = \sqrt{\boldsymbol{C}_{\mathbf{y},\mathbf{x}}^{\mathrm{H}}\boldsymbol{P}_{\mathcal{M}'}\boldsymbol{C}_{\mathbf{y},\mathbf{x}}}. \tag{B.29}$$

On the other hand, if we multiply Equation (B.27) by $\boldsymbol{\Psi}^{\mathrm{H}}\boldsymbol{M}'^{,\mathrm{H}}\boldsymbol{M}'$ on the left-hand side and exploit again the fact that $\boldsymbol{\Psi}^{\mathrm{H}}\boldsymbol{M}'^{,\mathrm{H}}\boldsymbol{M}'\boldsymbol{\Psi} = \boldsymbol{I}_M$, it follows

$$\tilde{\boldsymbol{\Lambda}} = \boldsymbol{\Psi}^{\mathrm{H}}\boldsymbol{M}'^{,\mathrm{H}}\boldsymbol{C}_{\mathbf{y},\mathbf{x}}. \tag{B.30}$$

Thus, at the optimum, we can replace the term $\boldsymbol{\Psi}^{\mathrm{H}}\boldsymbol{M}'^{,\mathrm{H}}\boldsymbol{C}_{\mathbf{y},\mathbf{x}}$ in Equation (B.23) by $\tilde{\boldsymbol{\Lambda}}$ as given in Equation (B.29). However, the maximum defined by Equation (B.23) can only be achieved by taking the non-negative eigenvalues of the general square root matrix which is the same as using the unique positive semidefinite square root matrix instead of the general square root matrix (cf. Appendix A.6). Thus, the subspace correlation measure can finally be written as

$$\rho\left(\mathcal{M}'\right) = \operatorname{tr}\left\{ \sqrt[\oplus]{C_{y,x}^{\mathrm{H}} P_{\mathcal{M}'} C_{y,x}} \right\}, \tag{B.31}$$

and the correlated subspace is the solution of the optimization

$$\left| \mathcal{M} = \operatorname*{argmax}_{\mathcal{M}' \in G(N,M)} \operatorname{tr}\left\{ \sqrt[\oplus]{C_{y,x}^{\mathrm{H}} P_{\mathcal{M}'} C_{y,x}} \right\}. \right| \tag{B.32}$$

In Equation (B.31), we omitted the real part operator because the square root matrix $\sqrt[\oplus]{C_{y,x}^{\mathrm{H}} P_{\mathcal{M}'} C_{y,x}}$ is by definition Hermitian and positive semidefinite (cf. Appendix A.6), thus, its trace is non-negative real-valued.

It remains to solve the optimization defined by Equation (B.32). With the parameterization of $P_{\mathcal{M}'} = \bar{M}' \bar{M}'^{,\mathrm{H}}$ where $\bar{M}' \in \mathbb{C}^{N \times M}$ and $\bar{M}'^{,\mathrm{H}} \bar{M}' = I_M$, the optimal parameter $\bar{M}$ whose columns span the correlated subspace $\mathcal{M}$, i.e., $\mathcal{M} = \operatorname{range}\{\bar{M}\}$, is obtained via the optimization

$$\bar{M} = \operatorname*{argmax}_{\bar{M}' \in \mathbb{C}^{N \times M}} \operatorname{tr}\left\{ \sqrt[\oplus]{C_{y,x}^{\mathrm{H}} \bar{M}' \bar{M}'^{,\mathrm{H}} C_{y,x}} \right\} \quad \text{s.t.} \quad \bar{M}'^{,\mathrm{H}} \bar{M}' = I_M. \tag{B.33}$$

With the singular value decomposition of $\bar{M}'^{,\mathrm{H}} C_{y,x} = U \Sigma V^{\mathrm{H}}$ where $U \in \mathbb{C}^{M \times M}$ as well as $V \in \mathbb{C}^{M \times M}$ are unitary matrices, and $\Sigma \in \mathbb{R}_{0,+}^{M \times M}$ is diagonal, it follows

$$\operatorname{tr}\left\{ \sqrt[\oplus]{C_{y,x}^{\mathrm{H}} \bar{M}' \bar{M}'^{,\mathrm{H}} C_{y,x}} \right\} = \operatorname{tr}\left\{ \sqrt[\oplus]{V \Sigma^2 V^{\mathrm{H}}} \right\} = \operatorname{tr}\left\{ \Sigma \right\}, \tag{B.34}$$

i.e., in Equation (B.33), the matrix $\bar{M}$ maximizes the sum of singular values of $\bar{M}'^{,\mathrm{H}} C_{y,x}$ over all $\bar{M}' \in \mathbb{C}^{N \times M}$. Since $U^{\mathrm{H}} \bar{M}'^{,\mathrm{H}} C_{y,x} V = U^{\mathrm{H}} U \Sigma V^{\mathrm{H}} V = \Sigma$, we get further[5]

$$\operatorname{tr}\left\{ \Sigma \right\} = \operatorname{Re}\left\{ \operatorname{tr}\left\{ V U^{\mathrm{H}} \bar{M}'^{,\mathrm{H}} C_{y,x} \right\} \right\} = \operatorname{Re}\left\{ \operatorname{tr}\left\{ \check{M}'^{,\mathrm{H}} C_{y,x} \right\} \right\}, \tag{B.35}$$

where $\check{M}' := \bar{M}' U V^{\mathrm{H}} \in \mathbb{C}^{N \times M}$ and the real part operator can be introduced because the sum of singular values is by definition real-valued. Due to the fact that $\bar{M}'^{,\mathrm{H}} \bar{M}' = I_M$, it holds also $\check{M}'^{,\mathrm{H}} \check{M}' = I_M$. Clearly, the matrix $U V^{\mathrm{H}}$ which transforms $\bar{M}'$ to $\check{M}'$ depends on $\bar{M}'$. However, since we are finally interested in the range space of $\bar{M}'$ and it holds that $\operatorname{range}\{\bar{M}'\} = \operatorname{range}\{\check{M}'\}$, the matrix

---

[5] Note that $\operatorname{tr}\{AB\} = \operatorname{tr}\{BA\}$ if $A \in \mathbb{C}^{m \times n}$ and $B \in \mathbb{C}^{n \times m}$.

$$\check{M} = \operatorname*{argmax}_{\check{M}' \in \mathbb{C}^{N \times M}} \operatorname{Re}\left\{ \operatorname{tr}\left\{ \check{M}'^{,\mathrm{H}} C_{y,x} \right\} \right\} \quad \text{s.t.} \quad \check{M}'^{,\mathrm{H}} \check{M}' = I_M, \tag{B.36}$$

is also a basis of the correlated subspace $\mathcal{M}$. Differentiating the corresponding Lagrangian function

$$L\left(\check{M}', \Lambda\right) = \operatorname{Re}\left\{ \operatorname{tr}\left\{ \check{M}'^{,\mathrm{H}} C_{y,x} - \left(\check{M}'^{,\mathrm{H}} \check{M}' - I_M \right) \Lambda \right\} \right\}, \tag{B.37}$$

$\Lambda \in \mathbb{C}^{M \times M}$, with respect to $\check{M}'^{,*}$ and setting the result to $\mathbf{0}_{N \times M}$ for $\check{M}' = \check{M}$, yields

$$\check{M}\left(\Lambda + \Lambda^{\mathrm{H}}\right) = C_{y,x}. \tag{B.38}$$

Again, we are only interested in the subspace spanned by the columns of $\check{M}$. Due to the fact that the Hermitian matrix $\Lambda + \Lambda^{\mathrm{H}} \in \mathbb{C}^{M \times M}$ does not change the subspace, it holds that $\operatorname{range}\{\check{M}\} = \operatorname{range}\{C_{y,x}\}$ and the correlated subspace is finally given by

$$\boxed{\mathcal{M} = \operatorname{range}\left\{C_{y,x}\right\}.} \tag{B.39}$$

C

Abbreviations and Acronyms

ARQ	Automatic Repeat reQuest
AV	Auxiliary Vector
AWGN	Additive White Gaussian Noise
BCD	Block Conjugate Direction
BCG	Block Conjugate Gradient
BCJR	Bahl–Cocke–Jelinek–Raviv
BER	Bit Error Rate
BGMRES	Block Generalized Minimal RESidual
Bi-CG	Bi-Conjugate Gradient
Bi-CGSTAB	Bi-Conjugate Gradient STABilized
bit	binary digit
BLR	Block Lanczos–Ruhe
BPSK	Binary Phase Shift Keying
BS	Backward Substitution
CDMA	Code Division Multiple-Access
cf.	confer (Latin, compare)
CF	Cholesky Factorization
CG	Conjugate Gradient
CME	Conditional Mean Estimator
CS	Cross-Spectral
CSI	Channel State Information
DFD	Decision Feedback Detector
DFE	Decision Feedback Equalization
DS	Direct Sequence
EDGE	Enhanced Data rates for GSM Evolution
e. g.	exempli gratia (Latin, for example)
et al.	et alii (Latin, and others)

EXIT	EXtrinsic Information Transfer
FEC	Forward Error Correction
FFT	Fast Fourier Transform
FIR	Finite Impulse Response
FLOP	FLoating point OPeration
FS	Forward Substitution
GMRES	Generalized Minimal RESidual
GPS	Global Positioning System
GSC	Generalized Sidelobe Canceler
GSM	Global Systems for Mobile communications
HI	Hermitian Inversion
i. e.	id est (Latin, that is)
i. i. e.	independent and identically distributed
INA	Institute for Numerical Analysis
ISI	InterSymbol Interference
KLT	Karhunen–Loève Transform
LLR	Log-Likelihood Ratio
LMS	Least Mean Squares
LS	Least Squares
LTI	Lower Triangular Inversion
LTP	Lower Triangular Product
MA	Multiple-Access
MAI	Multiple-Access Interference
MAP	Maximum A Posteriori
MI	Mutual Information
MIMO	Multiple-Input Multiple-Output
ML	Maximum Likelihood
MMSE	Minimum Mean Square Error
M-PSK	M-ary Phase Shift Keying
M-QAM	M-ary Quadrature Amplitude Modulation
MSE	Mean Square Error
MSMWF	MultiStage Matrix Wiener Filter
MSVWF	MultiStage Vector Wiener Filter
MSWF	MultiStage Wiener Filter
MUSIC	MUltiple SIgnal Classification
MVDR	Minimum Variance Distortionless Response
MWF	Matrix Wiener Filter
OVSF	Orthogonal Variable Spreading Factor
PC	Principal Component
QPSK	QuadriPhase Shift Keying

RC	Reduced-Complexity
RCMWF	Reduced-Complexity Matrix Wiener Filter
RLS	Recursive Least Squares
SIMO	Single-Input Multiple-Output
SNR	Signal-to-Noise Ratio
TV	Time-Variant
UCLA	University of California, Los Angeles
VCG	Vector Conjugate Gradient
VCS	Vector Cross-Spectral
viz.	videlicet (Latin, namely)
VPC	Vector Principal Component
VWF	Vector Wiener Filter
WCDMA	Wideband Code Division Multiple-Access
WF	Wiener Filter

References

1. M. Abramowitz and I. A. Stegun, *Handbook of Mathematical Functions With Formulas, Graphs, and Mathematical Tables*, Dover, 1972.

2. H. Akaike, "Block Toeplitz Matrix Inversion," *SIAM Journal on Applied Mathematics*, vol. 24, no. 2, pp. 234–241, March 1973.

3. S. P. Applebaum and D. J. Chapman, "Adaptive Arrays with Main Beam Constraints," *IEEE Transactions on Antennas and Propagation*, vol. AP-24, no. 5, pp. 650–662, September 1976.

4. W. E. Arnoldi, "The Principle of Minimized Iterations in the Solution of the Matrix Eigenvalue Problem," *Quarterly of Applied Mathematics*, vol. 9, no. 1, pp. 17–29, January 1951.

5. M. E. Austin, "Decision-Feedback Equalization for Digital Communication over Dispersive Channels," Technical Report 437, MIT Lincoln Lab, Lexington, MA, August 1967.

6. O. Axelsson, "Solution of Linear Systems of Equations: Iterative Methods," in *Sparse Matrix Techniques (Lecture Notes in Mathematics)*, V. A. Barker, Ed., vol. 572, pp. 1–51. Springer, 1977.

7. L. R. Bahl, J. Cocke, F. Jelinek, and J. Raviv, "Optimal Decoding of Linear Codes for Minimizing Symbol Error Rate," *IEEE Transactions on Information Theory*, vol. IT-20, no. 2, pp. 284–287, March 1974.

8. G. Bauch and N. Al-Dhahir, "Reduced-Complexity Space-Time Turbo-Equalization for Frequency-Selective MIMO Channels," *IEEE Transactions on Wireless Communications*, vol. 1, no. 4, pp. 819–828, October 2002.

9. P. Beckmann, *Orthogonal Polynomials for Engineers and Physicists*, The Golem Press, 1973.

10. W. Bell, "Signal Extraction for Nonstationary Time Series," *The Annals of Statistics*, vol. 12, no. 2, pp. 646–664, June 1984.

11. Benoit, "Note sur une méthode de résolution des équations normales provenant de l'application de la méthode des moindres carrés à un système d'équations linéaires en nombre inférieur à celui des inconnues – Application de la méthode à la résolution d'un système defini d'équations linéaires, Procédé du Commandant Cholesky (French, Note on a Method for Solving Normal Equations Occuring at the Application of the Least Squares Method to a System of Linear Equations Whose Number is Smaller than the Number of Unknowns – Application of the Method to Solve the Definite System of Linear Equations,

Approach of Commandant Cholesky)," *Bulletin Géodésique (French, Geodetical Bulletin)*, vol. 2, pp. 67–77, April/May/June 1924.

12. C. Berrou, A. Glavieux, and P. Thitimajshima, "Near Shannon Limit Error-Correcting Coding and Decoding: Turbo-Codes (1)," *IEEE International Conference on Communications*, pp. 1064–1070, May 1993.

13. W. M. Boothby, *An Introduction to Differentiable Manifolds and Riemannian Geometry*, Elsevier, 2002.

14. M. Botsch, G. Dietl, and W. Utschick, "Iterative Multi-User Detection Using Reduced-Complexity Equalization," in *Proceedings of the 4th International Symposium on Turbo Codes & Related Topics and the 6th International ITG-Conference on Source and Channel Coding (TURBO-CODING-2006)*, April 2006.

15. J. Brehmer, G. Dietl, M. Joham, and W. Utschick, "Reduced-Complexity Linear and Nonlinear Precoding for Frequency-Selective MIMO Channels," in *Proceedings of the 2004 IEEE 60th Vehicular Technology Conference (VTC 2004-Fall)*, September 2004, vol. 5, pp. 3684–3688.

16. J. Brehmer, M. Joham, G. Dietl, and W. Utschick, "Reduced Complexity Transmit Wiener Filter Based On a Krylov Subspace Multi-Stage Decomposition," in *Proceedings of the 2004 IEEE International Conference on Acoustics, Speech, and Signal Processing (ICASSP '04)*, May 2004, vol. 4, pp. 805–808.

17. J. W. Brewer, "Kronecker Products and Matrix Calculus in System Theory," *IEEE Transactions on Circuits and Systems*, vol. CAS-25, no. 9, pp. 772–781, September 1978.

18. C. Brezinski, "André Louis Cholesky," *Bulletin de la Société Mathématique de Belgique (French, Bulletin of the Mathematical Society of Belgium)*, pp. 45–50, December 1996, Supplement to volume 3, Numerical Analysis, A Numerical Analysis Conference in Honour of Jean Meinguet.

19. S. ten Brink, "Designing Iterative Decoding Schemes with the Extrinsic Information Transfer Chart," *AEÜ International Journal of Electronics and Communications*, vol. 54, no. 6, pp. 389–398, November 2000.

20. S. ten Brink, "Convergence Behavior of Iteratively Decoded Parallel Concatenated Codes," *IEEE Transactions on Communications*, vol. 49, no. 10, pp. 1727–1737, October 2001.

21. S. ten Brink, J. Speidel, and R.-H. Yan, "Iterative Demapping and Decoding for Multilevel Modulation," in *Proceedings of the Global Telecommunications Conference (GLOBECOM '98)*, November 1998, vol. 1, pp. 579–584.

22. S. ten Brink, J. Speidel, and R.-H. Yan, "Iterative Demapping for QPSK Modulation," *Electronics Letters*, vol. 34, no. 15, pp. 1459–1460, July 1998.

23. S. Burykh and K. Abed-Meraim, "Reduced-Rank Adaptive Filtering Using Krylov Subspace," *EURASIP Journal on Applied Signal Processing*, vol. 12, pp. 1387–1400, 2002.

24. K. A. Byerly and R. A. Roberts, "Output Power Based Partial Adaptive Array Design," in *Proceedings of the 23rd Asilomar Conference on Signals, Systems & Computers*, October 1989, vol. 2, pp. 576–580.

25. J. Capon, "High-Resolution Frequency-Wavenumber Spectrum Analysis," *Proceedings of the IEEE*, vol. 57, no. 8, pp. 1408–1418, August 1969.

26. R. H. Chan and M. K. Ng, "Conjugate Gradient Methods for Toeplitz Systems," *SIAM Review*, vol. 38, no. 3, pp. 427–482, September 1996.

27. P. S. Chang and A. N. Willson, Jr., "Analysis of Conjugate Gradient Algorithms for Adaptive Filtering," *IEEE Transactions on Signal Processing*, vol. 48, no. 2, pp. 409–418, February 2000.

28. P. L. Chebyshev, "Sur les questions de minima qui se rattachent à la représentation approximative des fonctions (French, On the Minima-Problem Occuring at the Approximate Representation of Functions)," *Mémoires de l'Académie Impériale des Sciences de Saint Pétersbourg. Sciences Mathématiques, Physiques et Naturelles. 1. Partie, Sciences Mathématiques et Physiques. 6. Série (French, Memoirs of the Imperial Academy of Sciences of Saint Petersburg. Mathematical, Physical, and Natural Sciences. 1. Part, Mathematical and Physical Sciences. 6. Series)*, vol. 7, pp. 199–291, 1859.

29. H. Chen, T. K. Sarkar, S. A. Dianat, and J. D. Brulé, "Adaptive Spectral Estimation by the Conjugate Gradient Method," *IEEE Transactions on Acoustics, Speech, and Signal Processing*, vol. ASSP-34, no. 2, pp. 272–284, April 1986.

30. W. Chen, U. Mitra, and P. Schniter, "On the Equivalence of Three Reduced Rank Linear Estimators with Applications to DS-CDMA," *IEEE Transactions on Information Theory*, vol. 48, no. 9, pp. 2609–2614, September 2002.

31. S. Choi, Y. U. Lee, and K. Hirasawa, "Real-Time Design of a Smart Antenna System Utilizing a Modified Conjugate Gradient Method for CDMA-Based Mobile Communications," in *Proceedings of the 1997 IEEE 47th Vehicular Technology Conference (VTC 1997-Spring)*, May 1997, pp. 183–187.

32. S. Chowdhury and M. D. Zoltowski, "Application of Conjugate Gradient Methods in MMSE Equalization for the Forward Link of DS-CDMA," in *Proceedings of the 2001 IEEE 54th Vehicular Technology Conference (VTC 2001-Fall)*, October 2001, pp. 2434–2438.

33. S. Chowdhury, M. D. Zoltowski, and J. S. Goldstein, "Reduced-Rank Adaptive MMSE Equalization for the Forward Link in High-Speed CDMA," in *Proceedings of the 43rd Midwest Symposium on Circuits and Systems*, August 2000, vol. 1, pp. 6–9.

34. P. Cifuentes, W. L. Myrick, S. Sud, J. S. Goldstein, and M. D. Zoltowski, "Reduced Rank Matrix Multistage Wiener Filter with Applications in MMSE Joint Multiuser Detection for DS-CDMA," in *Proceedings of the 2002 IEEE International Conference on Acoustics, Speech, and Signal Processing (ICASSP '02)*, May 2002, vol. 3, pp. 2605–2608.

35. B.-I. Clasen, "Sur une nouvelle méthode de résolution des équations linéaires et sur l'application de cette méthode au calcul des déterminants (French, On a New Method for Solving Systems of Linear Equations and Its Application to the Computation of Determinants)," *Annales de la Société Scientifique de Bruxelles (French, Annals of the Scientific Society of Brussels)*, vol. 12, no. 2, pp. 251–281, 1888.

36. P. Concus, G. H. Golub, and D. P. O'Leary, "A Generalized Conjugate Gradient Method for the Numerical Solution of Elliptic Partial Differential Equations," in *Sparse Matrix Computations*, J. R. Bunch and D. J. Rose, Eds., pp. 309–332. Academic Press, 1976.

37. J. W. Cooley, "How the FFT Gained Acceptance," in *A History of Scientific Computing*, S. G. Nash, Ed., pp. 133–140. ACM Press, Addison-Wesley, 1990.

38. J. W. Cooley and J. W. Tukey, "An Algorithm for the Machine Calculation of Complex Fourier Series," *Mathematics of Computation*, vol. 19, no. 2, pp. 297–301, April 1965.

39. J. W. Daniel, "The Conjugate Gradient Method for Linear and Nonlinear Operator Equations," *SIAM Journal on Numerical Analysis*, vol. 4, no. 1, pp. 10–26, 1967.

40. S. Darlington, "Linear Least-Squares Smoothing and Prediction, With Applications," *Bell Systems Technical Journal*, vol. 37, pp. 1221–1294, September 1958.

41. A. Dejonghe and L. Vandendorpe, "A Comparison of Bit and Symbol Interleaving in MMSE Turbo-Equalization," in *Proceedings of the 2003 IEEE International Conference on Communications (ICC 2003)*, May 2003, vol. 4, pp. 2928–2932.

42. G. Dietl, "Conjugate Gradient Implementation of Multi-Stage Nested Wiener Filter for Reduced Dimension Processing," Diploma thesis, Munich University of Technology, May 2001.

43. G. Dietl, M. Botsch, F. A. Dietrich, and W. Utschick, "Robust and Reduced-Rank Matrix Wiener Filter Based on the Block Conjugate Gradient Algorithm," in *Proceedings of the 6th IEEE Workshop on Signal Processing Advances in Wireless Communications (SPAWC 2005)*, June 2005, pp. 555–559.

44. G. Dietl and P. Breun, "Introduction to Reduced-Rank Matrix Wiener Filters," Technical Report TUM-LNS-TR-04-02, Munich University of Technology, April 2004.

45. G. Dietl, P. Breun, and W. Utschick, "Block Krylov Methods in Time-Dispersive MIMO Systems," in *Proceedings of the 2004 IEEE International Conference on Communications (ICC 2004)*, June 2004, vol. 5, pp. 2683–2688.

46. G. Dietl, I. Groh, and W. Utschick, "Low-Complexity MMSE Receivers Based on Weighted Matrix Polynomials in Frequency-Selective MIMO Systems," in *Proceedings of the 8th International Symposium on Signal Processing and Its Applications (ISSPA 2005)*, August 2005, vol. 2, pp. 699–702.

47. G. Dietl, M. Joham, P. Kreuter, J. Brehmer, and W. Utschick, "A Krylov Subspace Multi-Stage Decomposition of the Transmit Wiener Filter," in *Proceedings of the ITG Workshop on Smart Antennas, 2004*, March 2004.

48. G. Dietl, M. Joham, W. Utschick, and J. A. Nossek, "Long-Term Krylov-Prefilter for Reduced-Dimension Receive Processing in DS-CDMA Systems," in *Proceedings of the 2002 IEEE 56th Vehicular Technology Conference (VTC 2002-Fall)*, September 2002, pp. 492–495.

49. G. Dietl, C. Mensing, and W. Utschick, "Iterative Detection Based on Reduced-Rank Equalization," in *Proceedings of the 2004 IEEE 60th Vehicular Technology Conference (VTC 2004-Fall)*, September 2004, vol. 3, pp. 1533–1537.

50. G. Dietl, C. Mensing, and W. Utschick, "Iterative Detection Based on Widely Linear Processing and Real-Valued Symbol Alphabets," in *Proceedings of the Eleventh European Wireless Conference 2005*, April 2005, vol. 1, pp. 74–78.

51. G. Dietl, C. Mensing, W. Utschick, J. A. Nossek, and M. D. Zoltowski, "Multi-Stage MMSE Decision Feedback Equalization for EDGE," in *Proceedings of the 2003 IEEE International Conference on Acoustics, Speech, and Signal Processing (ICASSP '03)*, April 2003, vol. 4, pp. 509–513.

52. G. Dietl and W. Utschick, "On Reduced-Rank Aproaches to Matrix Wiener Filters in MIMO Systems," in *Proceedings of the 3rd IEEE International Symposium on Signal Processing and Information Technology (ISSPIT 2003)*, December 2003.

53. G. Dietl and W. Utschick, "Low-Complexity Equalizers – Rank Versus Order Reduction," in *Proceedings of the 7th IEEE Workshop on Signal Processing Advances in Wireless Communications (SPAWC 2006)*, 2006.

54. G. Dietl and W. Utschick, "MMSE Turbo Equalisation for Real-Valued Symbols," *European Transactions on Telecommunications*, vol. 17, no. 3, pp. 351–359, May/June 2006.

55. G. Dietl and W. Utschick, "Complexity Reduction of Iterative Receivers Using Low-Rank Equalization," *IEEE Transactions on Signal Processing*, vol. 55, no. 3, pp. 1035–1046, March 2007.

56. G. Dietl, M. D. Zoltowski, and M. Joham, "Recursive Reduced-Rank Adaptive Equalization for Wireless Communications," in *Proceedings of SPIE – The International Society for Optical Engineering*, April 2001, vol. 4395, pp. 16–27.

57. G. Dietl, M. D. Zoltowski, and M. Joham, "Reduced-Rank Equalization for EDGE via Conjugate Gradient Implementation of Multi-Stage Nested Wiener Filter," in *Proceedings of the 2001 IEEE 54th Vehicular Technology Conference (VTC 2001-Fall)*, October 2001, pp. 1912–1916.

58. F. Dietrich, G. Dietl, M. Joham, and W. Utschick, "Robust and Reduced Rank Space-Time Decision Feedback Equalization," in *Smart Antennas – State of the Art*, EURASIP Book Series on Signal Processing & Communications, chapter 10, part I: Receiver. Hindawi Publishing Corporation, 2005.

59. P. Ding, M. D. Zoltowski, and M. Firmoff, "Preconditioned Conjugate Gradient Based Fast Computation of Indirect Decision Feedback Equalizer," in *Proceedings of the 2004 IEEE International Conference on Acoustics, Speech, and Signal Processing (ICASSP '04)*, May 2004, vol. 4, pp. 1001–1004.

60. B. Dong and X. Wang, "Sampling-Based Soft Equalization for Frequency-Selective MIMO Channels," *IEEE Transactions on Communications*, vol. 53, no. 2, pp. 278–288, February 2005.

61. C. Douillard, M. Jézéquel, C. Berrou, A. Picart, P. Didier, and A. Glavieux, "Iterative Correction of Intersymbol Interference: Turbo-Equalization," *European Transactions on Telecommunications*, vol. 6, no. 5, pp. 507–511, September/October 1995.

62. A. Duel-Hallen, "A Family of Multiuser Decision-Feedback Detectors for Asynchronous Code-Division Multiple-Access Channels," *IEEE Transactions on Communications*, vol. 43, no. 2/3/4, pp. 421–434, February/March/April 1995.

63. C. Dumard and T. Zemen, "Double Krylov Subspace Approximation for Low Complexity Iterative Multi-User Decoding and Time-Variant Channel Estimation," in *Proceedings of the 6th IEEE Workshop on Signal Processing Advances in Wireless Communications (SPAWC 2005)*, June 2005, pp. 328–332.

64. C. Eckart and G. Young, "The Approximation of One Matrix by Another of Lower Rank," *Psychometrika*, vol. 1, no. 3, pp. 211–218, September 1936.

65. P. Elias, "Coding for Noisy Channels," in *IRE WESCON Convention Record, Part 4*, 1955, pp. 37–47.

66. J. Erhel and F. Guyomarc'h, "An Augmented Conjugate Gradient Method for Solving Consecutive Symmetric Positive Definite Linear Systems," *SIAM Journal on Matrix Analysis and Applications*, vol. 21, no. 4, pp. 1279–1299, 2000.

67. V. T. Ermolayev and A. G. Flaksman, "Signal Processing in Adaptive Arrays Using Power Basis," *International Journal of Electronics*, vol. 75, no. 4, pp. 753–765, 1993.

68. V. Faber and T. Manteuffel, "Necessary and Sufficient Conditions for the Existence of a Conjugate Gradient Method," *SIAM Journal on Numerical Analysis*, vol. 21, no. 2, pp. 352–362, April 1984.

69. R. Fletcher, "Conjugate Gradient Methods for Indefinite Systems," in *Proceedings of the Dundee Conference on Numerical Analysis, 1975, Numerical Analysis (Lecture Notes in Mathematics)*, G. A. Watson, Ed. 1976, vol. 506, pp. 73–89, Springer.

70. G. D. Forney, Jr., "The Viterbi Algorithm," *Proceedings of the IEEE*, vol. 61, no. 3, pp. 268–278, March 1973.

71. G. D. Forney, Jr., "Convolutional Codes II: Maximum Likelihood Decoding," *Information Control*, vol. 25, no. 3, pp. 222–266, July 1974.

72. L. Fox, H. D. Huskey, and J. H. Wilkinson, "Notes on the Solution of Algebraic Linear Simultaneous Equations," *The Quarterly Journal of Mechanics and Applied Mathematics*, vol. 1, pp. 149–173, 1948.

73. Z. Fu and E. M. Dowling, "Conjugate Gradient Projection Subspace Tracking," *IEEE Transactions on Signal Processing*, vol. 45, no. 6, pp. 1664–1668, June 1997.

74. C. F. Gauss, "Disquisitio de elementis ellipticis Palladis ex oppositionibus annorum 1803, 1804, 1805, 1807, 1808, 1809 (Latin, Disquisition on the Elliptical Elements of Pallas Based on the Oppositions in the Years 1803, 1804, 1805, 1807, 1808, and 1809)," in *Werke (German, Works)*, vol. 6, pp. 1–24. Königliche Gesellschaft der Wissenschaften zu Göttingen (German, The Royal Society of Sciences in Göttingen), 1874, Originally published in 1810.

75. C. F. Gauss, "Theoria interpolationis methodo nova tractata (Latin, Interpolation Theory Based on a Newly Derived Method)," in *Werke (German, Works)*, vol. 3, pp. 256–327. Königliche Gesellschaft der Wissenschaften zu Göttingen (German, The Royal Society of Sciences in Göttingen), 1876.

76. C. F. Gauss, "Gauss an Gerling (German, Gauss to Gerlin). Göttingen, December 26, 1823," in *Werke (German, Works)*, vol. 9, pp. 278–281. Königliche Gesellschaft der Wissenschaften zu Göttingen (German, The Royal Society of Sciences in Göttingen), B. G. Teubner, 1903.

77. W. Gautschi, "On Generating Orthogonal Polynomials," *SIAM Journal on Scientific and Statistical Computing*, vol. 3, no. 3, pp. 289–317, 1982.

78. A. Gersho and T. L. Lim, "Adaptive Cancellation of Intersymbol Interference for Data Transmission," *Bell Systems Technical Journal*, vol. 60, no. 11, pp. 1997–2021, Nov. 1981.

79. T. R. Giallorenzi and S. G. Wilson, "Multiuser ML Sequence Estimator for Convolutionally Coded Asynchronous DS-CDMA Systems," *IEEE Transactions on Communications*, vol. 44, no. 8, pp. 997–1008, August 1996.

80. T. R. Giallorenzi and S. G. Wilson, "Suboptimum Multiuser Receivers for Convolutionally Coded Asynchronous DS-CDMA Systems," *IEEE Transactions on Communications*, vol. 44, no. 9, pp. 1183–1196, September 1996.

81. K. S. Gilhousen, I. M. Jacobs, R. Padovani, A. J. Viterbi, L. A. Weaver, Jr., and C. E. Wheatley, III, "On the Capacity of a Cellular CDMA System," *IEEE Transactions on Vehicular Technology*, vol. 40, no. 2, pp. 303–312, May 1991.

82. A. Glavieux, C. Laot, and J. Labat, "Turbo Equalization over a Frequency Selective Channel," in *Proceedings of the International Symposium on Turbo Codes and Related Topics*, September 1997, pp. 96–102.

83. J. S. Goldstein, J. R. Guerci, and I. S. Reed, "An Optimal Generalized Theory of Signal Representation," in *Proceedings of the 1999 IEEE International Conference on Acoustics, Speech, and Signal Processing (ICASSP '99)*, March 1999, vol. 3, pp. 1357–1360.

84. J. S. Goldstein and I. S. Reed, "A New Method of Wiener Filtering and Its Application to Interference Mitigation for Communications," in *Proceedings of the IEEE Military Communications Conference (MILCOM 1997)*, November 1997, vol. 3, pp. 1087–1091.

85. J. S. Goldstein and I. S. Reed, "Reduced-Rank Adaptive Filtering," *IEEE Transactions on Signal Processing*, vol. 45, no. 2, pp. 492–496, February 1997.

86. J. S. Goldstein and I. S. Reed, "Subspace Selection for Partially Adaptive Sensor Array Processing," *IEEE Transactions on Aerospace and Electronic Systems*, vol. 33, no. 2, pp. 539–543, April 1997.

87. J. S. Goldstein and I. S. Reed, "Theory of Partially Adaptive Radar," *IEEE Transactions on Aerospace and Electronic Systems*, vol. 33, no. 4, pp. 1309–1325, October 1997.

88. J. S. Goldstein, I. S. Reed, D. E. Dudgeon, and J. R. Guerci, "A Multistage Matrix Wiener Filter for Subspace Detection," in *Proceedings of the IT Workshop on Detection, Estimation, Classification, and Imaging*, February 1999.

89. J. S. Goldstein, I. S. Reed, and L. L. Scharf, "A Multistage Representation of the Wiener Filter Based on Orthogonal Projections," *IEEE Transactions on Information Theory*, vol. 44, no. 7, pp. 2943–2959, November 1998.

90. J. S. Goldstein, I. S. Reed, L. L. Scharf, and J. A. Tague, "A Low-Complexity Implementation of Adaptive Wiener Filters," in *Proceedings of the 31st Asilomar Conference on Signals, Systems & Computers*, November 1997, vol. 1, pp. 770–774.

91. J. S. Goldstein, I. S. Reed, and P. A. Zulch, "Multistage Partially Adaptive STAP CFAR Detection Algorithm," *IEEE Transactions on Aerospace and Electronic Systems*, vol. 35, no. 2, pp. 645–661, April 1999.

92. G. H. Golub and D. P. O'Leary, "Some History of the Conjugate Gradient and Lanczos Algorithms: 1948–1976," *SIAM Review*, vol. 31, no. 1, pp. 50–102, March 1989.

93. G. H. Golub and R. Underwood, "The Block Lanczos Method for Computing Eigenvalues," in *Mathematical Software III*, J. R. Rice, Ed., pp. 361–377, Academic Press, 1977.

94. G. H. Golub and C. F. Van Loan, *Matrix Computations*, Johns Hopkins University Press, 1996.

95. A. Gorokhov and M. van Dijk, "Optimized Labeling Maps for Bit-Interleaved Transmission with Turbo Demodulation," in *Proceedings of the 2001 IEEE 53rd Vehicular Technology Conference (VTC 2001-Spring)*, May 2001, vol. 2, pp. 1459–1463.

96. A. Greenbaum and Z. Strakoš, "Predicting the Behavior of Finite Precision Lanczos and Conjugate Gradient Computations," *SIAM Journal on Matrix Analysis and Applications*, vol. 13, no. 1, pp. 121–137, January 1992.

97. T. N. E. Greville, "The Pseudoinverse of a Rectangular or Singular Matrix and Its Application to the Solution of Systems of Linear Equations," *SIAM Review*, vol. 1, no. 1, pp. 38–43, January 1959.

98. L. J. Griffiths and C. W. Jim, "An Alternative Approach to Linearly Constrained Adaptive Beamforming," *IEEE Transactions on Antennas and Propagation*, vol. AP-30, no. 1, pp. 27–34, January 1982.

99. C. W. Groetsch, *The Theory of Tikhonov Regularization for Fredholm Equations of the First Kind*, Research Notes in Mathematics; 105. Pitman, 1984.

100. I. Groh and G. Dietl, "Reduced-Rank MIMO Receivers Based on Krylov Subspaces and Random Matrices," Technical Report TUM-LNS-TR-05-01, Munich University of Technology, April 2005.

101. I. Groh, G. Dietl, and W. Utschick, "Preconditioned and Rank-Flexible Block Conjugate Gradient Implementations of MIMO Wiener Decision Feedback Equalizers," in *Proceedings of the 2006 IEEE International Conference on Acoustics, Speech, and Signal Processing (ICASSP '06)*, May 2006, vol. 4, pp. 441–444.

102. J. R. Guerci, J. S. Goldstein, and I. S. Reed, "Optimal and Adaptive Reduced-Rank STAP," *IEEE Transactions on Aerospace and Electronic Systems*, vol. 36, no. 2, pp. 647–663, April 2000.

103. D. Guo, S. Shamai (Shitz), and Sergio Verdú, "Mutual Information and Minimum Mean-Square Error in Gaussian Channels," *IEEE Transactions on Information Theory*, vol. 51, no. 4, pp. 1261–1282, April 2005.

104. J. Hadamard, *Lectures on Cauchy's Problem in Linear Partial Differential Equations*, Yale University Press, 1923.

105. J. Hagenauer, "Forward Error Correcting for CDMA Systems," in *Proceedings of the 4th International Symposium on Spread Spectrum Techniques and Applications (ISSSTA 1996)*, September 1996, vol. 2, pp. 566–569.

106. J. Hagenauer, "The Turbo Principle: Tutorial Introduction and State of the Art," in *Proceedings of the 2nd International Symposium on Turbo Codes & Related Topics*, September 1997, pp. 1–11.

107. J. Hagenauer, E. Offer, and L. Papke, "Iterative Decoding of Binary Block and Convolutional Codes," *IEEE Transactions on Information Theory*, vol. 42, no. 2, pp. 429–445, March 1996.

108. M. Hanke, *Conjugate Gradient Type Methods for Ill-Posed Problems*, Longman, 1995.

109. E. J. Hannan, "Measurement of a Wandering Signal Amid Noise," *Journal of Applied Probability*, vol. 4, no. 1, pp. 90–102, April 1967.

110. C. Hansen, *Rank-Deficient and Discrete Ill-Posed Problems*, SIAM, 1998.

111. S. Haykin and M. Moher, *Modern Wireless Communications*, Prentice Hall, 2005.

112. S. S. Haykin, *Adaptive Filter Theory*, Prentice Hall, 1986.

113. C. Heegard and S. B. Wicker, *Turbo Coding*, Kluwer, 1999.

114. M. T. Heideman, D. H. Johnson, and C. S. Burrus, "Gauss and the History of the Fast Fourier Transform," *IEEE Acoustics, Speech, and Signal Processing (ASSP) Magazine*, vol. 1, no. 4, pp. 14–21, October 1984.

115. M. R. Hestenes, "Pseudoinverses and Conjugate Gradients," *Communications of the ACM*, vol. 18, no. 1, pp. 40–43, 1975.

116. M. R. Hestenes, "Conjugacy and Gradients," in *A History of Scientific Computing*, S. G. Nash, Ed., pp. 167–179. ACM Press, Addison-Wesley, 1990.

117. M. R. Hestenes and E. Stiefel, "Methods of Conjugate Gradients for Solving Linear Systems," *Journal of Research of the National Bureau of Standards*, vol. 49, no. 6, pp. 409–436, December 1952.

118. H. Heuser and H. Wolf, *Algebra, Funktionalanalysis und Codierung (German, Algebra, Functional Analysis, and Coding)*, B. G. Teubner, 1986.

119. J. D. Hiemstra and J. S. Goldstein, "Robust Rank Selection for the Multistage Wiener Filter," in *Proceedings of the 2002 IEEE International Conference on Acoustics, Speech, and Signal Processing (ICASSP '02)*, May 2002, vol. 3, pp. 2929–2932.

120. M. L. Honig and J. S. Goldstein, "Adaptive Reduced-Rank Residual Correlation Algorithms for DS-CDMA Interference Suppression," in *Proceedings of the 32nd Asilomar Conference on Signals, Systems & Computers*, November 1998, vol. 2, pp. 1106–1110.

121. M. L. Honig and J. S. Goldstein, "Adaptive Reduced-Rank Interference Suppression Based on the Multistage Wiener Filter," *IEEE Transactions on Communications*, vol. 50, no. 6, pp. 986–994, June 2002.

122. M. L. Honig and R. Ratasuk, "Large-System Performance of Iterative Multiuser Decision-Feedback Detection," *IEEE Transactions on Communications*, vol. 51, no. 8, pp. 1368–1377, August 2003.

123. M. L. Honig and Y. Sun, "Performance of Iterative Multiuser Decision-Feedback Receivers," in *Proceedings of the 2001 IEEE Information Theory Workshop (ITW 2001)*, September 2001, pp. 30–32.

124. M. L. Honig, V. Tripathi, and Y. Sun, "Adaptive Decision Feedback Turbo Equalization," in *Proceedings of the International Symposium on Information Theory (ISIT 2002)*, June/July 2002, p. 413.

125. M. L. Honig and M. K. Tsatsanis, "Adaptive Techniques for Multiuser CDMA Receivers," *IEEE Signal Processing Magazine*, vol. 17, no. 3, pp. 49–61, May 2000.

126. M. L. Honig, G. Woodward, and P. D. Alexander, "Adaptive Multiuser Parallel-Decision-Feedback with Iterative Decoding," in *Proceedings of the International Symposium on Information Theory (ISIT 2000)*, June 2000, p. 335.

127. M. L. Honig, G. K. Woodward, and Y. Sun, "Adaptive Iterative Multiuser Decision Feedback Detection," *IEEE Transactions on Wireless Communications*, vol. 3, no. 2, pp. 477–485, March 2004.

128. M. L. Honig and W. Xiao, "Large System Performance of Reduced-Rank Linear Filters," in *Proceedings of the 37th Annual Allerton Conference on Communications, Systems, and Computing*, October 1999.

129. M. L. Honig and W. Xiao, "Adaptive Reduced-Rank Interference Suppression with Adaptive Rank Selection," in *Proceedings of the IEEE Military Communications Conference (MILCOM 2000)*, October 2000, vol. 2, pp. 747–751.

130. M. L. Honig and W. Xiao, "Performance of Reduced-Rank Linear Interference Suppression," *IEEE Transactions on Information Theory*, vol. 47, no. 5, pp. 1928–1946, July 2001.

131. R. A. Horn and C. R. Johnson, *Matrix Analysis*, Cambridge University Press, 1993.

132. H. Hotelling, "Analysis of a Complex of Statistical Variables into Principal Components," *Journal of Educational Psychology*, vol. 24, no. 6/7, pp. 417–441, 498–520, September/October 1933.

133. A. S. Householder, *The Theory of Matrices in Numerical Analysis*, Dover, 1964.

134. C. Hu and I. S. Reed, "Space-Time Adaptive Reduced-Rank Multistage Wiener Filtering for Asynchronous DS-CDMA," *IEEE Transactions on Signal Processing*, vol. 52, no. 7, pp. 1862–1877, July 2004.

135. Y. Hua, M. Nikpur, and P. Stoica, "Optimal Reduced-Rank Estimation and Filtering," *IEEE Transactions on Signal Processing*, vol. 49, no. 3, pp. 457–469, March 2001.

136. R. Hunger, "Floating Point Operations in Matrix-Vector Calculus," Technical Report TUM-LNS-TR-05-05, Munich University of Technology, 2005.

137. B. R. Hunt, "Deconvolution of Linear Systems by Constrained Regression and Its Relationship to the Wiener Theory," *IEEE Transactions on Automatic Control*, vol. 17, no. 5, pp. 703–705, October 1972.

138. C. G. J. Jacobi, "Ueber [sic] eine neue Auflösungsart der bei der Methode der kleinsten Quadrate vorkommenden lineären [sic] Gleichungen (German, On a New Method for Solving Linear Equations Occuring at the Least Squares Method)," *Astronomische Nachrichten (German, Astronomical News)*, vol. 22, no. 523, pp. 297–306, 1845.

139. C. G. J. Jacobi, "Über ein leichtes Verfahren die in der Theorie der Säcularstörungen [sic] vorkommenden Gleichungen numerisch aufzulösen (German, On a Simple Approach of Solving Numerically the Equations Occuring at the Theory of Secular Perturbance)," *Journal für die reine und angewandte Mathematik (German, Journal for Pure and Applied Mathematics, Crelle's Journal)*, vol. 30, pp. 51–94, 1846.

140. M. Joham, Y. Sun, M. D. Zoltowski, M. Honig, and J. S. Goldstein, "A New Backward Recursion for the Multi-Stage Nested Wiener Filter Employing Krylov Subspace Methods," in *Proceedings of the IEEE Military Communications Conference (MILCOM 2001)*, October 2001, vol. 2, pp. 28–31.

141. M. Joham and M. D. Zoltowski, "Interpretation of the Multi-Stage Nested Wiener Filter in the Krylov Subspace Framework," Technical Report TUM-LNS-TR-00-6, Munich University of Technology, November 2000, Also: Technical Report TR-ECE-00-51, Purdue University.

142. W. Jordan, *Handbuch der Vermessungskunde (German, Handbook of Surveying)*, Verlag der J. B. Metzler'schen Buchhandlung, 1888.

143. T. Kailath, A. H. Sayed, and B. Hassibi, *Linear Estimation*, Prentice Hall, 1999.

144. S. Kangshen, J. N. Crossley, and A. W.-C. Lun, *The Nine Chapters on the Mathematical Art*, chapter 8: Rectangular Arrays, pp. 386–438, Oxford University Press, 1999.

145. S. Kaniel, "Estimates for Some Computational Techniques in Linear Algebra," *Mathematics of Computation*, vol. 22, pp. 369–378, 1966.

146. A. Kansal, S. N. Batalama, and D. A. Pados, "Adaptive Maximum SINR RAKE Filtering for DS-CDMA Multipath Fading Channels," *IEEE Journal on Selected Areas in Communications*, vol. 16, no. 9, December 1998.

147. K. Karhunen, "Zur Spektraltheorie stochastischer Prozesse (German, On the Spectral Theory of Stochastic Processes)," *Suomalaisen Tiedeakatemian Toimituksia, Annales Academiæ Scientiarum Fennicæ. Serja A, I. Mathematica-Physica (Finish, Latin, Annals of the Finish Academy of Sciences. Series A, I. Mathematics-Physics)*, vol. 34, pp. 1–7, 1946.

148. P. Karttunen and R. Baghaie, "Conjugate Gradient Based Signal Subspace Mobile User Tracking," in *Proceedings of the 1999 IEEE 49th Vehicular Technology Conference (VTC 1999-Spring)*, May 1999, pp. 1172–1176.

149. W. Karush, "Convergence of a Method of Solving Linear Problems," *Proceedings of the American Mathematical Society*, vol. 3, no. 6, pp. 839–851, December 1952.

150. S. M. Kay, *Fundamentals of Statistical Signal Processing: Estimation Theory*, Prentice Hall, 1993.

151. S. M. Kay, *Fundamentals of Statistical Signal Processing: Detection Theory*, Prentice Hall, 1998.

152. A. N. Kolmogorov, "Sur l'interpolation et extrapolation des suites stationnaires (French, On the Interpolation and Extrapolation of Stationary Sequences)," *Comptes Rendus Hebdomadaires des Séances de l'Académie des Sciences (French, Proceedings of the Weekly Sessions of the Academy of Sciences)*, vol. 208, pp. 2043–2045, 1939.

153. A. N. Kolmogorov, "Интерполирование и экстраполирование стационарных случайных последовательностей (Russian, Interpolation and Extrapolation of Stationary Random Sequences)," Известия академии наук СССР, Серия математическая *(Russian, Bulletin of the Academy of Sciences of the USSR, Mathematical Series)*, vol. 5, pp. 3–14, 1941.

154. T. P. Krauss, M. D. Zoltowski, and G. Leus, "Simple MMSE Equalizers for CDMA Downlink to Restore Chip Sequence: Comparison to Zero Forcing and RAKE," in *Proceedings of the 2006 IEEE International Conference on Acoustics, Speech, and Signal Processing (ICASSP '00)*, June 2000, vol. 5, pp. 2865–2868.

155. H. Krim and M. Viberg, "Two Decades of Array Signal Processing Research," *IEEE Signal Processing Magazine*, vol. 13, no. 4, pp. 67–94, July 1996.

156. C. Lanczos, "An Iteration Method for the Solution of the Eigenvalue Problem of Linear Differential and Integral Operators," *Journal of Research of the National Bureau of Standards*, vol. 45, no. 4, pp. 255–282, October 1950.

157. C. Lanczos, "Solution of Systems of Linear Equations by Minimized Iterations," *Journal of Research of the National Bureau of Standards*, vol. 49, no. 1, pp. 33–53, July 1952.

158. C. Laot, R. Le Bidan, and D. Leroux, "Low-Complexity MMSE Turbo Equalization: A Possible Solution for EDGE," *IEEE Transactions on Wireless Communications*, vol. 4, no. 3, pp. 965–974, May 2005.

159. C. Laot, A. Glavieux, and J. Labat, "Turbo Equalization: Adaptive Equalization and Channel Decoding Jointly Optimized," *IEEE Journal on Selected Areas in Communications*, vol. 19, no. 9, pp. 1744–1752, September 2001.

160. K. Li and X. Wang, "EXIT Chart Analysis of Turbo Multiuser Detection," *IEEE Transactions on Wireless Communications*, vol. 4, no. 1, pp. 300–311, January 2005.

161. L. Li, A. M. Tulino, and S. Verdú, "Design of Reduced-Rank MMSE Multiuser Detectors Using Random Matrix Methods," *IEEE Transactions on Information Theory*, vol. 50, no. 6, pp. 986–1008, June 2004.

162. P. Li, W. Utschick, and J. A. Nossek, "A New DS-CDMA Interference Cancellation Scheme for Space-Time Rake Processing," in *Proceedings of the European Wireless 1999 & ITG Mobile Communications*, October 1999, pp. 385–389.

163. S. Lin and D. J. Costello, Jr., *Error Control Coding*, Prentice Hall, 2004.

164. M. Loève, "Fonctions aléatoires du second ordre (French, Random Functions of Second Order)," in *Processus stochastiques et mouvement brownien (French, Stochastic Processes and the Brownian Motion)*, Paul Lévy, Ed., pp. 299–352. Gauthier-Villars, 1948, Partly published in *Comptes Rendus de l'Académie des Sciences (French, Proceedings of the Academy of Sciences)*, vol. 220, pp. 295,380–382, 1945, vol. 222, pp. 469,628,942, 1946.

165. D. G. Luenberger, *Introduction to Linear and Nonlinear Programming*, Addison-Wesley, 1965.

166. M. Moher, "An Iterative Multiuser Decoder for Near-Capacity Communications," *IEEE Transactions on Communications*, vol. 46, no. 7, pp. 870–880, July 1998.

167. E. H. Moore, "On the Reciprocal of the General Algebraic Matrix," *Bulletin of the American Mathematical Society*, vol. 26, pp. 394–395, 1920.

168. E. H. Moore and R. W. Barnard, *General Analysis*, vol. I of *Memoirs of the American Philosophical Society*, American Philosophical Society, 1935.

169. S. Moshavi, E. G. Kanterakis, and D. L. Schilling, "Multistage Linear Receivers for DS-CDMA Systems," *International Journal of Wireless Information Networks*, vol. 3, no. 1, pp. 1–17, 1996.

170. M. S. Mueller and J. Salz, "A Unified Theory of Data-Aided Equalization," *Bell Systems Technical Journal*, vol. 60, no. 9, pp. 2023–2038, Nov. 1981.

171. R. J. Muirhead, *Aspects of Multivariate Statistical Theory*, Wiley, 1982.

172. R. R. Müller and S. Verdú, "Design and Analysis of Low-Complexity Interference Mitigation on Vector Channels," *IEEE Journal on Selected Areas in Communications*, vol. 19, no. 8, pp. 1429–1441, August 2001.

173. W. L. Myrick, M. D. Zoltowski, and J. S. Goldstein, "Anti-Jam Space-Time Preprocessor for GPS Based on Multistage Nested Wiener Filter," in *Proceedings of the IEEE Military Communications Conference (MILCOM 1999)*, October/November 1999, vol. 1, pp. 675–681.

174. W. L. Myrick, M. D. Zoltowski, and J. S. Goldstein, "GPS Jammer Suppression with Low-Sample Support Using Reduced-Rank Power Minimization," in *Proceedings of the 10th IEEE Workshop on Statistical Signal and Array Processing (SSAP 2000)*, August 2000, pp. 514–518.

175. D. P. O'Leary, "The Block Conjugate Gradient Algorithm and Related Methods," *Linear Algebra and Its Applications*, vol. 29, pp. 293–322, 1980.

176. H. Omori, T. Asai, and T. Matsumoto, "A Matched Filter Approximation for SC/MMSE Iterative Equalizers," *IEEE Communications Letters*, vol. 5, no. 7, pp. 310–312, July 2001.

177. R. Otnes and M. Tüchler, "EXIT Chart Analysis Applied to Adaptive Turbo Equalization," in *Proceedings of the Nordic Signal Processing Symposium*, October 2002.

178. D. A. Pados and S. N. Batalama, "Low-Complexity Blind Detection of DS/CDMA Signals: Auxiliary-Vector Receivers," *IEEE Transactions on Communications*, vol. 45, no. 12, pp. 1586–1594, December 1997.

179. D. A. Pados and S. N. Batalama, "Joint Space-Time Auxiliary-Vector Filtering for DS/CDMA Systems with Antenna Arrays," *IEEE Transactions on Communications*, vol. 47, no. 9, pp. 1406–1415, September 1999.

180. D. A. Pados and G. N. Karystinos, "An Iterative Algorithm for the Computation of the MVDR Filter," *IEEE Transactions on Signal Processing*, vol. 49, no. 2, pp. 290–300, February 2001.

181. D. A. Pados, F. J. Lombardo, and S. N. Batalama, "Auxiliary-Vector Filters and Adaptive Steering for DS/CDMA Single-User Detection," *IEEE Transactions on Vehicular Technology*, vol. 48, no. 6, November 1999.

182. C. C. Paige and M. A. Saunders, "LSQR: An Algorithm for Sparse Linear Equations and Sparse Least Squares," *ACM Transactions on Mathematical Software*, vol. 8, no. 1, pp. 43–71, March 1982.

183. A. Papoulis, *Probability, Random Variables, and Stochastic Processes*, McGraw-Hill, 1991.

184. R. Penrose, "A Generalized Inverse for Matrices," *Mathematical Proceedings of the Philosophical Society*, vol. 51, pp. 406–413, 1955.

185. B. Picinbono and P. Chevalier, "Widely Linear Estimation with Complex Data," *IEEE Transactions on Signal Processing*, vol. 43, no. 8, pp. 2030–2033, August 1995.

186. R. L. Pickholtz, D. L. Schilling, and L. B. Milstein, "Theory of Spread-Spectrum Communications – A Tutorial," *IEEE Transactions on Communications*, vol. COM-30, no. 5, pp. 855–884, May 1982.

187. R. L. Plackett, "Some Theorems in Least Squares," *Biometrika*, vol. 37, no. 1/2, pp. 149–157, June 1950.

188. H. V. Poor, *An Introduction to Signal Detection and Estimation*, Springer, 1988.

189. H. V. Poor and S. Verdú, "Probability of Error in MMSE Multiuser Detection," *IEEE Transactions on Information Theory*, vol. 43, no. 3, pp. 858–871, May 1997.

190. B. Porat and B. Friedlander, "Blind Equalization of Digital Communications Channels Using High-Order Moments," *IEEE Transactions on Signal Processing*, vol. 39, pp. 522–526, February 1991.

191. R. Prasad and T. Ojanperä, "An Overview of CDMA Evolution Towards Wideband CDMA," *IEEE Communications Surveys & Tutorials*, vol. 1, no. 1, pp. 2–29, 1998, http://www.comsoc.org/pubs/surveys.

192. R. Price and P. E. Green, Jr., "A Communication Technique for Multipath Channels," *Proceedings of the IRE*, vol. 46, no. 3, pp. 555–570, March 1958.

193. J. G. Proakis, *Digital Communications*, McGraw-Hill, 1995.

194. H. Qian and S. N. Batalama, "Data Record-Based Criteria for the Selection of an Auxiliary Vector Estimator of the MMSE/MVDR Filter," *IEEE Transactions on Communications*, vol. 51, no. 10, pp. 1700–1708, October 2003.

195. V. Ramon and L. Vandendorpe, "Predicting the Performance and Convergence Behavior of a Turbo-Equalization Scheme," in *Proceedings of the 2005 IEEE International Conference on Acoustics, Speech, and Signal Processing (ICASSP '05)*, March 2005.

196. M. C. Reed, C. B. Schlegel, P. D. Alexander, and J. A. Asenstorfer, "Iterative Multiuser Detection for CDMA with FEC: Near-Single-User Performance," *IEEE Transactions on Communications*, vol. 46, no. 12, pp. 1693–1699, December 1998.

197. J. K. Reid, "The Use of Conjugate Gradients for Systems of Equations Processing 'Property A'," *SIAM Journal on Numerical Analysis*, vol. 9, no. 2, pp. 325–332, June 1972.

198. D. Reynolds and X. Wang, "Low-Complexity Turbo-Equalization for Diversity Channels," *Signal Processing*, vol. 81, no. 5, pp. 989–995, 2001.

199. D. C. Ricks and J. S. Goldstein, "Efficient Architectures for Implementing Adaptive Algorithms," in *Proceedings of the 2000 Antenna Applications Symposium*, September 2000, pp. 29–41.

200. A. Ruhe, "Implementation Aspects of Band Lanczos Algorithms for Computation of Eigenvalues of Large Sparse Symmetric Matrices," *Mathematics of Computation*, vol. 33, no. 146, pp. 680–687, April 1979.

201. Y. Saad, "On the Lanczos Method for Solving Symmetric Linear Systems with Several Right-Hand Sides," *Mathematics of Computation*, vol. 48, no. 178, pp. 651–662, April 1987.

202. Y. Saad, *Numerical Methods for Large Eigenvalue Problems*, Halstead Press – out of print, 1992.

203. Y. Saad, "Analysis of Augmented Krylov Subspace Methods," *SIAM Journal on Matrix Analysis and Applications*, vol. 18, no. 2, pp. 435–449, April 1997.

204. Y. Saad, *Iterative Methods for Sparse Linear Systems*, SIAM, 2003.

205. Y. Saad and M. H. Schultz, "GMRES: A Generalized Minimal Residual Algorithm for Solving Nonsymmetric Linear Systems," *SIAM Journal on Scientific and Statistical Computing*, vol. 7, no. 3, pp. 856–869, July 1986.

206. Y. Saad and H. A. van der Vorst, "Iterative Solution of Linear Systems in the 20th Century," *Journal of Computational and Applied Mathematics*, vol. 123, no. 1–2, pp. 1–33, November 2000.

207. M. Sadkane, "Block-Arnoldi and Davidson Methods for Unsymmetric Large Eigenvalue Problems," *Numerische Mathematik (German, Numerical Mathematics)*, vol. 64, pp. 195–212, 1993.

208. M. Sadkane, "A Block Arnoldi-Chebyshev Method for Computing the Leading Eigenpairs of Large Sparse Unsymmetric Matrices," *Numerische Mathematik (German, Numerical Mathematics)*, vol. 64, pp. 181–193, 1993.

209. E. L. Santos and M. D. Zoltowski, "On Low Rank MVDR Beamforming Using the Conjugate Gradient Algorithm," in *Proceedings of the 2004 IEEE International Conference on Acoustics, Speech, and Signal Processing (ICASSP '04)*, May 2004, vol. 2, pp. 173–176.

210. L. L. Scharf, *Statistical Signal Processing*, Addison-Wesley, 1991.

211. L. L. Scharf, "The SVD and Reduced Rank Signal Processing," *Signal Processing*, vol. 25, no. 2, pp. 113–133, 1991.

212. L. L. Scharf, L. T. McWhorter, E. K. P. Chong, J. S. Goldstein, and M. D. Zoltowski, "Algebraic Equivalence of Conjugate Gradient Direction and Multistage Wiener Filters," in *Proceedings of the 11th Annual Workshop on Adaptive Sensor Array Processing (ASAP)*, March 2003, pp. 388–392.

213. L. L. Scharf and D. W. Tufts, "Rank Reduction for Modeling Stationary Signals," *IEEE Transactions on Acoustics, Speech, and Signal Processing*, vol. ASSP-35, no. 3, pp. 350–355, March 1987.

214. R. O. Schmidt, "Multiple Emitter Location and Signal Parameter Estimation," *IEEE Transactions on Antennas and Propagation*, vol. AP-34, no. 3, pp. 276–280, March 1986, Originally published in *Proceedings of the Rome Air Development Center (RADC) Spectrum Estimation Workshop*, October 1979.

215. M. K. Schneider and A. S. Willsky, "Krylov Subspace Estimation," *SIAM Journal on Scientific Computing*, vol. 22, no. 5, pp. 1840–1864, 2001.

216. R. A. Scholtz, "The Origins of Spread-Spectrum Communications," *IEEE Transactions on Communications*, vol. COM-30, no. 5, pp. 822–854, May 1982.

217. F. Schreckenbach, N. Görtz, J. Hagenauer, and G. Bauch, "Optimization of Symbol Mappings for Bit-Interleaved Coded Modulation with Iterative Decoding," *IEEE Communications Letters*, vol. 7, no. 12, pp. 593–595, December 2003.

218. H. R. Schwarz, H. Rutishauser, and E. Stiefel, *Numerik symmetrischer Matrizen (German, Numerics of Symmetric Matrices*, B. G. Teubner, 1968.

219. J. A. Scott, "An Arnoldi Code for Computing Selected Eigenvalues of Sparse, Real, Unsymmetric Matrices," *ACM Transactions on Mathematical Software*, vol. 21, no. 4, pp. 432–475, December 1995.

220. L. Seidel, "Ueber [sic] ein Verfahren, die Gleichungen, auf welche die Methode der kleinsten Quadrate führt, sowie lineäre [sic] Gleichungen überhaupt, durch successive [sic] Annäherung aufzulösen (German, On an Approach for Solving Equations Obtained From the Least Squares Method or Linear Equations in General by Successive Approximation)," *Abhandlungen der mathematisch-physikalischen Classe [sic] der königlich bayerischen Akademie der Wissenschaften (German, Essay of the Mathematical-Physical Class of the Royal Bavarian Academy of Sciences)*, vol. 11, pp. 81–108, 1874.

221. G. M. A. Sessler and F. K. Jondral, "Low Complexity Polynomial Expansion Multiuser Detector with Strong Near-Far Resistance," in *Proceedings of the 2002 IEEE 56th Vehicular Technology Conference (VTC 2002-Fall)*, September 2002, vol. 3, pp. 1869–1872.

222. G. M. A. Sessler and F. K. Jondral, "Low Complexity Polynomial Expansion Multiuser Detector for CDMA Systems," *IEEE Transactions on Vehicular Technology*, vol. 54, no. 4, pp. 1379–1391, July 2005.

223. V. Simoncini and E. Gallopoulos, "An Iterative Method for Nonsymmetric Systems with Multiple Right-Hand Sides," *SIAM Journal on Scientific Computing*, vol. 16, no. 4, pp. 917–933, July 1995.

224. V. Simoncini and E. Gallopoulos, "Convergence Properties of Block GMRES and Matrix Polynomials," *Linear Algebra and Its Applications*, vol. 247, pp. 97–119, 1996.

225. A. van der Sluis and H. A. van der Vorst, "The Rate of Convergence of Conjugate Gradients," *Numerische Mathematik (German, Numerical Mathematics)*, vol. 48, pp. 543–560, 1986.

226. H. Stark and J. W. Woods, *Probability, Random Processes, and Estimation Theory for Engineers*, Prentice Hall, 1986.

227. Z. Strakoš, "On the Real Convergence Rate of the Conjugate Gradient Method," *Linear Algebra and Its Applications*, vol. 154–156, pp. 535–549, 1991.

228. G. Strang, "A Proposal for Toeplitz Matrix Calculations," *Studies in Applied Mathematics*, vol. 74, no. 2, pp. 171–176, 1986.

229. S. S. Sud and J. S. Goldstein, "A Low Complexity Receiver for Space-Time Coded CDMA Systems," in *Proceedings of the 2002 IEEE International Conference on Acoustics, Speech, and Signal Processing (ICASSP '02)*, May 2002, vol. 3, pp. 2397–2400.

230. Y. Sun and M. L. Honig, "Performance of Reduced-Rank Equalization," in *Proceedings of the International Symposium on Information Theory (ISIT 2002)*, June/July 2002.

231. Y. Sun, M. L. Honig, and V. Tripathi, "Adaptive, Iterative, Reduced-Rank Equalization for MIMO Channels," in *Proceedings of the IEEE Military Communications Conference (MILCOM 2002)*, 2002, vol. 2, pp. 1029–1033.

232. Y. Sun, V. Tripathi, and M. L. Honig, "Adaptive Turbo Reduced-Rank Equalization for MIMO Channels," *IEEE Transactions on Wireless Communications*, vol. 4, no. 6, pp. 2789–2800, November 2005.

233. A. Thorpe and L. Scharf, "Reduced Rank Methods for Solving Least Squares Problems," in *Proceedings of the 23rd Asilomar Conference on Signals, Systems & Computers*, October 1989, vol. 2, pp. 609–613.

234. A. N. Tikhonov, "Solution of Incorrectly Formulated Problems and the Regularization Method," *Soviet Mathematics Doklady*, vol. 4, pp. 1035–1038, 1963.

235. A. N. Tikhonov and V. Y. Arsenin, *Solutions of Ill-Posed Problems*, Winston, 1977.

236. J. Todd, "The Prehistory and Early History of Computation at the U.S. National Bureau of Standards," in *A History of Scientific Computing*, S. G. Nash, Ed., pp. 251–268. ACM Press, Addison-Wesley, 1990.

237. L. N. Trefethen and D. Bau, III, *Numerical Linear Algebra*, SIAM, 1997.

238. W. F. Trench, "An Algorithm for the Inversion of Finite Toeplitz Matrices," *Journal of the Society for Industrial and Applied Mathematics*, vol. 12, no. 3, pp. 515–522, September 1964.

239. M. Tüchler, R. Koetter, and A. Singer, "Turbo Equalization: Principles and New Results," *IEEE Transactions on Communications*, vol. 50, no. 5, pp. 754–767, May 2002.

240. M. Tüchler, A. Singer, and R. Koetter, "Minimum Mean Squared Error Equalization Using A-priori Information," *IEEE Transactions on Signal Processing*, vol. 50, no. 3, pp. 673–683, March 2002.

241. D. W. Tufts and R. Kumaresan, "Singular Value Decomposition and Improved Frequency Estimation Using Linear Prediction," *IEEE Transactions on Acoustics, Speech, and Signal Processing*, vol. ASSP-30, no. 4, pp. 671–675, August 1982.

242. D. W. Tufts, R. Kumaresan, and I. Kirsteins, "Data Adaptive Signal Estimation by Singular Value Decomposition of a Data Matrix," *Proceedings of the IEEE*, vol. 70, no. 6, pp. 684–685, June 1982.

243. G. Ungerboeck, "Channel Coding with Multilevel/Phase Signals," *IEEE Transactions on Information Theory*, vol. 28, no. 1, pp. 55–67, January 1982.

244. H. L. Van Trees, *Optimum Array Processing, Part IV of Detection, Estimation, and Modulation Theory*, Wiley, 2002.

245. M. K. Varanasi and T. Guess, "Optimum Decision Feedback Multiuser Equalization with Successive Decoding Achieves the Total Capacity of the Gaussian Multiple-Access Channel," in *Proceedings of the 31st Asilomar Conference on Signals, Systems & Computers*, November 1997, vol. 2, pp. 1405–1409.

246. S. Verdú, "Minimum Probability of Error for Asynchronous Gaussian Multiple-Access Channels," *IEEE Transactions on Information Theory*, vol. IT-32, no. 1, pp. 85–96, January 1986.

247. S. Verdú, *Multiuser Detection*, Cambridge University Press, 1998.

248. A. J. Viterbi, "Error Bounds for Convolutional Codes and an Asymptotically Optimum Decoding Algorithm," *IEEE Transactions on Information Theory*, vol. IT-13, no. 2, pp. 260–269, April 1967.

249. V. V. Voevodin, "The Question of Non-Self-Adjoint Extension of the Conjugate Gradients Method is Closed," *USSR Computational Mathematics and Mathematical Physics*, vol. 23, no. 2, pp. 143–144, 1983.

250. D. V. Voiculescu, K. J. Dykema, and A. Nica, *Free Random Variables*, American Mathematical Society, 1992.

251. H. A. van der Vorst, "Bi-CGSTAB: A Fast and Smoothly Converging Variant of Bi-CG for the Solution of Nonsymmmetric Linear Systems," *SIAM Journal on Scientific and Statistical Computing*, vol. 13, no. 2, pp. 631–644, March 1992.

252. H. A. van der Vorst, "Krylov Subspace Iteration," *Computing in Science & Engineering*, vol. 2, no. 1, pp. 32–37, January/February 2000.

253. H. F. Walker, "Implementation of the GMRES Method Using Householder Transformations," *SIAM Journal on Scientific Computing*, vol. 9, no. 1, pp. 152–163, 1988.

254. X. Wang and H. V. Poor, "Iterative (Turbo) Soft Interference Cancellation and Decoding for Coded CDMA," *IEEE Transactions on Communications*, vol. 47, no. 7, pp. 1046–1061, July 1999.

255. G. A. Watson, "An Algorithm for the Inversion of Block Matrices of Toeplitz Form," *Journal of the Association for Computing Machinery*, vol. 20, no. 3, pp. 409–415, July 1973.

256. M. E. Weippert, J. D. Hiemstra, J. S. Goldstein, and M. D. Zoltowski, "Insights from the Relationship Between the Multistage Wiener Filter and the Method of Conjugate Gradients," in *Proceedings of the Sensor Array and Multichannel Signal Processing Workshop*, August 2002, pp. 388–392.

257. B. Widrow and M. E. Hoff, Jr., "Adaptive Switching Circuits," in *IRE WESCON Convention Record, Part 4*, 1960, pp. 96–104.

258. B. Widrow, P. E. Mantey, L. J. Griffiths, and B. B. Goode, "Adaptive Antenna Systems," *Proceedings of the IEEE*, vol. 55, no. 12, pp. 2143–2159, December 1967.

259. N. Wiener, *Extrapolation, Interpolation, and Smoothing of Stationary Time Series*, MIT Press, 1949.

260. N. Wiener and E. Hopf, "Über eine Klasse singulärer Integralgleichungen (German, On a Class of Singular Integral Equations)," *Sitzungsberichte der Königlich Preussischen [sic] Akademie der Wissenschaften zu Berlin (German, Proceedings of the Royal Prussian Academy of Sciences in Berlin)*, pp. 696–706, 1931.

261. G. Woodward and M. L. Honig, "Performance of Adaptive Iterative Multiuser Parallel Decision Feedback with Different Code Rates," in *Proceedings of the 2001 IEEE International Conference on Communications (ICC 2001)*, June 2001, vol. 3, pp. 852–856.

262. G. Woodward, R. Ratasuk, and M. L. Honig, "Multistage Multiuser Decision Feedback Detection for DS-CDMA," in *Proceedings of the 1999 IEEE International Conference on Communications (ICC 1999)*, June 1999, vol. 1, pp. 68–72.

263. G. Woodward, R. Ratasuk, M. L. Honig, and P. D. Rapajic, "Minimum Mean-Squared Error Multiuser Decision-Feedback Detectors for DS-CDMA," *IEEE Transactions on Communications*, vol. 50, no. 12, pp. 2104–2112, December 2002.

264. W. Xiao and M. L. Honig, "Convergence Analysis of Adaptive Full-Rank and Multi-Stage Reduced-Rank Interference Suppression," in *Proceedings of the 2000 Conference on Information Sciences and Systems*, March 2000, vol. 1, pp. WP2-6–WP2 11.

265. W. Xiao and M. L. Honig, "Large System Transient Analysis of Adaptive Least Squares Filtering," *IEEE Transactions on Information Theory*, vol. 51, no. 7, pp. 2447–2474, July 2005.

266. X. Yang, T. K. Sarkar, and E. Arvas, "A Survey of Conjugate Gradient Algorithms for Solution of Extreme Eigen-Problems of a Symmetric Matrix," *IEEE Transactions on Acoustics, Speech, and Signal Processing*, vol. 37, no. 10, pp. 1550–1556, October 1989.

267. D. M. Young, "A Historical Review of Iterative Methods," in *A History of Scientific Computing*, S. G. Nash, Ed., pp. 180–194. ACM Press, Addison-Wesley, 1990.

268. N. Young, *An Introduction to Hilbert Space*, Cambridge University Press, 1988.

269. M. Zoltowski and E. Santos, "Matrix Conjugate Gradients for the Generation of High-Resolution Spectograms," in *Proceedings of the 37st Asilomar Conference on Signals, Systems & Computers*, November 2003, vol. 2, pp. 1843–1847.

270. M. D. Zoltowski, W. J. Hillery, S. Özen, and M. Fimoff, "Conjugate-Gradient-Based Decision Feedback Equalization with Structured Channel Estimation for Digital Television," in *Proceedings of SPIE – The International Society for Optical Engineering*, April 2002, vol. 4740.

271. M. D. Zoltowski, T. P. Krauss, and S. Chowdhury, "Chip-Level MMSE Equalization for High-Speed Synchronous CDMA in Frequency Selective Multipath," in *Proceedings of SPIE – The International Society for Optical Engineering*, April 2000, vol. 4045, pp. 187–197.

272. P. A. Zulch, J. S. Goldstein, J. R. Guerci, and I. S. Reed, "Comparison of Reduced-Rank Signal Processing Techniques," in *Proceedings of the 32nd Asilomar Conference on Signals, Systems & Computers*, November 1998, vol. 1, pp. 421–425.

Index

GPSR Compliance

The European Union's (EU) General Product Safety Regulation (GPSR) is a set of rules that requires consumer products to be safe and our obligations to ensure this.

If you have any concerns about our products, you can contact us on

ProductSafety@springernature.com

In case Publisher is established outside the EU, the EU authorized representative is:

Springer Nature Customer Service Center GmbH
Europaplatz 3
69115 Heidelberg, Germany